Smart Grid Technologies and Applications

Smart Grid Technologies and Applications

Mohammad Ashique Azad

ARCLER
PRESS

www.arclerpress.com

Smart Grid Technologies and Applications

Mohammad Ashique Azad

Arcler Press
4164 Lakeshore Road
Burlington, ON L7L 1A4
Canada
www.arclerpress.com
Email: orders@arclereducation.com

ISBN: 978-1-77956-967-7 (Hardcover)

ABOUT THE AUTHOR

Mr. Mohammad Ashique Azad serves as Assistant Professor in the Department of Computer Science and Engineering at K.L. University, Vijayawada, Andhra Pradesh, India, where he combines academic excellence with cutting-edge research in quantum technologies. He is currently pursuing his doctoral research (PhD) in Quantum Computing from the National Institute of Technology (NIT), Jamshedpur, positioning himself at the forefront of this transformative field. His research expertise encompasses Quantum Computing with specialized focus on Quantum Machine Learning (QML), Quantum Cryptography, Quantum Sensing, and Quantum Convolutional Neural Networks (QCNN). With over five years of distinguished teaching experience across premier engineering institutions and universities in India, Mr. Azad has cultivated a reputation for academic rigor and innovative pedagogy. His unique multidisciplinary foundation spanning B.Sc (Zoology), B.Tech and M.Tech in Computer Science and Engineering, MBA (Human Resource Management), PGDEMA, and Diploma in Quantum Computing enables him to approach complex research challenges with diverse perspectives. He has established an impressive scholarly portfolio with numerous publications in prestigious national and international journals and conferences, complemented by multiple patents, books and book chapters that underscore his ability to translate theoretical insights into practical innovations, making significant contributions to the advancement of quantum technologies and computational science.

TABLE OF CONTENTS

LIST OF FIGURES

PREFACE

Transformation in the power system is one of the defining engineering problems of the 21st century. The traditional one-way flow of power-based grids of electricity, which have been fine-tuned to deliver power from central generation to consumers, are under unprecedented stress due to rising demand, integration of renewables, and the need for resilience against outages. Climate change and sustainability imperatives have heightened the need for cleaner and more efficient power systems at the same time. This has led to smart grids—adaptive, intelligent networks that use advanced communication, sensing, and control technologies to revolutionize the generation, delivery, and utilization of electricity. By harnessing the combination of energy engineering and information technology, smart grids can change the way societies generate, distribute, and utilize electricity.

Aside from bolstering reliability, smart grids herald a paradigm shift toward active consumer participation and decentralized power generation. With smart meters, demand-side management programs, microgrids, and electric vehicle integration, the customer is not a passive receiver but an active player in managing energy. Renewable sources such as wind and solar, which were once seen as variable and intermittent, are increasingly being accessed with smart grid technologies that level out their power through storage and sophisticated control. But with this interconnectivity bring issues: cybersecurity weaknesses, regulatory gaps, and the issue of integrating heterogeneous technologies. Solutions to such risks and opportunities require a nuanced appreciation of the technical, operational, and policy considerations that underpin modern smart grid systems.

Smart Grid Technologies and Applications has been conceived as an encyclopedic introduction to both the theoretical underpinnings and the pragmatic applications of smart grids. The book covers the most important building blocks of smart grid infrastructure, such as advanced metering, communications, renewables integration, storage, and cybersecurity. It offers readers a rational sequence from the fundamentals to leading-edge applications, focusing not only on the technological developments but also on the systemic and social implications of rolling out smart grids. Each chapter is composed to meet both academic and professional audiences by combining conceptual soundness, applied illustrations, and forward-looking analysis.

Chapter 1 is a background for smart grids and how they evolved from conventional power systems. It presents drivers to change, such as energy sustainability, consumer empowerment, and grid reliability. The chapter also sets a context by presenting a contrast between traditional and smart grids, with the shift from centralized to distributed intelligence.

Chapter 2 deals with smart grid architecture, describing the tiered structure of generation, transmission, distribution, and consumption. It describes how physical infrastructure gets combined with digital communication and control systems. The chapter lays great stress on interoperability among diverse components, from substations and sensors to cloud platforms, to facilitate seamless operation across the grid.

Chapter 3 examines smart metering and advanced metering infrastructure (AMI). It describes how smart meters take real-time measurements of use, provide two-way communication, and support dynamic pricing plans. The chapter also addresses the application of AMI to support demand response, outage detection, and energy efficiency participation by customers.

Chapter 4 provides explanations on smart grid communication technology, recognizing that reliable exchange of data is critical to their functionality. Chapter 4 covers wired and wireless communication technologies including fiber optics, power line communication, Wi-Fi, and cellular networks. The chapter further establishes the significance of communication protocols, latency needs, and cybersecurity concerns in guaranteeing credible communication channels.

Chapter 5 presents demand-side management (DSM), demonstrating how consumers can play an active role in managing demand and supply. Chapter 5 addresses load shifting, peak shaving, and demand response programs, demonstrating the role of pricing signals and automated control in increasing efficiency. The chapter also examines the role played by behavioral change and consumer incentives in successful DSM implementation.

Chapter 6 deals with the technical and operational challenges of connecting wind, solar, and other variable resources to the grid, addressing the methods of forecasting, the grid stabilization strategies, and the value of flexible resources such as storage and demand response. The successful integration of renewables at the national and local community levels is put forward with case studies.

Chapter 7 discusses microgrids and distributed energy resources (DERs) and how they can build grid resilience and local autonomy. It is regarding the design, control, and operation of microgrids and how small wind turbines, community energy storage, and rooftop solar can be incorporated into them. It emphasizes the role that microgrids have in both energy access in off-grid regions and resilience in cities.

Chapter 8 covers energy storage technology, the backbone of variability management and ensuring reliability. The chapter covers batteries, pumped hydro, compressed air, flywheels, and newer technologies such as hydrogen storage. The chapter recognizes metrics for performance such as efficiency, lifetime, and scalability and equates them to grid applications ranging from frequency regulation to renewable smoothing.

Chapter 9 is on electric vehicles and smart charging. It discusses the dual role of electric vehicles as energy consumers and potential storage resources through vehicle-to-grid (V2G) systems. Charging infrastructure, scheduling, and grid impacts are examined, including peak load reduction and renewable integration support through smart charging strategies.

Chapter 10 addresses cybersecurity in smart grids, a field of urgent importance given the digital integration of systems. It identifies vulnerabilities in communication networks, metering infrastructure, and control centers along with potential cyber-attacks such as denial-of-service attacks and data breaches. The chapter identifies defense mechanisms such as encryption, intrusion detection, and compliance to ensure the resilience of smart grids to cyber threats.

Electrical engineering, power systems, computer science, and information technology students will view it as an structured study guide that integrates technical knowledge with real-world applications. Researchers will like how it weaves together practice and theory, especially in renewable integration and cybersecurity. Utility, energy industry, and tech company professionals can leverage its insights towards planning, operating, and optimizing smart grid systems. Policy and regulators will benefit from its method to standards, communication infrastructures, and social nature of consumer involvement. Weaving engineering basics, emerging technologies, and pragmatic methods together, this book is a comprehensive handbook for learning about and shaping the destiny of power systems.

CHAPTER 1

INTRODUCTION TO SMART GRIDS

Contents

INTRODUCTION

The emergence of smart grids is a crucial step in the production, distribution, and consumption of electricity. In the past, the power grid was based on a one-way system of communication, whereby power utilities provided energy without consumers' feedback in real time. With increased demand for electricity, bringing renewable resources into play, and encouraging energy efficiency, an outdated system like that has been insufficient. With the help of advanced information and communication technologies (ICT), smart grids aim to enhance operating reliability and flexibility, and become a realistic way to meet the energy challenges of today. As Borghetti et al. (2013) state, to become a smarter grid does not only fix the limitations of traditional infrastructure but also entails new technological and management paradigms for efficient energy consumption and production. Smart grid technology is founded at its core on making near real-time communication between various segments of the grid, i.e., generation, transmission, distribution, and consumption points. This enables energy use and integration of various sources of energy required in supporting the complexities of today's electricity use. As defined by Chin et al. (2011), such intelligent networks call for dynamic communication protocols in order to be interoperable and secure in responding to the challenge of privacy in data management.

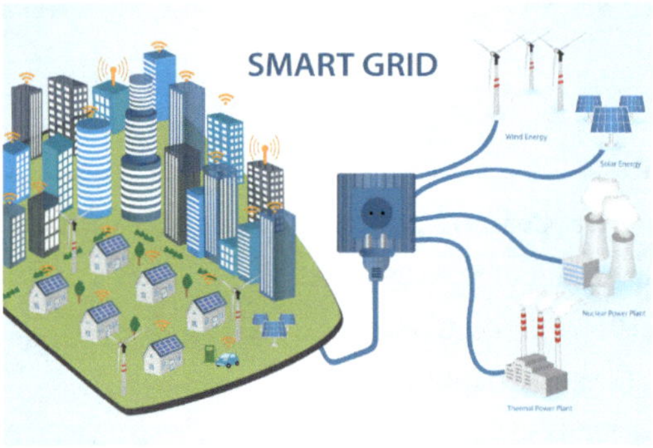

Figure 1.1. Smart grid system overview.

Source: https://media.licdn.com/dms/image/v2/D5612AQF8pCeytX4eGg/article-cover_image-shrink_720_1280/article-cover_image-shrink_720_1280/0/1703133572218?e=2147483647&v=beta&t=wcE3Asi4IqpCnU93aCljyf5XHC7lpVTSN2VbJEJlAuA.

The coordinated efforts at standardization undertaken in Europe offer an applicable case study of how technological innovation and regulation can collaborate in yielding a complete smart grid platform that is more efficient and sustainable. As the world is confronted with climate change and energy security challenges, the application of smart grids provides a multi-faceted solution to these intimidating issues. Not only do the systems provide a platform for the integration of renewable energy but also allow for consumer engagement through the application of smart metering technologies that empower individuals to drive their energy consumption. Furthermore, emphasis on the smart grid's security, as indicated in Chin et al. (2011), maintains the network's stability to absorb any cyber-attacks or disruption. In summary, the evolution of the smart grid is an inevitable step in the right direction, central to creating an adaptive energy environment responsive to contemporary society's requirements while setting the stage for a better future (Figure 1.1).

Definition of Smart Grids

Smart grids notion reflects a paradigm shift in the traditional electrical grid structure, characterized by the use of advanced communication, automation, and control technology. Unlike legacy grids, which rely on a one-way supply of electricity and information, smart grids facilitate two-way communication between end-users and utility providers. This bidirectionality facilitates dynamic energy resource management so that consumers can take an active role in energy conservation and demand response. With higher penetration of smart meters and IoT devices, household and commercial installations are being integrated into the more responsive network, resulting in increased energy efficiency and reliability of energy supply. Moreover, smart grids leverage emerging technologies to enhance operating flexibility and security, an aspect that qualifies them to play a central role in fulfilling contemporary energy demands. In particular, deployment of machine learning algorithms offers forecasted analytics for the optimization of energy supply during peak periods to reduce the likelihood of blackouts and enhance system robustness. Furthermore, the connection of renewable energy sources such as solar and wind power necessitates a flexible grid with which to connect to the variable inputs easily. Following recent research, the shift towards a metaverse-like environment for energy systems accounts for the need for advanced communications infrastructure akin to that described in innovation such as Integrated Sensing and Communications, whose purpose is to advance the interconnectivity of networks (Liu et al., 2022). Lastly, the definition of smart grids must incorporate their utilization for improving sustainability and environmental stewardship. Moving towards a smart grid structure is not just about technological innovation but policy reform that optimizes energy efficiency and reduces carbon footprint. With the development of smart grids, they support decarbonization through providing avenues of enhanced energy management and integration of distributed energy resources. The dynamic between smart grids and renewable energy technologies is a vital field of study, suggested by the ongoing controversies in the literature regarding the need for integrated frameworks that balance improvements in performance with sustainability goals. A sophisticated understanding of their different characteristics is essential in order to overcome the inherent issue of scalability and interoperability in the modern grid technologies (Wang et al., 2022) (Figure 1.2).

Figure 1.2. Smart grid.

Source: https://electricalacademia.com/wp-content/uploads/2017/11/smart-grid.jpg.

Importance of Smart Grids in Modern Energy Systems

The transition to more efficient energy systems is largely dependent on the application of smart grids, which facilitate a high integration of renewable sources of energy. With the integration of sophisticated technologies, smart grids make it possible for the integration of clean energy such as solar and wind power into the conventional energy mix to reduce the reliance on fossil fuel. This synergy is essential in addressing the effects of climate change as it increases overall energy efficiency and encourages environmentally friendly consumption. In addition, smart grids also offer consumers instant feedback on their consumption, which allows them to make informed decisions on consumption behavior. The openness of such interconnectedness is central to enabling energy conservation and efficiency in resource use, so that smart grids are not merely a technological innovation but also an energizer of behavioral change in energy consumption both at the personal and societal levels (Motlagh et al., 2020). The Internet of Things (IoT) is critical in smart grids because it provides the platform for improved grid functions. IoT technologies enhance the interconnection of various devices and systems within the energy network, allowing for more efficient monitoring and control over energy supply and demand. This interconnectedness significantly adds to the dynamism of the energy system to responses in supply and demand, with increased integration of intermittent renewable supplies. In addition, advancements in IoT platforms and hardware are driving predictive analytics innovations that enable grid operators to make better predictions about demand trends. This not only optimizes the operational efficiency but also avoids power grid collapse, giving more reliability and resiliency in the delivery of energy. IoT and smart grids are a quintessential example of innovations in new-generation energy systems with vast potential for growth in energy management (Shafique et al., 2020). Lastly, smart grid construction serves to enhance energy security with enhanced resilience to cyber as well as physical disruptions. As energy consumption increases and sustainability concerns turn issue-based, the guarantee of a secure energy supply is of prime consideration. Advanced security features, such as encryption and decentralized control, are built into smart grids to guard against weaknesses. Moreover, they allow utilities to respond smartly to disturbances or outages, using real-time data to quickly redirect operations and reconnect power. Such adaptive capabilities are vital in an atmosphere where energy infrastructure becomes increasingly vulnerable to sabotage by hostile groups. Thus, smart grids not only improve energy infrastructure but also ensure a more robust and secure energy environment, which is imperative to future energy sustainability and resilience (Motlagh et al., 2020) (Figure 1.3).

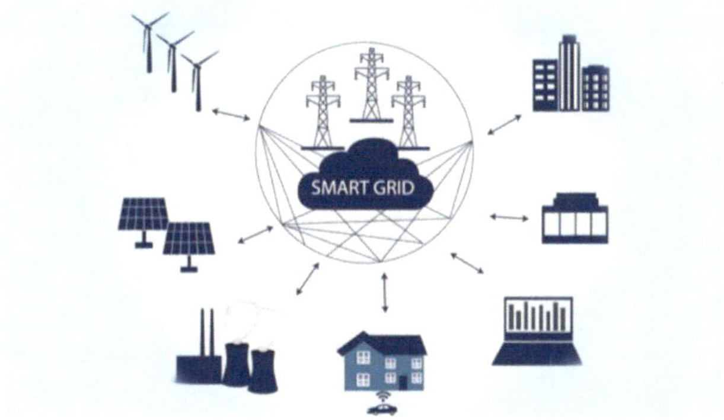

Figure 1.3. The role of smart grids in modern energy management.

Source: https://media.licdn.com/dms/image/v2/D5612AQE9Ul7Rv3akqw/article-cover_image-shrink_600_2000/article-cover_image-shrink_600_2000/0/1726849779116?e=2147483647&v=beta&t=KEgbKYEvno3uZHwKZAgj6Alk70Jg6cjbXO8V-SW_yDY.

1.1. HISTORICAL CONTEXT

The evolution of smart grids starts with the increasing need for an efficient and dependable electric power system, which became more urgent over the last few decades. Previous transformations towards industrialization made utilities modify their operations to manage heightened consumer expectations and unstable energy demands. This concept of smart grids emerged from these problems as a move to integrate sophisticated technologies in power distribution networks. Smart grids, characterized by two-way communication, real-time monitoring, and automatic control, are set to revolutionize the power industry deeply, according to analysts. They exist not simply as a response to technological advances but as a strategic initiative intended to address environmental challenges and carbon-neutral targets, compelling meaningful political and financial investment at the government level (Axon et al., 2014). In various geopolitical contexts, the imperative for smart grid implementation converges with defining moments regarding energy policy and environmental awareness. The agenda of reducing carbon emissions has been transformed from a faddy agenda to a necessary aspect of government plans, especially for the European Union and the United States.

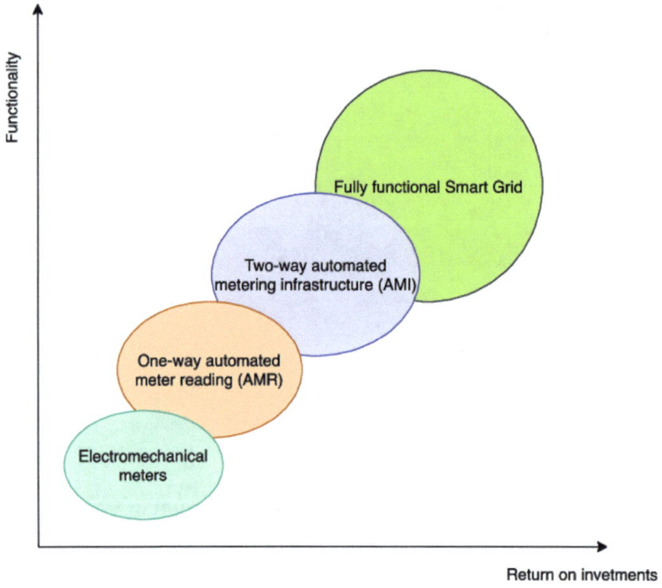

Figure 1.4. Evolution of smart grid.

Source: https://www.researchgate.net/profile/Akhil-Joseph-5/publication/346853333/figure/fig2/AS:967012410994689@1607565133514/Evolution-of-smart-grid-Source-21.png.

The smart grid has thus become linked to a transition towards a clean energy era. As countries report their climate plans, integration with a smart grid is done as a column in order to achieve these guidelines. Regardless of this, the transition is complex, involving public opinion, regulatory frameworks, and technological infrastructure as interlinked barriers needing integrated understanding and engagement by different stakeholders (Axon et al., 2014). The background of Argentina's electrical grid history illustrates the tension between existing infrastructure and the transition to smart technology. At present, as the electrical network operates at near-full capacity, the operating constraints highlight the necessity for modernization. Despite the probable advantages of smart grid technologies, including increased efficiency in managing resources and increased longevity of equipment, the deployment has been held back by cost largely. The application of old systems in the earlier years has slowed down the process. Instances, like data analytics application in monitoring transformers, illustrate the reality of step-by-step progression needed to take advantage of proven technologies in smart grid settings. Resolution of such historical issues is central to maximizing the potential of smart grids in the Argentinian situation (Figure 1.4).

1.1.1. Evolution of Electrical Grids

Evolution of electrical grids has witnessed dramatic change over the past century, transforming from traditional, centralized paradigms to more complex, decentralized ones. Early electrical grids were primarily one-way systems of electricity transmission from generators to consumers. However, with increasing demand for energy, along with technological advancements, there was a need for more interactive infrastructure. Smart grid technology is one such major innovation in this transition, wherein digital communication and advanced sensors are utilized to improve grid control and performance. Smart grids allow stakeholders to optimize energy use by providing a two-way interface for communication between consumers and utilities, making it possible to integrate renewable energy resources, which is a big step towards sustainable and resilient energy systems. As there is a steady rise in demand for reliable and efficient energy, the role of digitalization for electrical grids becomes increasingly clear. The convergence of smart technologies enables access to record levels of data, which assists utilities in real-time diagnostics of issues and enhancing service reliability. The digital revolution supports the linkage of energy systems with United Nations Sustainable Development Goals (SDGs), by promoting sustainable behavior and fair access to energy resources. The possible benefits of these digital innovations are several, addressing significant problems such as sustainable food production, access to drinking water, and renewable energy generation. Smart grids, in this regard, mark a new era when energy systems can, hopefully, improve society's well-being and facilitate sustainability in the environment, being of vital relevance in climate change issues and biodiversity (Mondéjar et al., 2021). Furthermore, with the rising tide of electric vehicles (EVs) as a worldwide auto trend, the development of electrical grids needs to adapt to accommodate the growing injection of EV charging networks. The development of next-generation battery technologies, such as lithium-ion batteries, is at the forefront of making this adaptation possible. While current battery technologies have enabled the practical application of EVs, further innovation is required to meet the emerging demands on performance, cost, and safety in this growing market (Zeng et al., 2019). The energy grid should be upgraded with comparable changes as these batteries become more prevalent so that it can support the additional loads and facilitate efficient energy distribution. Thus, electrical grid development is still intertwined with technology developments in battery storage and electric transport toward a greener and more sustainable energy future (Figure 1.5).

Figure 1.5. History of the electrical grid.

Source: https://dynamic-assets.hitachienergy.com/is/image/hitachiabb-powergrids/hitachi-energy-transmission-towers-2436x810-2:21-9?wid=1200&hei=514&fmt=webp-alpha&fit=crop%2C1.

1.1.2. Key Technological Advancements Leading to Smart Grids

The evolution of smart grids is based on some significant technological advances that bring more energy management and distribution efficiency. Among such new technologies, renewable energy technology advances, specifically solar energy systems, have received significant attention. Solar power is central to the quest for a sustainable, resilient energy future by offering a decentralized source of power that can be simply integrated into existing grids. The focus on solar technologies not only answers increasing worldwide energy demand but also aids in climate change mitigation. Proof exists in the diversified benefits of solar energy implementations across various climates, necessitating expert methodology in its implementation. With adequate information about and adaptation to environmental conditions, nations, including such a nation as Nigeria, are capable of capitalizing on solar energy access simultaneously with economic growth and upholding environmental integrity, which makes it a corner stone of intelligent grid construction (Izuka et al., 2023). Digitalization is also a significant technology innovation transforming the smart grid environment. The integration of digital technologies, including the Internet of Things (IoT), artificial intelligence (AI), and advanced metering infrastructure, has facilitated the availability and analysis of real-time data, thereby transforming the way grids operate. Through proper planning and management, these digital technologies enhance the quality of forecasting and provide real-time response to variations in consumer demand. These innovations not only improve the reliability and efficiency of power delivery but also create an environment that is conducive to trading and competition in the energy market. Systematic review of over 54,000 academic articles reaffirms how such digital technologies are part of the inherent system that improves the operational parameters and regulatory compliance of energy markets. While smarter grids evolve, defeating challenges such as data integration and system interoperability remains central to effectively harnessing their potential (Aghahadi et al., 2024). Other than digitalization and renewable technologies, innovations in battery storage systems have also served as determinants in transitioning toward smarter grids. These systems counteract the intermittency of renewable resources by allowing for the storage and release of excess energy generated during peak production periods. These functions provide a more flexible and predictable supply of energy, ensuring that consumer demands are met at all times despite generation volatility. With the advancement in battery technology, where there are increased energy densities and increased rates of charging, economic viability for large-scale storage solutions becomes more apparent. The efficiency of renewable energy integration is not only improved through such advancements but also helps improve grid stability at peak demand. Therefore, the complementarity of energy storage solutions and other emerging technologies represents a notable step toward fully functional smart grids.

1.1.3. Regulatory Changes and Their Impact

Implementation of smart grid necessitates far-reaching regulatory changes with a powerful impact on the production and consumption of energy. Present energy policy is strongly advanced for the traditional grid paradigm, which comes to limit the innovations necessary for smart infrastructure. The regulatory regimes must change to support decentralized production like integrating renewable energies into the grid. For instance, regulations should promote the deployment of smart metering technologies and demand response programs that promote energy efficiency and customer choice. While shifting regulations to include such innovations, governments may also stimulate competition in markets, thus reducing costs and dependability of services. As pointed out in the literature, the pursuit of energy sustainability emphasizes the role of legislative frameworks in accordance with technological developments so that smart grids can operate optimally within a rapidly changing energy landscape (Agupugo et al., 2025). Additionally, the funding climate for smart grid technology heavily relies on facilitating regulatory frameworks. Regulatory reforms can create a conducive investment climate that motivates public and private stakeholders to invest in smart grid projects. Policymakers can build incentives such as tax credits, financing arrangements, and subsidies aimed at smart grid installation and integration of renewable energy. These systems can motivate stakeholders such as energy firms and financial institutions to invest in innovative technology that results in a stronger energy system. Additionally, with an accelerated move towards the use of renewable energy sources, government regulatory policies must also deal with possible barriers to market investment. By creating conducive financing and investment environments, governments can promote economic security by way of diversification of the energy base, thus promoting sustainable growth and decreasing dependence on the fossil fuel sources (Ivanova et al., 2025). The impacts of policy reforms on the development of smart grids are varied, not only impacting the technological landscape but also the environmental and social effects. With increasingly liberalizing regulations, smart grid infrastructures can maximize the efficiency of energy transmission, resulting in lower greenhouse emissions and higher environmental sustainability. Furthermore, the application of smart grids facilitates equitable access to energy, promoting social justice by providing equal access to renewable energy resources for underprivileged communities. Enhanced regulatory systems will simplify the process of forming microgrids in remote communities, allowing control of energy needs by local communities. Therefore, as the world grapples with the urgent issues of climate change and energy security, regulations underpinning smart grid development will play an imperative role in designing a future that is in tandem with sustainability objectives and also promotes inclusive economic growth, demonstrating the revolutionary power of strategic regulatory changes (Agupugo et al., 2025; Ivanova et al., 2025).

1.2. COMPONENTS OF SMART GRIDS

Smart grid elements constitute an diverse group of technologies and infrastructure that enable real-time data management and communication in the energy sector. At the heart of making smart grids work are the integration of sensor networks and communications infrastructures that enable operation across levels of generation, transmission, distribution, and consumption. Such integration enables information flow smoothly, allowing utilities to optimize energy distribution and enhance the grid's reliability. The connectivity of these components serves a useful purpose in the overall operation of energy efficiency, since grid managers can respond more rapidly to fluctuating levels of demand and supply. Additionally, the development of standardized communication protocols is necessary to enable interoperability between various technologies in the smart grid architecture, thus enabling unproblematic working among various manufacturers and service providers (Chin et al., 2011).

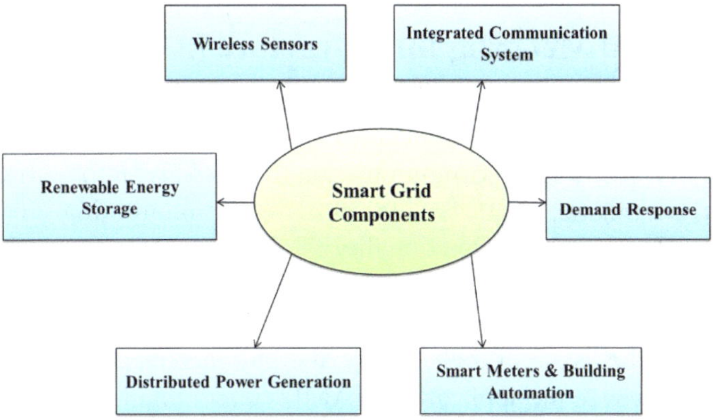

Figure 1.6. Diagram of smart grid components.

Source: https://www.stet-review.org/articles/stet/full_html/2024/01/stet20240131/stet20240131-fig2.jpg.

Another important aspect of smart grids is the implementation of Smart Home Energy Management Systems (SHEMS). SHEMS empower customers to be proactive in energy usage and efficiency by managing house energy consumption according to real-time data. SHEMS help make smart choices, assisting users in monitoring their energy consumption patterns and aid demand-side management strategies. Despite technological capacity demonstrated in SHEMS, existing literature reports a tendency to prioritize technical solutions over user-centered approaches that would add more to energy-saving procedures. The limitation highlights the imperative for transdisciplinary studies that integrate technological and human-centered behavioral aspects in order to realize the maximum capacity of SHEMS in household energy management (McIlvennie et al., 2020). Besides, smart grids utilize the deployment of advanced metering infrastructure (AMI) as a crucial component for enabling two-way communication between consumers and utilities. AMI infrastructure consists of data management systems, communication networks, and smart meters, which provide utilities with detailed data on energy consumption patterns. This kind of data enables more accurate billing, enhances demand response capacity, and makes integration of renewable resources easier. The installation of AMI also allows for more effective detection and response to outages, reducing restoration times and generating customer satisfaction. However, AMI rollout is contentious with regard to privacy and security, and stringent measures must be implemented to protect consumer data. Balancing the benefits of smart metering with privacy concerns is essential to establish trust and ensure the success of smart grid technologies (Chin et al., 2011) (Figure 1.6).

1.2.1. Advanced Metering Infrastructure (AMI)

AMI is the key component of the smart grid infrastructure, which assists in improving the communication link between electricity suppliers and consumers. By integrating communication technology into current energy distribution paradigms, AMI facilitates real-time monitoring of electricity consumption. This two-way functionality facilitates more accurate billing and better demand-side management programs, leading to a more efficient electric grid. The deployment of AMI systems supports consumers in making intelligent decisions about the usage of their energy, thereby increasing energy efficiency and savings. However, the application of AMI is not problem-free when it comes to cybersecurity threats that need proactive interventions to mitigate associated risks (Barros et al., 2025). While the huge benefits presented by Advanced Metering Infrastructure are significant, the application of networked systems has serious security threats. The AMI systems infrastructure is susceptible to a variety of cyber-attacks, including unauthorized access and data modification, that can undermine the integrity of the energy management system. Attack scenarios can lead to large-scale service outages with catastrophic consequences affecting not just utility companies but also consumers who rely on continuous energy provisioning. Therefore, the exploration of the vulnerabilities that are inherent to AMI, including hardware and communications vulnerabilities, is vital towards making the smart grid operations sustainable and reliable. The increased sophistication of cyber-attacks proves the need for proper security solutions that target intrusion detection and risk avoidance measures, thereby enabling the AMI system to work securely in the overall smart grid setting (Barros et al., 2025). To enhance the security stance of Advanced Metering Infrastructure, there is a need for a strong strategy that emphasizes cooperation among policymakers, technology vendors, and stakeholders. Effective cybersecurity governance can facilitate the development of best practice frameworks tailored to AMI systems to cover gaps in system vulnerabilities and regulatory requirements. Proactive risk management is paramount while integrating next-generation technologies such as artificial intelligence and machine learning for strong threat detection. Second, the use of blockchain technology to authenticate data will ensure consumer data integrity in the AMI system. As power grids evolve, the enhancement of the security environment surrounding AMI will become key to safeguarding smart grids from the issues posed by an increasingly digital world, eventually culminating in energy sustainability and resilience (Ahmed et al., 2025) (Figure 1.7).

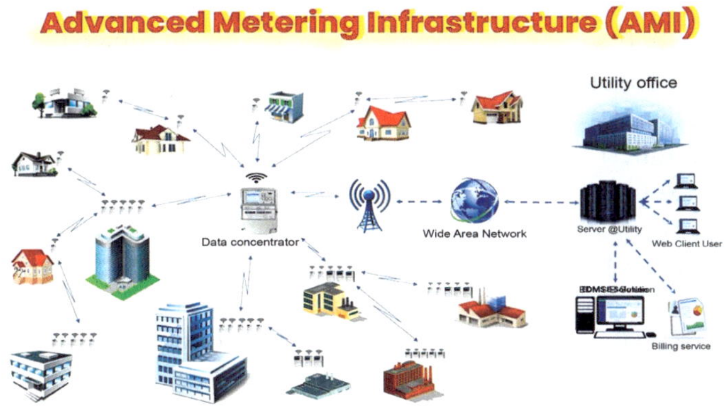

Figure 1.7. Understanding advanced metering infrastructure (AMI) in smart grid system.

Source: https://i.ytimg.com/vi/Lby1sDHpXKo/maxresdefault.jpg.

1.2.2. Communication Networks

Communication network advancement is key in the process of conceptualizing and running smart grids, an upgraded electricity infrastructure that seeks to improve the reliability and efficiency of energy supply. Traditional electricity infrastructures have not met modern needs, and the necessity has therefore come to improve towards smarter systems equipped with Information and Communication Technologies (ICT). Interconnection of these technologies allows real-time sharing of data, facilitating better observation and management of energy distribution. Based on recent studies, the smart grid concept is an indicator of the ability of ICT to evolve, making infrastructure required to sustain alternative sources of energy and demand-side management possible (Borghetti et al., 2013). Thus, communication networks are not merely an operational effectiveness cornerstone but also a means of resolving regulatory and safety issues inherent in the modern energy systems. Smart grids' most significant characteristic is grounded in their application of near-real-time communication capabilities across various parts of the grid, ranging from generation points through transmission points, distribution points, to consumption points. This dynamic engagement allows for maximizing energy consumption and enhancing the grid's overall performance. Addressing barriers to unfettered communication entails problems such as interoperability and future-proof technology applied in communication networks. Addressing these issues has broader implications for smart metering and good practices in energy management, which necessitate collective effort in developing uniform standards and protocols. The discussion of communications research is employed to suggest the potential of existing networking technologies to enable energy management, ultimately translating into lower operational costs and improved service delivery (Chin et al., 2011). It is imperative that the stakeholders in the energy sector promote such advances to enjoy the fullest benefits of smart grid infrastructure. Additionally, the integration of sophisticated communication networks into smart grids puts into focus important aspects of security and privacy. With the systems becoming more interconnected, there can be vulnerabilities that can lead to operational disturbances or security breaches. It becomes important to construct secure protocols with resilience, particularly in the wake of reliance on near-real-time data exchange among diverse grid entities. Regulatory regimes adapt in the face of the challenge posed by emerging threats to address apprehension regarding the safety of digital infrastructures. Effective communication networks not only ensure smooth functioning but also assist in developing consumers' and regulatory bodies' trust. The advance setup of security systems will thus ensure sustained expansion and implementation of smart grids as an efficient response to modern energy management (Borghetti et al., 2013; Chin et al., 2011) (Figure 1.8).

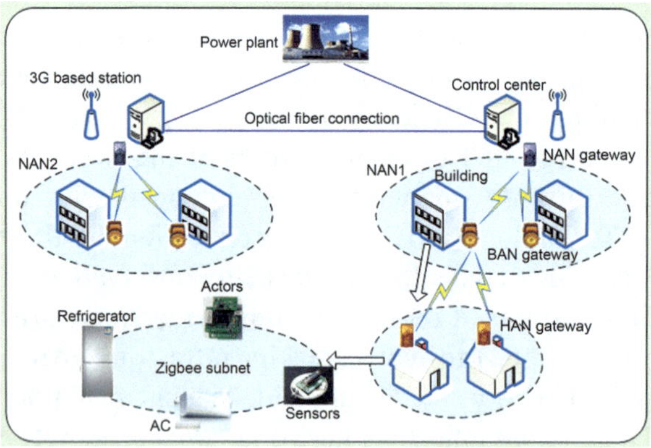

Figure 1.8. Communication architecture of smart grid.

Source: https://www.researchgate.net/profile/Jiafu-Wan/publication/264846553/figure/fig3/AS:669000623222793@1536513586538/Communication-architecture-of-smart-grid.ppm.

1.2.3. Distributed Energy Resources (DER)

DER are increasingly being regarded as the most important building blocks of smart grids, as they are heterogeneous collections of technologies that not only generate electricity but also store and regulate electricity close to where it is being used. DER coupling increases energy systems' resilience and efficiency by decentralizing energy production, thereby decreasing the dependence on traditional, centralized power plants. This shift facilitates a more sustainable energy ecosystem, in harmony with global initiatives to limit carbon emissions but also to accommodate growing needs for renewable energy.

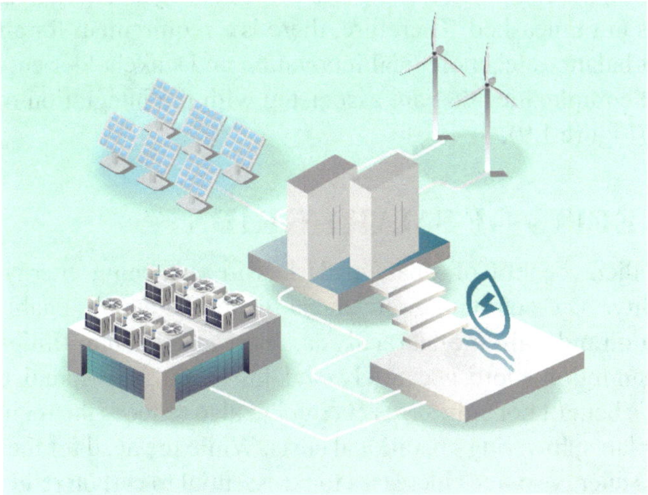

Figure 1.9. Distributed energy resources (DER).

Source: https://www.glaciergrid.com/hubfs/DER%20image4.png.

Moreover, the possibility of DER installations such as solar panels and battery storage units to stand alone or parallel to the grid facilitates a revolutionary approach to energy management, significantly departed from conventional supply paradigms. The integration of DER with smart grids brings in a myriad of technical and regulatory problems to be addressed to realize their maximum potential benefits. Effective operational paradigms must be ensured to provide seamless interface between distributed resources and existing grid infrastructure already in place. This means not only the technological capability but also the economic and regulatory means of making integration happen. As noted, numerous initiatives in the United States and Europe have emerged for developing approaches and standards to measure DER contributions to the general grid performance (Steven et al., 2012). For maximum performance of such resources, stakeholders must prioritize data sharing and collaborative platforms to facilitate comprehensive understanding and analysis of deployments in different locations. Also, public acceptance and engagement are essential to the large-scale deployment of DER in smart grids. As these resources often involve consumer engagement—e.g., community solar programs or demand response programs—data privacy, safety, and energy supply reliability concerns become a top priority. In addition, the integration of smart technologies, including advanced metering and communication systems, is essential to facilitate two-way interaction between consumers and utilities (Axon et al., 2014). Absent encouraging the symbiotic association among industry stakeholders, regulators, and consumers, the full potential of DER in supporting an efficient and resilient smart grid is not unleashed. Therefore, there is a requirement for an integrated strategy that balances technological innovation and stakeholder engagement to navigate the complexities that are associated with the integration of DER into smart grids (Figure 1.9).

1.3. BENEFITS OF SMART GRIDS

The most salient benefit of smart grids lies in enhancing energy efficiency and reliability. By combining advanced technologies to enable two-way communication and real-time data analysis, smart grids enable utilities to monitor energy consumption more accurately and immediately respond to shifts in demand. This benefit not only saves energy but also reduces the requirement for new power plants, lowering operational costs. While the need for the integration of renewable energy sources increases to be essential to carbon reduction, smart grids ensure the necessary infrastructure for properly managing the intermittent supply of energy. Balancing efficient supply management with consumer engagement brings energy demand and supply into a better harmonization, paving the way for the future of more sustainable energy consumption (Axon et al., 2014). Besides, smart grids also offer greater energy security, which is critical at a period of rising geopolitical threats and climate change impacts. By facilitating the use of distributed resources such as solar panels and wind turbines, smart grids lower reliance on traditional sources of power that are likely to be susceptible to market volatility or foreign influence. In the context of the European Union (EU), such innovations are essential to seek energy independence and offset the threats of energy imports. Intelligent grids with enhanced technologies built-in enable better risk handling and streamlining the energy supply chain, thereby pre-empting any probable disruptions and stabilizing the energy environment. Thus, the strategic deployment of smart grids can rebalance energy security relations by increasing self-sufficiency and energy system resilience (Abrhám et al., 2018). Finally, smart grid installation can create huge economic benefits through the creation of jobs and technological innovation. As smart grid technologies gain greater demand across various sectors, more opportunities for employment arise in areas such as engineering, information technology, and energy management. With this shift not only does economic development occur but also a workforce intelligent enough to cope with the rapidly evolving face of energy arises. Furthermore, smart grid-related innovation stimulates competition among energy suppliers, fostering innovation in service delivery and cost savings to consumers. Smart grids eventually result in a more dynamic energy sector, which benefits the economy and society as a whole, by revealing new business models and new avenues of investment (Axon et al., 2014) (Figure 1.10).

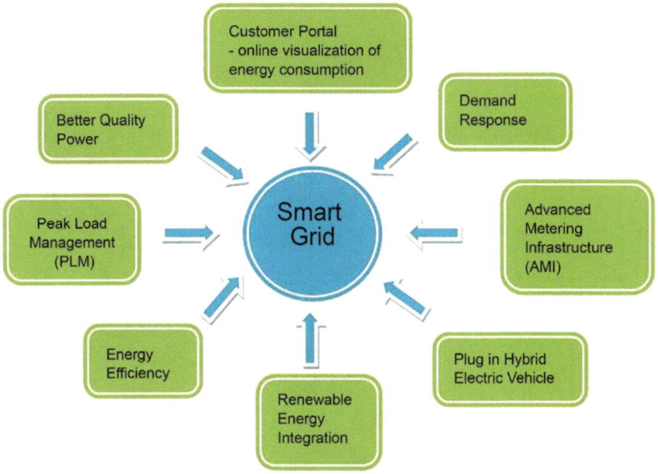

Figure 1.10. Benefits of smart grid technology.

Source: https://www.researchgate.net/profile/Thokozani-Shongwe/publication/330257532/figure/fig1/AS:1136182335414272@1647898386668/Benefits-of-Smart-Grid-Technology-44.jpg.

1.3.1. Enhanced Energy Efficiency

Greater energy efficiency is a key sector of smart grid technologies, whereby the application of Information and Communication Technologies (ICT) facilitates more responsive and dynamic energy systems. Embracing these advancements in contemporary cities leads to greater operational capacity within the energy field, specifically electrical stability and distribution. Smart grids leverage ICT to enhance communications security and optimize energy generation and consumption, thereby resulting in a more sustainable energy environment. For instance, new paradigms of demand response allow prosumer communities—

parties or entities generating electricity for consumption by themselves—to participate actively in load balancing and peak shaving, thereby reducing energy demand during times of high consumption. This collaborative framework, not only enhances efficiency but also allows for consumers to engage more actively in their energy consumption pattern, paving the way towards collective reduction in greenhouse gases and shift towards renewable resources (Bonetto et al., 2017). Also, the implementation of Smart Home Energy Management Technologies (SHEMS) identifies tremendous scope for enhanced energy efficiency at the domestic level. These technologies enable residents to observe and manage their energy use prudently, thereby contributing to demand-side management initiatives.

Figure 1.11. Enhanced energy efficiency.

Source: https://cdnbbsr.s3waas.gov.in/s3c6335734dbc0b1ded766421cfc611750/uploads/2021/03/2021030180.png.

Although initial research focuses primarily on the technology and control aspects of SHEMS, increasing need is felt to understand the behavior of users to these systems. A holistic approach must incorporate technical aspects as well as the driving factors of energy usage behind it. By integrating user-focused strategies into already implemented techno-focused ones, SHEMS can trigger high-value participation, enhance energy conservation practices, and support peak demand management. This three-pronged strategy is key in unlocking the maximum potential of home energy savings because it aligns technology with behavior to enhance overall energy efficiency (McIlvennie et al., 2020). Furthermore, the union of big data analytics with intelligent grid infrastructures makes possible compelling opportunities for improved energy efficiency. Processing of gigantic data sets enables utilities to predict patterns of demand and identify usage behaviors, which in turn make more insightful decisions for the distribution of energy possible. This level of visibility facilitates greater operational efficiency since energy providers can adjust their supply dynamically in real-time, responding to demand variability and managing resources. Moreover, big data has the potential to boost user engagement by providing actionable insights for energy usage, hence promoting energy-efficient practices among users. As people become more comfortable with the use of these technologies, they will continue using data insights to adjust their lifestyles towards sustainability. Thus, the convergence of data analytics and smart grids not only enhances energy efficiency but also allows consumers to take charge of their energy usage, contributing towards a more resilient and green power system (Bonetto et al., 2017; McIlvennie et al., 2020) (Figure 1.11).

1.3.2. Improved Reliability and Resilience

The arrival of smart grids significantly enhances the reliability and resilience of electrical infrastructure in modern societies. With the integration of new communication technology into traditional energy infrastructure, smart grids enable real-time control and monitoring of energy transmission. Synchronization of this nature optimizes energy distribution, allowing for quick response to volatility of demand as well as potential disruptions. With growing vulnerabilities in the aging power grid, the ability to anticipate and insulate against these risks is more critical than ever. Robust security controls need to be implemented in order to provide operational integrity, considering the complex interplay of system elements and current cyber threats. The multi-dimension concept of security addressed in Hauser et al. (2018) highlights the effectiveness of good decision-making mechanisms in smart grids as a means to provide greater resilience against system disruptions. Microgrids offer themselves as a key component in the smart grid architecture, ushering in increased reliability and community resilience. Microgrids are increasingly sought after in the United States, particularly as the trend in policies is in their favor for construction. By being half-independent from the main grid, microgrids are able to efficiently control energy production and consumption, something that is highly beneficial with instances of outages or emergencies. Local control not only decouples dependence on traditional infrastructures but also provides energy stability to selected communities. The technical improvements in microgrids, as well as novel techniques of control and market interactions, also guarantee the microgrids' reliability. As explained in Bao et al. (2018), the experience of learning from successful U.S. demonstration projects reveals an educated path for other countries to emulate these resilience-building practices in their power infrastructure. The resilience of smart grids is also enhanced by applying game-theoretic models to review and address security concerns. These models provide informative insights into strategic interactions that occur among the elements of the grid and potential attackers. By carrying out an exhaustive analysis of these interactions, actors are able to develop improved security measures and response strategies. Deployment of cross-layer resilient controllers as well as secure network routing protocols is critical in the mitigation of vulnerabilities, particularly in the data communication and networking layers. Fighting these multi-layered security issues, as outlined in Hauser et al. (2018), not only enhances the operational reliability of smart grids but also makes them more secure against external attacks. Through continuous innovative advancements and adoption of advanced methodologies, smart grid resilience can significantly be enhanced, leading to an energy future that is more reliable and secure.

1.3.3. Environmental Impact and Sustainability

The integration of smart grids in cities has great potential to encourage sustainability and reduce ecological footprints. Smart grids apply advanced digital technologies in order to optimize energy distribution and consumption, leading to increased efficiency and reduced greenhouse gas emissions. Improved monitoring functions enable real-time data capture that assists in shaping energy consumption patterns, allowing more nuanced demand response measures. This action reduces dependence on fossil fuels and the ecological footprint of traditional energy production. Smart city transition continues to be a prime example of this potential, with efficient utilization of resources such as water and electricity achievable in integrated networks. However, of utmost concern is careful consideration of the life cycle analysis of the technologies deployed in smart grids to ensure their environmental benefit outweighs any negative impacts (Berville et al., 2025). While the potential advantages of smart grids cannot be denied, the environmental problems caused by them cannot be ignored. The extensive application of information and communication technology (ICT) inevitably leads to high energy consumption, particularly that of data centers and network facilities. These centers, unless operated sustainably, can be one of the biggest carbon emitters, negating the expected effects of moving towards smart grids. Among those measures is the use of edge computing and efficient use of data centers. Such measures are central to mitigating such impacts and enabling sustainable behaviors. Also, the environmental risk posed by electronic waste and life cycle totals of smart grid components call for anticipatory design and policy-making. Designing healthy constructs that enable circular economy ideas will make sure that smart grid technology achieves its sustainability potential (Berville et al., 2025). Air quality is yet another significant consideration which is impacted by smart grids since they have the potential to directly help reduce dangerous emissions from industry, transport, and other mechanisms. Big plants and powerhouses have been classically primary sources of air pollution, releasing nitrogen and carbon-containing compounds, and particulate matter that are harmful to health and the ecosystem. By facilitating the use of renewable resources and improved energy efficiency, smart grids make it possible to decrease emissions in the energy industry. Improved management of grids also makes it possible to use electric vehicles (EVs), which can also decrease air pollutants in urban areas. As urban cities progress towards smarter infrastructural systems, it is essential that not only are the benefits of smart grids being harnessed but also their net impacts upon air quality with great caution, particularly within sensitive populations. It is necessary to address these interconnected issues to establish sustainable solutions in urban cities.

1.4. CHALLENGES IN IMPLEMENTATION

Smart grid implementation is plagued by a myriad of challenges that significantly hinder progress towards an advanced energy grid. Most prominent among them is the staggering cost of building and upgrading existing infrastructure. A transition from traditional grids to smart grids not only involves capital investment but also equally periodic long-term financial obligations extending several decades into the future. These investments have to be well-planned to meet both current and future needs, which makes decision-making processes cumbersome. The complexity is also heightened by the requirement to have a holistic approach to understanding intertwined variables like environmental effects and social consequences. Therefore, the stakeholders must recognize that fair sharing of benefits and costs among various participants is necessary in the reduction of uncertainties and encouraging investments in this dynamic environment (Hulya et al., 2014). Another serious challenge lies in the technological integration of dissimilar energy resources into the smart grid structure. The greater penetration of intermittent renewable energy sources, such as wind and solar, causes serious operation-related problems. Current transmission and distribution infrastructure may not be well-equipped to manage the variability of such renewable sources, creating potential reliability issues. Moreover, incorporating newer technology is demanding data management and analytical tools, which are resource-intensive and may require new skill sets within the workforce. Unless these technology hurdles are overcome, the potential of smart grids will never be fully tapped into, limiting their ability to enhance energy efficiency and stability (ELENA et al., 2011). Regulatory and policy frameworks are also tremendous obstacles to efficient realization of smart grids. Various regions have varying policies that have the ability to bring about market fragmentation in the energy sector, hindering collaboration and standardization. This regulatory landscape must evolve to support innovation while maintaining stability in energy supply and pricing. Furthermore, stakeholders must engage with various communities and societal groups to address public concerns regarding privacy, security, and economic equity. Effective communication and collaboration among all parties involved are vital for surmounting these challenges. With the energy ecosystem still under development, the need for stable policies that consider emerging technologies along with demands of the market grows more pressing (Hulya et al., 2014) (Figure 1.12).

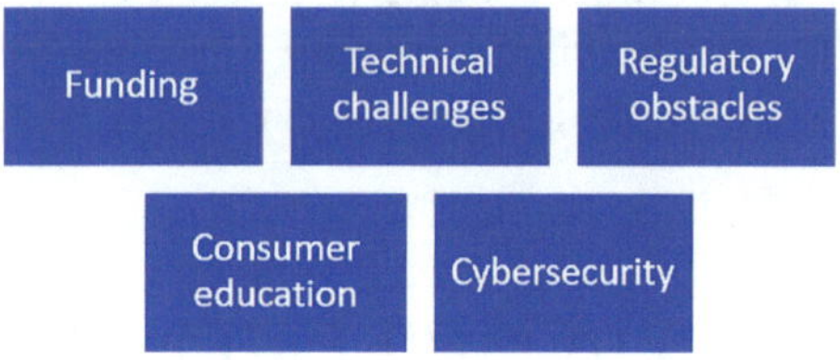

Figure 1.12. The challenges of implantation of smart grid.

Source: https://www.researchgate.net/publication/375175638/figure/fig2/AS:11431281202483659@1698859032526/The-challenges-of-implantation-of-smart-grid.png.

1.4.1. High Initial Costs

The transition to smart grids from traditional electrical grids has numerous advantages, but the astronomical initial cost of such a transition poses serious challenges. Upgrading existing infrastructure to accommodate smart technologies is costly, both in terms of hardware and software. Utilities must not only purchase more advanced metering systems and communication equipment but also invest in training their staff to manage the new technologies effectively. The cost can put off most companies from embracing the shift to smart grid since they put current expenditures against distant future benefits. Thus, the upfront expense is a major barrier that requires strategic planning, finance sources, and maybe government subsidy to encourage take-up and overcome resistance to change. In addition, the technicity involved in introducing smart grid components only serves to increase the cost factor. As already noted, the current electricity network infrastructure simply no longer accommodates present requirements due to the integration of alternative sources of power and forming energy-saving strategies (Borghetti et al., 2013). The transformation requires drastic modifications to the grid, which in turn may result in higher costs. For instance, the integration of renewable energy like solar and wind power into the smart grid framework requires significant modifications in the existing systems to achieve reliability and efficiency. This synchronization is not only a technological effort but also one of finance since utilities have to contend with increased capital investment in green infrastructure under turbulent energy prices. Therefore, wise cost-benefit estimation becomes imperative for policymakers to justify investment against potential future savings and environmental benefits. Life cycle cost analysis also provides an entire view of the financial influence of smart grid investments. An evaluation of both the traditional and smart grid infrastructures' overall costs can reveal the potential benefits of adopting advanced technology despite initial expenses (Rodríguez Jara et al., 2016). Through analysis of energy consumption, emissions, and operating efficiency, utilities can have a better idea about the economic future of transitioning to smart grids. These calculations often demonstrate the manner in which initial costs can be recouped through lower operational costs and increased reliability of provision in the long term. As such financial analysis becomes increasingly influential among stakeholders like regulators and service engineers across the energy industry, it is able to help encourage policies or incentives toward adoption of smart grid, and in turn contribute toward an evolution of sustainable practices for managing energy.

1.4.2. Cybersecurity Risks

The use of smart grids in modern energy grids has largely enhanced efficiency and dependability of operations but simultaneously represents a credible cybersecurity threat. The ability of cybercriminals to attack weaknesses in infrastructure is exemplified by incidents such as the Ukraine cyberattack that compromised substation networks and supervisory control and data acquisition (SCADA) systems. These attacks pinpoint a key concern: the likelihood of coordinated attacks provoking mass-scale disruptions and cascading failures on interconnected grids. The problem of maintaining these networks in operation—achieving functionality while ensuring security—calls for an ongoing framework with real-time monitoring and anomaly detection, ultimately intended to fortify cyberinfrastructure from sophisticated threats (Ginter et al., 2018). As there is a growing industrial Internet of Things (IoT) within smart grids, there is enhanced security threat, and thus a comprehensive risk assessment paradigm becomes absolutely essential. Cyber-attacks such as targeted distributed denial of service (DDoS) illustrate critical vulnerabilities, highlighting the need for regulation controls such as the European Union's NIS Directive and GDPR. Both these are a paradigm shift that necessitate robust security practices for interconnected industrial systems. As expected, the shift from offline to online infrastructures calls for new risk management models that account for the temporal properties of security threats. These concerns need to be addressed in order to prevent vulnerabilities from becoming institutionalized in the new smart energy supply chain, best practices remaining ahead of the game with technological advancement (Accenture Technology et al., 2018). Furthermore, the possible consequences of cybersecurity attacks extend beyond regional operational outage; they affect public faith in smart grid technology's reliability. In-depth knowledge of infrastructure complexity is necessary because convergence of various systems from power generation to grid management broadens the potential attack space. Stakeholders will have to prioritize cybersecurity investment and devise end-to-end strategies that not only address protection of existing assets but also utilize lessons from past attacks. Open dialogue among the regulatory institutions, industry players, and researchers will ensure that the danger presented by the cyber vulnerabilities is addressed without stifling an innovative-friendly environment. By establishing the discussion channel, the energy industry can better forecast impending challenges and stay away from the danger posed by attackers (Accenture Technology et al., 2018; Ginter et al., 2018).

1.4.3. Regulatory and Policy Barriers

The operation and deployment of smart grids are severely hampered by an overwhelming number of policy and regulatory barriers that hinder them from being deployed and functioning. One key setback is the existing regime of regulation, which always lags behind technological development. Policy documents developed for traditional energy infrastructure might not be able to capture the decentralized and high-tech nature of smart grid technologies. Therefore, the technology firms and energy utilities are confronted with how they can ride over these ageing rules, which inadvertently stifle innovation and disrupt collaboration. For instance, transparency in algorithms and data protection can bring about rigid compliance costs and redirect resources from essential technological innovation, thereby reducing operational efficiency. Closing these regulatory gaps is vital in order to facilitate a robust platform for developing smart grids because it will enable regulatory adaptation to the new operational conditions brought about by innovative energy solutions. Moreover, the convergence of smart grids and artificial intelligence introduces another level of complexity, specifically in complying with European Union law. The evolving regulatory landscape, particularly those regulations that pertain to technology openness, data privacy, and cybersecurity, is a challenging environment for the energy industry players. These regulations, such as the General Data Protection Regulation (GDPR) and the EU Cyber Resilience Act, are intensive in terms of requirements that can limit the installation of AI-driven smart grid systems. While such regulations are necessary for consumer protection and responsibility, they could act as bottlenecks for AI solution development and deployment, which is imperative to enhance grid efficiency and reliability. The fine balance of regulatory objectives and the need for innovation is critical to safeguard the shaping of these policies that drive technological advancement in the smart grid industry, rather than suppressing it, in order to realize its full potential (Jørgensen et al., 2025). In addition, harmonizing the tenets of Circular Economy (CE) into smart grid networks is a paradigm of sustainable energy practice but one fraught with immense regulatory hurdles. Current policy frameworks are usually not well designed to enable such innovations that ensure not only energy efficiency but also sustainability through waste reduction and resource recovery. The hurdles encompass lacking incentivization of distributed energy systems and insufficient overall policy action to stimulate circular practices. Smart grid advancement relies on collaborative models that encourage public-private partnership and innovation and manage financial risks in research and development. Through calls for targeted policy reform, e.g., tax credits to renewable technologies and partnerships to facilitate knowledge transfer, CE-smart grid program convergence can greatly improve sustainability objectives and ultimately ease the transition to sustainable, low-carbon energy infrastructures (Munonye, 2025).

1.5. CASE STUDIES

Case studies form the core of advancing the understanding of smart grids, particularly by illustrating the application of theoretical models into real-world application. Through detailed analysis of various deployments, scientists can ascertain the success of smart grid technologies and innovations in specific contexts. For example, considering an assessment of regional smart grid pilot programs, it can be seen that there are differences in public acceptance of technology integration, e.g., smart meters or smart communication systems. Such case studies highlight the intricacies of smart grid deployment, magnifying the interrelationships of supply and demand dynamics, regulatory issues, and technology advancements, as mentioned in Axon et al. (2014). These studies capture rich experiences and present a better perspective on the way smart grids could be optimized to allow stakeholders to develop precise interventions that are capable of optimizing efficiency of operations as well as consumers' engagement. Besides, the integration of Information and Communication Technologies (ICT) into case studies of smart grids reveals huge enhancements in energy supply and management. Not only do these technologies enhance the stability and reliability of power supply but also enable new paradigms of energy consumption, such as demand response programs that empower consumers with the authority to balance loads of energy. In certain projects where domestic dwellings are prosumers, selling energy while managing consumption optimally, data collected from these programs enrich the vocabulary of energy efficiency and sustainability. Analysis of these case studies brings out the multifaceted effects of ICT integration, as noted in Bonetto et al. (2017). Through the presentation of various collaborative efforts between consumers, industries, and regulators, these case studies illustrate the revolutionizing ability that smart grids have to provide for both local societies and the whole energy industry. Lastly, socio-technical aspects uncovered in smart grid case studies play a very important role to public acceptability and broader implications of technology take-up. Data privacy, cybersecurity, and perceived smart technology risks are generally tested in case analyses, resulting in debate on regulatory frameworks needed and consumer education. For instance, resistance case studies on smart meter installation reveal issues with data privacy and management that need to be addressed. Such facts highlight the importance of stakeholder engagement and communication in enabling the uptake of smart grid technology. For Axon et al. (2014), societal attitudes and technological progress are determinative factors in making smart grid projects successful. Therefore, careful case studies need to be done in order to develop a well-balanced strategy to implement smart grids on the basis of both technological potential and social needs.

1.5.1. Successful Smart Grid Implementations Worldwide

The global push for successful smart grid deployments has received a lot of attention as countries strive to upgrade their energy infrastructures. A good example is the integration of hybrid energy systems which use solar thermal collectors, photovoltaic modules, and natural gas internal combustion engines. A comparative study of various configurations for various climatic regions of Spain shows the possible primary energy saving and emission reductions that these systems can achieve. By using advanced software like TRNSYS 17 for calculations and optimization, regulators, manufacturers, and energy service engineers are provided with useful insights about the efficiency of smart grids. The emphasis on life cycle cost analyses also provides comparative analysis of different energy configurations, displaying the viability of employing smart grid technologies in varying environmental conditions. Successful deployment is thus strongly interlinked with the prudent assessment of appropriate technological designs for particular geographical conditions towards more sustainable energy options. Additionally, smart grids play a fundamental role in addressing the increasing global demand for renewable energy, fueled primarily by environmental and energy security requirements. While wind and solar energy are significant growth opportunities, traditional power transmission and distribution networks generally lack the capability to accommodate high volumes of variable renewable sources in a worthwhile way. The issue is mobilizing sizeable investments that are needed for retrofitting these energy systems to support smart power grids. A holistic approach must be pursued in order to meet the interdependent challenges on physical, environmental, cyber, social, economic, and policy fronts. This heavy planning will not only minimize uncertainties but also foster conducive investment climates. With nations seeking to upgrade their energy systems, the successful deployment of smart grids will rely on determining and resolving the complexities involved in these investments, ultimately realizing a greener and more efficient energy future (Hulya et al., 2014). Socioeconomic impacts of smart grid programs extend beyond their financial aspects; they revolutionize the very way energy infrastructures are considered and utilized by societies. Smart grid development workshops have emphasized the need to assess benefits that extend beyond externalities and social factors. For example, stakeholder engagement across different industries can enhance public involvement and willingness toward renewable programs. In addition, when countries deploy smart grid technologies, they must factor in equitable allocation of short-term costs and long-term benefits among stakeholders. These are at the core of considerations for investment in smart grid technologies. Through offering the concurrent possibility of cost-benefit analysis and interaction among diverse entities, these projects can promote the impact of smart grids on the energy sector as well as on society as a whole (Hulya et al., 2014) (Figure 1.13).

Figure 1.13. Successful smart grid projects: Collaborations between utilities, vendors, and real-world use cases.

Source: https://media.licdn.com/dms/image/v2/D4D12AQFPUOR98GuECg/article-cover_image-shrink_720_1280/article-cover_image-shrink_720_1280/0/1729846543128?e=2147483647&v=beta&t=OGzVj1JrRpya6F0DH1N_YV8Ic_Yhzh-cQAVCw4vd551U.

1.5.2. Lessons Learned from Failed Projects

The studies of smart grid projects have revealed useful lessons on the real-world challenges and pitfalls involved in deploying sophisticated power systems. Lessons derived from unsuccessful projects provide compelling reasons for strong planning and stakeholder involvement. In most instances, unsuccessful projects occurred due to insufficient stakeholder input, causing a disparity between community needs and project goals.

Furthermore, inadequate risk assessment processes can destroy even the most technologically advanced projects. As discussed in Steven et al. (2012), evaluation of past smart grid projects needs to have a common understanding of terminologies and methodologies because such clarity facilitates improved stakeholder coordination and communication. By analyzing failed projects critically, policymakers and practitioners can build frameworks that focus on transparency and effective communication and hence lead to more successful and sustainable deployments of smart grids. Furthermore, the importance of sound assessment methodologies cannot be overstated in smart grid development. Experience has learned us that without a structured assessment framework, projects get lost direction and eventually fail to meet their desired outcomes. For example, Anastopoulou et al. (2018) observes that though technology innovation can offer personalized learning environments, effective utilization of such technologies in practice goes without checks.

This analogy emphasizes rigorous testing of intelligent grid technologies to establish their true application potential in practical scenarios and effectiveness. Emphasizing collection and sharing of data between different regions, as in Steven et al. (2012), can do much to enhance combined knowledge and support informed decision-making. Ultimately, lessons learned from previous failures in smart grid projects shed light on pathways to more evidence-based strategies that draw on both technological advancement and intensive evaluation procedures.

Additionally, similarities in problems in smart grid deployments worldwide spell a compelling motivation for international collaboration. Cross-border efforts, including those enabled by the US-EU Energy Council, facilitate the exchange of lessons learned, best practices, and experience across contexts. The misfortunes of past projects not only provide warning examples but also motivating factors for developing more effective collaborative frameworks. Facilitating communication and collaboration among nations can make it easier to share both successes and failures, thus enhancing the success rate of future projects. As is explored in Steven et al. (2012), the distillation of lessons from unsuccessful smart grid projects can inform policy initiatives for a higher good, and investments in smart technology can yield concrete, equitable dividends. By a culture of shared learning and collective accountability, stakeholders may effectively spearhead the evolution of wiser, more resilient energy infrastructures.

1.5.3. Comparative Analysis of Different Approaches

The evolution of smart grids represents a paradigm shift in energy management and distribution that is marked by the intersection of sophisticated technologies and smart systems. Comparative analysis of the different approaches towards the evolution of smart grids brings into focus the requirement to blend multiple technologies so that future energy demands can be effectively met. EPRIs IntelliGrid initiative and the European SmartGrids Technology Platform both address the necessity of creating an extensible framework utilizing a set of advanced technical solutions and developing a toolbox of proved technical solutions essential to creating next-generation energy infrastructure. Not only does this joint effort facilitate superior decision-making capabilities, but it also promotes extensible, flexible architectures. In addition, the performance of such intelligent systems should be evaluated as much as to provide real-time solutions that adapt to the individual requirements of a changing energy situation as presented in the current literature (Catterson et al., 2010). Energy optimization is a critical issue in the case of smart grids, particularly for household applications where user comfort and energy utilization is of prime concern. Studies indicate that different optimization strategies have been formulated, all aiming to reduce energy usage without reducing the satisfaction of users. Comparison among these different approaches is employed to highlight the complexity of creating efficient algorithms capable of adapting dynamically to real-time user behavior without diminishing energy efficiency. There is still a broad literature gap in the application of advanced optimization methods particularly tailored for energy consumption in smart homes. This gap calls for further research on cutting-edge theoretical and practical models that economically integrate the new optimization tendencies and their scales on deployments of smart grids (Fayaz et al., 2019). Additionally, the incorporation of intelligent systems into smart grid architecture is critical in solving both energy distribution and optimization. The application of multi-agent systems, for example, increases the flexibility of smart grids in their operation by facilitating decentralized decision-making processes that can be responsive to changing energy demands and system conditions. These systems not only enhance the efficiency and reliability of the energy networks but also play a critical role in anticipating energy needs and balancing consumption patterns. With progress in the area, concerted efforts of researchers and practitioners in developing new solutions will be crucial in actualizing an adaptable and sustainable energy system. Thus, comparative evaluation of different approaches to integrate intelligence in smart grids is a critical component of unleashing their potential to enhance energy efficiency and customer satisfaction (Catterson et al., 2010).

1.6. FUTURE TRENDS

The future for smart grids is becoming increasingly characterized by the integration of newer technologies that have the potential to make energy more efficient and sustainable. As society transitions to a low-carbon energy system, smart grids take center stage in this process, facilitating essential functions like two-way communication between producers and consumers of energy. Past research is interested in these technologies, noting that improved management of intermittent supply as well as system resilience are important. Simple definitions and theoretical models of smart grids remain unestablished, however, and future work will have to close this gap. As institutions and industries collaborate to make this transition happen, an intersitutional understanding of these basic ideas is necessary to advance the discussion around smart grid technologies (Axon et al., 2014).

Figure 1.14. Future of smart grids.

Source: https://www.azocleantech.com/image-handler/ts/20250312105227/ri/950/src/images/Article_Images/ImageForArticle_1954_17417911454567121.jpg.

Beyond the technology, socio-economic drivers behind the deployment of smart grids are profound. Political and financial backing by institutions such as the European Union and the UK government is a sign of collective agreement regarding the promise of smart grids as job-creating game-changers, resulting in reduced energy costs. Furthermore, smart grids are as much dependent on technology as on public acceptance of the same. Data safety, privacy, and security concerns of relevance pose the significant issues to be overcome. These multifaceted fears necessitate a collaborative approach among the industry actors, regulators, and consumers beyond merely the technological capacity to encompass the general implications of smart grid deployment (Axon et al., 2014). For the future, scenario techniques provide a helpful method to predict and anticipate evolving interactions among various participants in the smart grid system. By incorporating economic, regulatory, and socio-technical factors, such scenarios can provide useful insights regarding likely paths for the development of smart grids past attempts at implementing such methods have been more toward policy formulation or economic growth; however, an analysis of greater depth must consider interdependent variables such as supply and demand dynamics, regulatory regimes, and emerging technologies. The embracement of a systems thinking mindset will be essential for policymakers and stakeholders as they continue to navigate the complex sphere of smart grid deployment, that energy distribution in the future is sensitive to sustainability goals and meets the demands of different stakeholders (Axon et al., 2014) (Figure 1.14).

1.6.1. Integration of Renewable Energy Sources

The transition to a sustainable energy system mainly relies on the integration of renewable energy sources (RES) into smart grids. Since communities tend to place more and more value on reducing carbon emissions, smart grids appear to be core infrastructure capable of coping with the fluctuating nature of renewable energy production. This integration not only calls for advanced technological frameworks but also demands accurate understanding of operational as well as theoretical terms in the energy industry. Public tolerance, regulatory frameworks, and between-industry cooperation are at the core of this transition. Multiple stakeholders, including consumers and policymakers, are encouraged by the potential economic benefits and job creation that exist within the creation of smart grids. However, issues such as cybersecurity and data privacy are important in addressing in order to enable mass deployment of these technologies, as evident in expectations during smart grid integration discourse (Axon et al., 2014). The technical issues of integrating renewable energy into smart grids cannot be overstated. Of particular note is the intermittency nature of solar and wind resources, which creates challenges that need steady energy storage systems (ESS). These systems are tasked with balancing supply and demand and consequently enhancing the reliability of the grid. Furthermore, the control and communication role of both RES and ESS result in a more complex grid architecture, which inadvertently makes it more susceptible to cybersecurity breaches. Research indicates a growing need to optimize the integration of RES within traditional power systems, with methods like OPF being utilized along with ESS scheduling software. Two-way optimization ensures efficient energy management while addressing threats from the transition towards a more dynamic grid relationship (Li et al., 2023). The incorporation of machine learning techniques, and for this scenario, reinforcement learning (RL), offers promising avenues for optimizing the incorporation of renewable energy into smart grids. By the use of real-time data, RL algorithms can reduce the interactions among system elements to optimize system efficiency. These algorithms not only address the scheduling challenges of RES variability but also help in the practice of cybersecurity, making them efficient tools in modern grid management. Real-world simulation studies from actual data provide ample information regarding the efficacy and limitations of such machine learning algorithms. As the energy sector keeps evolving, understanding the interplay among RES, ESS, and sophisticated optimization tools becomes imperative for achieving the highest possible effectiveness from smart grids, ultimately leading the way toward a greener energy future (Li et al., 2023) (Figure 1.15).

Figure 1.15. Renewable integration - Energy system.

Source: https://iea.imgix.net/ff09f6fa-b39c-4eaa-9eb4-7cba813e36f9/shutterstock_2425853399.jpg?auto=compress%2Cformat&fit=min&h=630&q=80&rect=0%2C400%2C6000%2C3377&w=1200.

1.6.2. Role of Artificial Intelligence (AI) and Big Data

Integration of AI and Big Data in smart grids has transformed the energy system operation to be more efficient and responsive. AI can handle vast amounts of data generated by consumers and utilities, allowing for predictive analytics that optimize energy consumption and distribution. For instance, machine learning approaches can identify patterns in consumption, allowing utility firms to make more accurate demand predictions. Consequently, this enhanced predictive capability is reflected in better resource allocation and energy efficiencies, breaking the bounds of conventional methods overwhelmed by the complexity of big data. The importance of the innovation is emphasized through the applications described in research, where big data analysis is at the center of smart grid functions, including microgrid control and utilization of renewable energy resources, to emphasize how AI and data-driven insights enable pre-emptive management of electricity networks (Gharehpetian et al., 2019). Also, the role of Big Data and AI extends to demand response systems, facilitating increased consumer involvement and engagement in energy management. The smart grid smart homes and businesses of today function more as prosumers, generating their own power while, by load balancing, decreasing peak demand stress. Integration of Information and Communication Technologies (ICT) facilitates these interactions, allowing for real-time monitoring and message exchange between consumers and energy companies. This interactive grid utilizes big data analysis to inform prosumers about their energy consumption patterns and promote energy-efficient behavior via value-based initiatives. These innovations not only promote more energy-efficient behaviors but also enhance grid stability, representing the necessary interdependence between technology and community participation in energy use (Bonetto et al., 2017). Lastly, the combination of Big Data and AI in smart grid systems is an integrated approach to handling energy whereby the innovation potential continues to grow. As IoT-based devices become more prevalent and cloud computing is even more accessible, the options for leveraging data in smart grids expand. Big data analytics helps decision-makers navigate various problems, including the identification of energy theft and the integration of distributed energy resources. In the future, attention to these technologies will be most likely to support the transition towards more resilient and adaptive energy infrastructures to cope with the demands of current society. With the evolving world, the promise of information privacy and cybersecurity will be the overriding aspect since stakeholders will have to comprehend the added complexity introduced by such innovative technologies further emphasizing the pivotal role of AI and Big Data in defining the future of energy systems (Gharehpetian et al., 2019).

1.6.3. Potential for Consumer Engagement and Empowerment

The evolution of smart grids offers an unmatched degree of consumer engagement and empowerment, fundamentally transforming the nature of energy consumption and management. By becoming active and playing the role of prosumers, consumers gain control over producing energy, storing it, and even distributing it to their communities. This transformation is especially noted in the creation of peer-to-peer (P2P) electricity markets, through which prosumers are directly able to trade electrical energy and investments. Such markets create a consumer-oriented environment through which individuals can choose how they obtain their energy. Therefore, consumers' involvement not only enhances their agency but also pushes traditional energy systems to transform in line with this new paradigm, fostering a more sustainable and efficient energy system as envisaged in recent works on future designs for smart grids (Baroche et al., 2018). Moreover, smart meter installation also plays a significant role in enhancing consumer empowerment through instant feedback on consumption. The Smart Meter Implementation Programme (SMIP) of the United Kingdom is one prime example of such a program, which aims to have smart meters and in-home display units installed in households. By enabling consumers to observe their consumption of energy, the SMIP fosters decision-making on energy usage habits that ultimately encourages increased conservation and savings. However, their deployment has been plagued by issues with distributional, procedural, and recognition justice. By using an energy justice frame to analyze these issues, there is potential to redress current injustices so that everyone benefits equally from better smart grid technology (Brockway et al., 2018). As smart grids evolve, consumer empowerment and participation have ceased to be a theory; they manifest in realistic strategies that are encouraging higher levels of participation and collaboration among customers. Community energy projects, particularly involving prosumers, are activated as the base of cooperative energy management and sustainability. This culture of collaboration can provide the basis for creative design of operations that keep communities in charge of their energy assets, thus fostering an enhanced degree of responsibility and ownership among customers. By encouraging such mutual undertakings alongside enabling policies and designs, the transition to responsive and resilient energy systems can be expedited. Lastly, when consumers are their own owners in smart grids, the connection between individual choices and collective implications becomes stronger, paving the way toward a more integrated and democratized energy future (Baroche et al., 2018).

SUMMARY

Having concluded the research into smart grids, it is imperative to highlight their revolutionary potential in modern energy scenarios. Smart grids are a significant innovation in the distribution and management of electricity using advanced technologies to enhance operating efficiency, reliability, and sustainability. The integration of renewable energy sources, along with smart energy management systems, has the potential to transform how electricity is both generated and consumed. As explained, new technologies such as energy storage systems and distributed energy generation are significant enablers in making this possible.

Not only do these advancements eliminate dependence on fossil fuels but they contribute immensely to the reduction of carbon emissions. Therefore, the meaning of smart grids extends far beyond utility enhancements and ushers in a new era in the generation of power focusing on energy conservation and environmental protection. In addition, the merge of smart grids with new technologies such as machine learning cannot be overlooked. Machine Learning (ML) provides real-time insights into energy consumption patterns, enabling predictive maintenance and optimizing energy supply. The potential of the future for lowering efficiency losses and improving system reliability using ML is a powerful argument for integrating it into smart grid systems. As put forth, such smart systems enable real-time data analysis, leading to more responsive and dynamic energy management plans. This harmonization of smart grids with ML technologies ultimately results in a more robust energy infrastructure that can face the challenges of volatile energy needs and increased penetration of renewable energy sources. Evidently, the future of energy management is very much intertwined with the innovation brought about by smart grids and associated technologies. In the future, continued research into smart grid technology can further promote opportunities in energy management and alleviate the inherent complexities of traditional energy systems. The study underscores the need for additional research and intersector collaboration so that maximum benefits can be derived from smart grids. With smart energy management systems, it is not hard to envision a future where energy security, environmental sustainability, and economic viability become collectively compatible. In line with this vision is research that points to the necessity of intersectoral collaboration in speeding up the adoption of innovative energy solutions. The way to sustainable energy systems will definitely involve embracing these innovations and making collective commitments to the transformative potential of smart grids, ultimately paving the way to a more efficient and green future (Husseini et al., 2024; Lewandowski, 2025).

The concept of smart grids is a revolutionary leap in energy management that integrates advanced technologies with existing electrical infrastructure. Characterized by shared communication between the consumers and providers, smart grids facilitate efficient consumption and distribution of energy. Some of its fundamental characteristics are real-time observation, decentralized energy resource integration, and demand response. Not only do these technologies ensure enhanced reliability and efficiency of the energy systems, but they also significantly contribute to a low-carbon energy future. But most of the key characteristics of smart grids are still defined in an inconsistent manner, and greater agreement is required in theoretical as well as applied terms. According to research, disagreements regarding the definitions and functions of smart grids can influence the trajectory of their deployment and integration into the current energy grid (Axon et al., 2014). Furthermore, political and financial assistance for smart grid initiatives demonstrates their worth in relation to energy sustainability. Governments and industries recognize that the deployment of smart grids can lead to reduced greenhouse gas emissions, job creation, and lower energy production and distribution expenses. Despite such benefits, the successful application of smart grids will largely depend on public acceptance of the technologies and risks involved, i.e., data privacy and cyber risks. Stakeholder dialogue between industries, regulators, and consumers would be crucial to manage such multifaceted aspects while determining acceptability and feasibility of smart grid innovations. Lastly, these elements in isolation depict the comprehensive setting in which smart grids will operate as they demonstrate the interconnectedness of social, economic, and technological elements (Axon et al., 2014). Together with operational schemes and popularity, smart grid plans must incorporate a sequence of sophisticated energy management strategies (EMSs) into their planning. With nations transitioning to decentralized and renewable sources of energy, the role of EMSs becomes even more significant, particularly in applications for hydrogen microgrids and other emerging technologies. The use of advanced control techniques, such as Model Predictive Control, can coordinate energy supply and demand in an optimal manner, thereby improving the performance of the system. Experimental studies of EMSs indicated significant differences in performance under different control policies, which could influence the efficiency of smart grid applications. The findings based on these experimental evaluations highlight the importance of enhancing energy management techniques in smart grids as they directly influence the stability and resilience of next-generation energy systems

REFERENCES

1. Abrhám, J., Britchenko, I., Garškaitė-Milvydienė, K., Jankovic, S., et al. (2018). Energy security issues in contemporary Europe. https://core.ac.uk/download/322492552.pdf (accessed on 30th September 2025).
2. Accenture Technology, Adams, A., Alperovitch, D., Anastasi, M., Anderson, R., Bilge, L., et al. (2018). Avoiding the internet of insecure industrial things. https://core.ac.uk/download/141198917.pdf (accessed on 30th September 2025).
3. Aghahadi, M., Berizzi, A., Bosisio, A., Forciniti, G., et al. (2024). Digitalization processes in distribution grids: A comprehensive review of strategies and challenges. https://core.ac.uk/download/614056229.pdf (accessed on 30th September 2025).
4. Agupugo, C. P., Tochukwu, M. F. C., Ogunmoye, K. A., Mosha, A. S., & Chuku, C. P. (2025). Energy sustainability, AI, and wireless electricity generation. *International Journal of Future Engineering Innovations*. https://www.semanticscholar.org/paper/3e631a9fc1553886498b64e43c15d519f269809f (accessed on 30th September 2025).
5. Ahmed, M., Kanago, J. A., Hemalatha, J. N., & Srivani, S. G. (2025). Advanced metering infrastructure survey on security: Vulnerabilities, attacks, and countermeasures for a 5kW PV microgrid. *2025 Emerging Technologies for Intelligent Systems (ETIS)*. https://www.semanticscholar.org/paper/0565a3628d35985480be773a99cbda1950fb637f (accessed on 30th September 2025).
6. Akinc, H., & Mirosevic, M. (2014). *Enlargement and Integration Action Activity Workshop on Costs, Benefits and Impact Assessment of Smart Grids for Europe and Beyond*. https://core.ac.uk/download/38628116.pdf (accessed on 30th September 2025).
7. Alecu, E., Colta, A., Fulli, G., Gangale, F., Giordano, V., Maschio, I., Mengolini, A. M., et al. (2011). *Smart Grid Projects in Europe– Lessons Learned and Current Developments*. https://core.ac.uk/download/38625657.pdf (accessed on 30th September 2025).
8. Anastopoulou, S., Holmes, W., Mavrikis, M., Schaumburg, H., et al. (2018). Technology-enhanced personalised learning: Untangling the evidence. https://core.ac.uk/download/161526173.pdf (accessed on 30th September 2025).
9. Axon, C., Balta-Ozkan, N., Cipcigan, L., Connor, P., et al. (2014). Scenarios for the development of smart grids in the UK: Literature review. https://core.ac.uk/download/155776124.pdf (accessed on 30th September 2025).

10. Bao, Y., Feng, W., Jin, M., Liu, X., et al. (2018). A review of microgrid development in the United States–A decade of progress on policies, demonstrations, controls, and software tools. https://core.ac.uk/download/323310623.pdf (accessed on 30th September 2025).

11. Baroche, T., Moret, F., Pinson, P., Soares, T., et al. (2018). Peer-to-peer and community-based markets: A comprehensive review. http://arxiv.org/abs/1810.09859 (accessed on 30th September 2025).

12. Barros, P., Agupugo, C. P., Ejichukwu, E., Hayden, M. D., & Ogunmoye, K. A. (2025). Smart grid security: Safeguarding sustainable energy systems from cyber threats. *World Journal of Advanced Research and Reviews*. https://www.semanticscholar.org/paper/ddfd828326f000eb0a98e753e835771c0d551db4 (accessed on 30th September 2025).

13. Berville, C., Croitoru, C., & Bode, F. (2025). Life cycle analysis in the context of smart cities. *E3S Web of Conferences*. https://www.semanticscholar.org/paper/6aa8fcdbbd30404e295c7cc85bb1fcf9c4b525f7 (accessed on 30th September 2025).

14. Bonetto, R., & Rossi, M. (2017). Smart grid for the smart city. https://core.ac.uk/download/84922332.pdf (accessed on 30th September 2025).

15. Bordons Alba, C., Guerra Macho, J. J., Rosa Iglesias, M. F., Valverde Isorna, A., et al. (2016). Energy management strategies in hydrogen smart-grids: A laboratory experience. https://core.ac.uk/download/288003657.pdf (accessed on 30th September 2025).

16. Borghetti, G., Menga, G., Procaccianti, G., Kok, J. K., Ardito, L., Morisio, M., Mo, Y., et al. (2013). Smart grid technologies in Europe: An overview. https://core.ac.uk/download/pdf/11431923.pdf (accessed on 30th September 2025).

17. Bossart, S., & Giordano, V. (2012). Assessing smart grid benefits and impacts: EU and U.S. initiatives. https://core.ac.uk/download/38627650.pdf (accessed on 30th September 2025).

18. Brockway, P., Foxon, T., Miller, J., Sorrell, S., et al. (2018). The United Kingdom smart meter rollout through an energy justice lens. https://core.ac.uk/download/217284947.pdf (accessed on 30th September 2025).

19. Catterson, V., Davidson, E., & McArthur, S. (2010). The role of intelligent systems in delivering the smart grid. https://core.ac.uk/download/9037661.pdf (accessed on 30th September 2025).

20. Chin, W. H., Efthymiou, C., Fan, Z., Gormus, A., et al. (2011). Smart grid communications: Overview of research challenges, solutions, and standardization activities. http://arxiv.org/abs/1112.3516 (accessed on 30th September 2025).

21. Ekins, P., & Watson, J. (2014). UK energy in a global context: Synthesis report. https://core.ac.uk/download/20391336.pdf (accessed on 30th September 2025).
22. El Husseini, F., Noura, H. N., & Vernier, F. (2024). Machine-learning-based smart energy management systems: A review. *2024 International Wireless Communications and Mobile Computing (IWCMC)*. https://www.semanticscholar.org/paper/2d31bf6cc9d9c47c16a45bf62c948392a09b0cc5 (accessed on 30th September 2025).
23. Fayaz, M., Lajis, A., Nasir, H., Shah, A., et al. (2019). A review on energy consumption optimization techniques in IoT-based smart building environments. https://core.ac.uk/download/300473894.pdf (accessed on 30th September 2025).
24. Gharehpetian, G. B., Marzband, M., Moradi, J., Nafisi, H., et al. (2019). Attributes of big data analytics for data-driven decision making in cyber-physical power systems. https://core.ac.uk/download/305121394.pdf (accessed on 30th September 2025).
25. Ginter, A., Ten, C. W., Vasilakos, A. V., Yamashita, K., et al. (2018). Impact assessment of hypothesized cyberattacks on interconnected bulk power systems. http://arxiv.org/abs/1801.01048 (accessed on 30th September 2025).
26. Gotay Sardiñas, J., Jimenez, V. A., Rodriguez, S. A., Will, J., et al. (2018). Adjustment of model parameters to estimate distribution transformers' remaining lifespan. https://core.ac.uk/download/224998148.pdf (accessed on 30th September 2025).
27. Hauser, C. H., Lorenz, D. H., Fudenberg, D., Santacana, E., Kuipers, F., Wu, F. F., Xue, G., et al. (2018). Multi-layer cyber-physical security and resilience for smart grid. http://arxiv.org/abs/1810.00282 (accessed on 30th September 2025).
28. Ivanova, T., Petrenko, K., Skorobogatova, N., Tymoshenko, N., & Mykytiuk, O. (2025). Investment activity in renewable energy as a component of economic security. *Theoretical and Applied Issues of Economics*. https://www.semanticscholar.org/paper/a0f07ae5c3e976110143e86bb40f7a9e5ef3c644 (accessed on 30th September 2025).
29. Ivanykina, T. V., Shkrabtak, N. V., & Praskova, J. (2025). Analysis of the impact of emissions from the Blagoveshchensk asphalt concrete plant "Nikko Cbd-100" JSC "Asphalt" on the atmospheric air and measures to protect it. *International Scientific and Practical Conference "Smart Cities and Sustainable Development of Regions"*

(SMARTGREENS 2024). https://www.semanticscholar.org/paper/a0b6e2ae6cf6b1e692ddb1faccfb31df8b6d7009 (accessed on 30th September 2025).

30. Izuka, U., Izuka, U., Lottu, O. A., Lottu, A., et al. (2023). Solar energy adaptation and efficiency across diverse Nigerian and global climates: A review of technological advancement. https://core.ac.uk/download/596254248.pdf (accessed on 30th September 2025).
31. Jørgensen, B., Gunasekaran, S., & Ma, Z. (2025). Impact of EU laws on AI adoption in smart grids: A review of regulatory barriers, technological challenges, and stakeholder benefits. *Energies*. https://www.semanticscholar.org/paper/d88076289bc92e535440c338e09afa4b44b0c984 (accessed on 30th September 2025).
32. Lewandowski, K. (2025). Modern energy sources for sustainable buildings: Innovations and energy efficiency in green construction. *Energies*. https://www.semanticscholar.org/paper/8863d9c375057e2b90f34d093fa2e40df3e0779f (accessed on 30th September 2025).
33. Li, Y. (2023). Achieving high renewable energy integration in smart grids with machine learning. https://core.ac.uk/download/588293905.pdf (accessed on 30th September 2025).
34. Liu, F., Cui, Y., Masouros, C., Xu, J., Han, T. X., Eldar, Y. C., & Buzzi, S. (2022). Integrated sensing and communications: Toward dual-functional wireless networks for 6G and beyond. *IEEE Journal on Selected Areas in Communications, 40*, 1728–1767. https://doi.org/10.1109/jsac.2022.3156632 (accessed on 30th September 2025).
35. McIlvennie, C., Pritoni, M., & Sanguinetti, A. (2020). Of impacts, agents, and functions: An interdisciplinary meta-review of smart home energy management systems research. https://core.ac.uk/download/323062198.pdf (accessed on 30th September 2025).
36. Mondéjar, M. E., Avtar, R., Baños Díaz, H. L., Dubey, R. K., Esteban, J., Gómez-Morales, A., Hallam, B., et al. (2021). Digitalization to achieve sustainable development goals: Steps towards a smart green planet. *The Science of The Total Environment, 794*, 148539–148539. https://doi.org/10.1016/j.scitotenv.2021.148539 (accessed on 30th September 2025).
37. Motlagh, N. H., Mohammadrezaei, M., Hunt, J. D., & Zakeri, B. (2020). Internet of Things (IoT) and the energy sector. *Energies, 13*, 494–494. https://doi.org/10.3390/en13020494 (accessed on 30th September 2025).
38. Munonye, W. C. (2025). Circular economy meets smart energy grids: Designing systems for resource optimization and carbon reduction.

Frontiers in Sustainability. https://www.semanticscholar.org/paper/8fee3a36f9caed22e4d55c89ac75e8b584e203c3 (accessed on 30th September 2025).

39. Rodríguez Jara, E. Á., Romero Rodríguez, L., Salmerón Lissén, J. M., Sánchez Ramos, J., et al. (2016). Analysis of the economic feasibility and reduction of a building's energy consumption and emissions when integrating hybrid solar thermal/PV/micro-CHP systems. https://core.ac.uk/download/288003155.pdf (accessed on 30th September 2025).
40. Shafique, K., Khawaja, B. A., Sabir, F., Qazi, S., & Mustaqim, M. (2020). Internet of Things (IoT) for next-generation smart systems: A review of current challenges, future trends and prospects for emerging 5G-IoT scenarios. *IEEE Access, 8*, 23022–23040. https://doi.org/10.1109/access.2020.2970118 (accessed on 30th September 2025).
41. Wang, Y., Su, Z., Zhang, N., Xing, R., Liu, D., Luan, T. H., & Shen, X. (2022). A survey on metaverse: Fundamentals, security, and privacy. *IEEE Communications Surveys & Tutorials, 25*, 319–352. https://doi.org/10.1109/comst.2022.3202047 (accessed on 30th September 2025).
42. Zeng, X., Li, M., Abd El-Hady, D., Alshitari, W., Al-Bogami, A. S., Lü, J., & Amine, K. (2019). Commercialization of lithium battery technologies for electric vehicles. *Advanced Energy Materials, 9*. https://doi.org/10.1002/aenm.201900161 (accessed on 30th September 2025).

CHAPTER 2

SMART GRID ARCHITECTURE

Contents

INTRODUCTION

The transformation of the electric grid from a traditional, centralized system to a decentralized and dynamic system is captured by the emergence of smart grid architecture. The necessity to adapt to the sophisticated energy landscape of the present with the adoption of renewables, growing consumer load, and aggressive energy conservation requirements is the impetus for the transformation. Traditional electricity systems have proven to be inadequate to address these problems, necessitating an infrastructure capable of supporting bi-directional energy transfer and communication of data in real time. Based on advanced Information and Communication Technologies (ICT), smart grids provide enhanced reliability, efficiency, and sustainability in the management of energy, which eventually leads to a more reliable electricity system that is able to support traditional as well as renewable sources of energy (Chin et al., 2011).

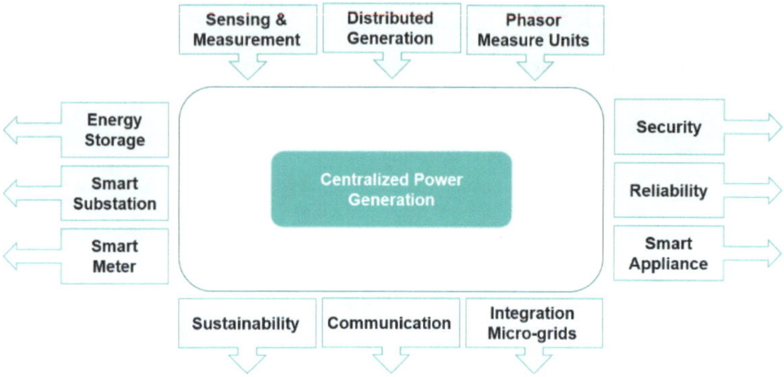

Figure 2.1. Architecture of smart grid system.

Source: https://stellarix.com/wp-content/uploads/2024/06/Architecture-of-smart-grid-system.png.

One of the basic attributes of smart grid architecture is that it depends on near-real-time communication between various components of the grid, including the generation, transmission, and distribution systems. Not only does it encourage energy optimization, but it also makes the grid more responsive to dynamic demands and supply levels. Moreover, the integration of smart metering technologies plays a crucial role in facilitating data analytics to maximize operational efficiency and developing feedback systems that empower consumers. This advanced communication infrastructure, though, presents critical concerns, especially in the areas of system security and privacy. It is imperative to develop solutions to overcome these concerns to establish an interoperable and future-proofed smart grid that can effectively manage energy resources while ensuring user data protection (Chin et al., 2011). In addition to the technology advancements, regulatory frameworks have a huge influence on determining the future of smart grid architecture. As countries are speeding up to become energy-independent and sustainable, governments are compelled to balance the impacts of regulations on the growth of smart grids. A comprehensive regulatory framework will not only facilitate the investment in innovative grid infrastructure, but also ensure the transition to smart grids is equitable and just. Harmonization of communication protocols and standards, which is common in the European Union, would be a best-practice model throughout the globe. With intersectoral cooperation, the smart grid can achieve its maximum potential, ultimately becoming a secure, efficient, and green energy environment (Borghetti et al., 2013) (Figure 2.1).

Definition of Smart Grid

The smart grid is a significant leap in the planning and management of electricity supply systems from the traditional ways to a sophisticated system that can handle the complexity of modern-day energy demands. Essentially, the smart grid merges cutting-edge digital communication technology and artificial intelligence to ensure an efficient and responsive distribution of energy. This progress facilitates two-way communication between consumers and utilities and enables real-time data transfer, maximizing electricity use and reliability. The adaptive nature of the smart grid, not only with the ability to empower consumers to monitor and adjust energy consumption but also to integrate renewable resources, addresses the sustainability imperatives of today's power grids. Technology integration of multi-agent systems is one of the most important mechanisms facilitating smart grid functionality. Multi-agent systems function through autonomous agents that can communicate and interact cooperatively without being centrally controlled in an effort to perform various tasks in the grid. In a way, multi-agent systems enable components to function with loose coupling, thereby enhancing flexibility and scalability within smart grid structures. Such configuration is particularly critical in managing the dynamic sides of power distribution, as the agents are in a better position to adapt to changing conditions—such as variable production from renewables or fluctuating demand—than traditional centralized models. As the above indicates, "the transformation of energy networks from passive to active systems requires the embedding of intelligence within the network" (Ai et al., 2012), which identifies how multi-agent technology contributes to the efficiency and effectiveness of operation of smart grids. Along with technological advancements, regulatory frameworks also need to play an important part in deciding the development of smart grids. The emerging electricity system must converge with contemporary energy policy, environmental regulations, and economic drives. For instance, the current necessity to integrate alternative energy as well as react to energy-saving programs necessitates a robust system of governance that fosters innovation while ensuring security and reliability. The "older electricity network infrastructure has been found to be insufficient" to serve these purposes, emphasizing the imperative of introducing smart grids (Borghetti et al., 2013). Thus, rigorous analysis of not only the technological aspects but also the regulatory environment is imperative to fully comprehend the definition and implication of smart grids, ultimately leading to an efficient and green energy future (Figure 2.2).

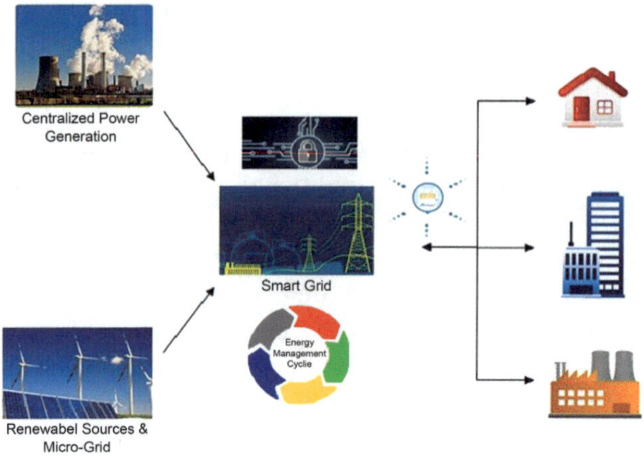

Figure 2.2. Working concept of smart grid.

Source: https://ars.els-cdn.com/content/image/1-s2.0-S2090447920301064-gr1_lrg.jpg.

Importance of Smart Grid in Modern Energy Systems

Energy system development in the face of challenges of the present times necessitates shifting from traditional infrastructures to smart grids. The traditional power infrastructure for electricity has proven inadequate in managing such complexities as the integration of alternative sources of power, increased electricity demand, and requirement for energy-saving initiatives. This shortcoming is one major reason why smart grids are necessary because they employ advanced technology in improving and simplifying electricity distribution. By utilizing Information and Communication Technologies (ICT) integration, smart grids improve the reliability and efficiency of energy systems, making it easy to include renewable energy sources in the grid. As the world's energy scenario continues to change, it is essential to move towards smart grids to ensure sustainability, improve system resilience, and avoid shortages of energy, thereby meeting the dynamic demands of the consumer and the environment (Borghetti et al., 2013). In addition to increasing efficiency, smart grids also improve security and robustness in energy networks. These characteristics are critical during an era characterized by increased cyber-attacks and the vulnerability of critical infrastructure. Smart grid architecture integrates resilient security elements designed to maintain operations and data integrity, which is critical to guaranteeing users' and stakeholders' trust. Game and decision theory application is a vital tool in mapping the interactions among various elements of a grid as well as approximating the threats that may be exploited by attackers. By applying layered security, smart grids are able to effectively predict as well as counter security attacks such that energy distribution is not affected even under the threat of attack. This proactive security strategy provides a robust platform for energy management in which diversified energy resources are brought together while the integrity of the grid is protected against any malicious behavior (Hauser et al., 2018). The intelligent architecture of the smart grid also enables better demand response, which strongly influences consumer behavior and energy usage patterns. By providing real-time data and intelligence, smart grids allow consumers to make rational decisions regarding their energy use, raising the level of participation in energy management. This entails not only maximizing the consumption of energy but also promoting energy-saving habits and devices among consumers. Smart grids further facilitate dynamic pricing systems, which can lead to consumer savings while promoting energy usage during off-peak times. This shift in usage patterns contributes to more stable energy prices and reduces the burden on the grid, ultimately enhancing the reliability of the energy supply. The synergy thereby achieved between the consumers and the grid is a crucial component of modern energy systems, and the importance of smart grids is thus becoming increasingly beyond doubt in striving for a sustainable and secure energy future.

Overview of Smart Grid Architecture

The Smart Grid (SG) architecture represents a paradigm shift away from traditional electrical systems, which facilitate the intricate two-way information flow between suppliers and consumers. The paradigm shift addresses such relevant issues as energy wastage, reliability, and growing electricity demand. Compared to traditional power grids, which operate primarily on the basis of a one-way flow of electricity, the Smart Grid utilizes advanced technologies and appliances that can facilitate real-time communication and data exchange. These developments support more efficient monitoring, analysis, and control over electricity generation, transmission, and usage processes. The incorporation of IoT technologies, emphasized in SG models, further enhances the architecture by deploying sensors, actuators, and intelligent meters along the energy infrastructure, encouraging connectivity and automation essential to modern energy management (Copeland et al., 2019). For the Smart Grid infrastructure, Information and Communication Technologies (ICT) plays an unavoidable role in delivering efficient and reliable network operations.

ICT integration is an enabler for deploying management and optimization solutions in various domains to overcome the limitations of the conventional electrical infrastructures. Smart Grids are designed to include alternative energy sources and adapt to shifting electricity demand, increasingly a valuable role in the era of green energy initiatives. Moreover, Smart Grids' ability to include renewable energy sources such as wind and solar power represents a significant advancement for the energy infrastructure. It is necessary to explore the regulatory frameworks that surround Smart Grid development, particularly in the context of environments such as that of the European Union, to examine how such frameworks would be capable of adapting and facilitating overall innovation in smart grid operations (Borghetti et al., 2013). An important feature of Smart Grid design is the focus on security and resilience against evolving cyber threats. With increasing digitalization of power grids, the potential dangers associated with device interconnectivity need robust cybersecurity measures. Proper security protocols need to be put in place not just to protect sensitive data but also to maintain supply continuity. Other research avenues under the SG architecture concentrate on bolstering these security measures as well as overcoming existing technical bottlenecks. Technology developers, regulatory bodies, and electrical engineers working in interdisciplinary teams will be vital to assist in creating the Smart Grids capabilities. These elements combined create a strong architecture that can anticipate ahead and adapt to emerging energy requirements and adopt novel technologies that improve service provision and consumer empowerment (Copeland et al., 2019) (Figure 2.3).

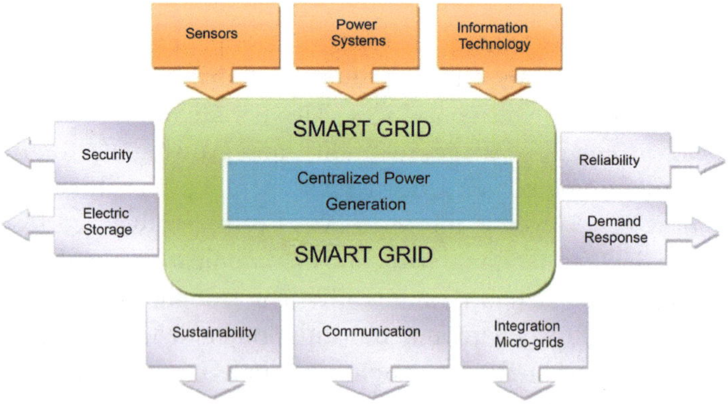

Figure 2.3. Overview of smart grid architecture.

Source: https://ars.els-cdn.com/content/image/1-s2.0-S2090447920301064-gr2_lrg.jpg.

2.1. COMPONENTS OF SMART GRID

One of the primary components of a smart grid is having advanced metering infrastructure (AMI) included, which greatly enhances the ability to monitor and manage energy in real-time. AMI employs smart meters that provide consumers and utilities with instant access to information on energy consumption, enabling energy efficiency and demand-side management. While traditional meters only record overall electricity usage, smart meters can measure electricity consumption on almost real-time levels, with consumers able to change their behavior based on real-time information. Upgrading to AMI is imperative, particularly in addressing the inadequacies of the existing electricity network infrastructure, which is struggling to keep up with existing issues such as fluctuating energy levels and the integration of renewable energy technologies (Borghetti et al., 2013). Therefore, AMI is a fundamental component of smart grid architecture because it provides consumers and utilities with the means to optimize energy consumption and increase the reliability of the grid overall. A second very important element of smart grid architecture is the assimilation of renewable energy sources and distributed generation. As society increasingly focuses on sustainability, the need to integrate alternatives such as solar, wind, and hydroelectric power becomes inevitable. The smart grid enables such generation sources to be directly interconnected to the power delivery network, enabling a model of decentralized power with increased resistance to outages. The model provides diversity to the power supply and assists in meeting environmental objectives by reducing the consumption of fossil fuels. Furthermore, the integration of such sources requires sophisticated management systems that can regulate the supply and demand as well as stability of the grid. The layered nature of smart grid security, highlighted in recent studies, reiterates the requirement for secure protocols to ensure the safe operation of the grid's distributed resources (Hauser et al., 2018). Hence, renewable integration is a critical factor that serves as a supporting element in the smart grids innovative and green system. Last but not least, the role of advanced communication technologies cannot be overemphasized in the components of a smart grid. These technologies facilitate seamless data transfer between smart grid devices like sensors, equipment, and control systems. Information and Communication Technologies (ICT) enhance the efficiency and flexibility of the energy networks, enabling utilities to monitor system performance, detect outages, and deal with emergencies on time. This interconnectedness also allows for greater security and resilience, as in new game-theoretic techniques that model the interaction of system elements, projecting possible vulnerabilities of the grid (Hauser et al., 2018). Communication is also needed in order to implement demand response programs that incentivize customers to reduce or shift energy use during peak hours. Thus, high-tech communication technology is the backbone of the smart grid, enabling it to solve today's energy problems while ensuring user satisfaction and engagement (Figure 2.4).

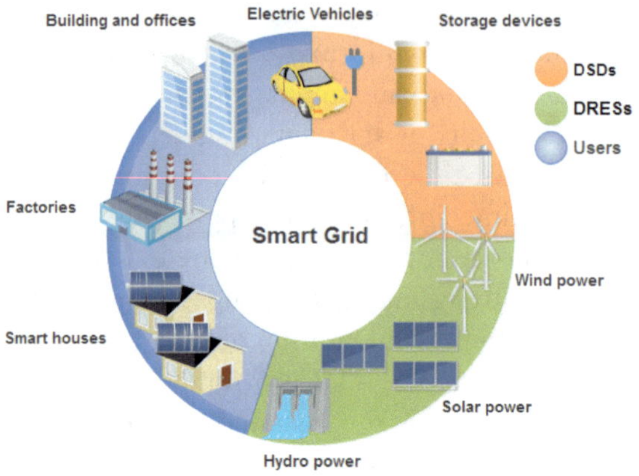

Figure 2.4. Smart grid components.

Source: https://www.researchgate.net/profile/S-Harous/publication/339267794/figure/fig1/AS:893967252082688@1590149810480/Smart-Grid-Components.ppm.

2.1.1. Smart Meters

The addition of smart meters in the structure of the smart grid is an important move toward improved management of energy. These smart metering systems provide real-time data regarding energy consumption patterns so that utilities can optimize energy distribution and reduce wastage. Smart meters replace traditional meters by embracing technologies such as cloud analytics and distributed energy resource management that not only streamline operations but also enable sustainability goals. According to recent studies, the integration of smart grid technologies into existing infrastructures can potentially deliver significant carbon footprint savings and improved operational performance. To implement these technologies, however, there are sophisticated strategies required to address problems such as data management and regulatory requirements, particularly in cities where infrastructure can be more complex (Bonepalli, 2025). Besides, technical architecture for the smart meter has also been evolving with blockchain technology usage. The breakthrough significantly enhances the security and efficiency of the energy transaction because it allows peer-to-peer (P2P) transactions in virtual power plant (VPP) systems. Protection of the transaction data and facilitation of smooth interaction between the grid and consumers, blockchain-based smart meters allow real-time observation of energy and effective load management. Consumer data measurement and analysis maximize grid responsiveness while creating opportunities for decentralized energy markets. Evidenced in recent studies is the observation that smart meters' ability to operate within a blockchain setting creates measurable benefits, including higher profit margins and improved energy balance between stakeholders, thereby creating a stronger grid (Kaif et al., 2025). The future of smart meters in the smart grid framework will also continue to evolve, particularly through advancement in machine learning and artificial intelligence. The two can potentially enhance predictive analysis so that there are more precise energy management and consumer engagement schemes. With enormous amounts of real-time data collected by smart meters, utilities can predict changes in energy demand and respond to ensure grid stability. Moreover, the adaptability of smart metering systems makes them adaptable to various regimes of regulation, thereby providing a widely usable remedy for the issue of energy distribution. As energy professionals strive towards enhanced operational efficiency and sustainability, subsequent research evaluating the long-term performance and cost advantage of these technologies will be valuable (Bonepalli, 2025; Kaif et al., 2025) (Figure 2.5).

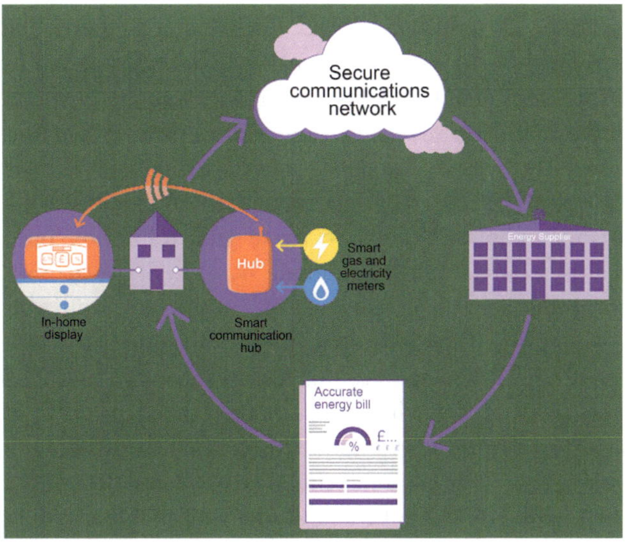

Figure 2.5. Basic flow chart of smart meters' operations.

Source: https://cdn-kabof.nitrocdn.com/nfjHFOYSVknMSLxSPIZUhvuJiK-cRMGGc/assets/images/optimized/rev-a4c25e5/2.bp.blogspot.com/-HPWH832m-1no/V1MQM8BC3sI/AAAAAAAABYY/-evCJDRf2yADNDrHAvPfSGJyp-WFTpLQewCLcB/s640/Smart+meter+process_Graphic_V2-01.png.

2.1.2. Advanced Sensors and Control Systems

The adoption of smart grid architecture represents a significant change in the management of electrical grids, particularly through the use of advanced sensors and control systems. These technologies enable the exchange of information bi-directionally, facilitating energy providers and consumers to communicate appropriately. Advanced sensing, by way of IoT devices, is a vital function for monitoring electricity usage in real-time, allowing data that is instrumental to optimizing energy distribution and utilization. Moreover, the capacity to gather and analyze great volumes of data allows it to identify trends and project energy requirements, thereby maximizing operational efficiency in the grid. Since the smart grid depends increasingly on such technologies, it engulfs conventional infrastructure by bringing in sophisticated interactivity that keeps pace with the dynamic face of contemporary energy systems, a great departure from the previous uni-directional electrical grids that were prevalent (Copeland et al., 2019). In practical uses, the role of advanced sensors extends beyond data gathering; they form part of control systems that regulate the passage of electricity along the smart grid. Sensor-driven real-time analytics, for instance, enable fast decision-making in load balancing and fault detection, all of which are necessary for grid stability. At UC Berkeley, one particular project on building a Building-to-Grid testbed demonstrated the efficacy of this technology. Through careful testing and installation of an array of devices, researchers illustrated how integrated metering and monitoring systems can optimize energy efficiency through dynamically adapting themselves to consumption patterns (Peffer et al., 2010). This scheme is a demonstration of how sensor technology advancements not only increase grid reliability but also help to transition towards a cleaner energy climate by reducing waste and increasing responsiveness towards consumer needs. The future of intelligent grid design largely relies on continued advancements in sensors and control systems, as they form the basis of sophisticated solutions to future issues. But the way to fully optimized smart grids is not without challenges. Data privacy, cybersecurity, and technological integration heterogeneity must be dealt with collaboratively between stakeholders. This also accentuates the need for ongoing research into IoT-enabled applications and frameworks that enable seamless interaction between disparate elements of the grid. Also, creating co-operation between higher education organizations, such as the UC Berkeley program, and industry players is necessary to enhance these technologies (Peffer et al., 2010). With changing patterns of energy consumption, enhancing sensor technology and connecting control systems efficiently will be of prime importance to realize the full capability of smart grids in ensuring better reliability, efficiency, and quick response to future energy needs (Copeland et al., 2019).

2.1.3. Communication Networks

The integration of communication networks into Smart Grid Architecture is a paradigm transition away from traditional power systems. Traditional grids experienced a one-way flow of information, and the ultimate outcome was inefficiencies such as wastage of energy and the inability to handle increasing consumer demand. Smart grids, however, facilitate bi-directional communication among service providers and consumers. This shift is significant to enhance reliability and security and promote more efficient energy consumption and management. As highlighted in literature today, Smart Grids rely on a combination of monitoring, analysis, and control devices distributed across the various components of the grid, which necessitates an assured communications infrastructure to manage the increased complexity effectively (Copeland et al., 2019). These advanced communication systems are the backbone and means for intelligent energy management, enabling real-time data sharing and decision-making. Internet of Things plays an important part in developing communication networks for Smart Grid systems. With IoT devices like sensors, actuators, and smart meters deployed inside them, integration of the grid elements becomes much more enhanced.

These systems support round-the-clock monitoring and data acquisition, which leads to increased efficiency and automation of energy management systems. IoT integration allows multiple network functions during the process of energy generation, transmission, and distribution. Smart device deployment requires scalable communication techniques that have the capability to efficiently communicate data, track energy loads, and control grid operations dynamically. Central to this endeavor is the provision of connectivity promising seamless interaction among diverse devices, thus addressing the difficulties brought about by the energy requirements of the current age (Copeland et al., 2019). This expansion towards IoT-enabled grids not only streamlines activities but also allows consumers actively to participate in energy management. Overcoming the hurdles involved in the embracement of effective communication networks in Smart Grids involves a tremendous amount of research and standardization. A critical aspect of this transformation is to render the different technologies and standards employed in the grid interoperable. The issues of communication in this case involve receiving data transferred securely and reliably while maintaining consumer privacy. Strategic integration of the existing networking technologies with the needs of energy management can pave the way for more integrated Smart Grid systems. Moreover, collaborative efforts in standardization, as they exist in the European context, highlight the importance of harmonized communication protocols, which will be vital in ensuring smart metering and energy distribution networks. By addressing these communications issues,

Smart Grid architectures in the future can optimize energy usage and enhance the overall resilience of the power infrastructure (Figure 2.6).

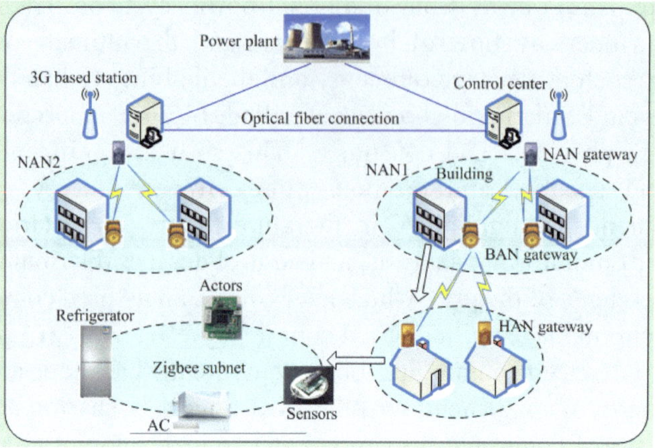

Figure 2.6. Communication architecture of smart grid.

Source: https://www.researchgate.net/profile/Jiafu-Wan/publication/264846553/figure/fig3/AS:669000623222793@1536513586538/Communication-architecture-of-smart-grid.ppm.

2.2. BENEFITS OF SMART GRID IMPLEMENTATION

Smart Grid implementation is a breakthrough step over traditional power systems, basically refining the efficiency and dependability of energy management. The most significant benefit is the facilitation of bi-directional energy flow, wherein not only central power generation but also distributed power generation due to renewable sources such as solar and wind is feasible. This shift enables energy producers, like household owners with solar panels, to feedback excess energy into the grid, thus supporting sustainability and less consumption of fossil fuels. Moreover, integration of smart technologies enables monitoring and control of power in real time, allowing utilities to tackle shifts in demand and supply in real time. These technologies lead to a more elastic grid that is capable of handling variable energy sources, ultimately yielding cheaper energy costs for consumers and improved overall system performance (Develder et al., 2010). Additionally, Smart Grid implementation significantly enhances data collection and usage, allowing for a more interactive relationship among service providers and consumers. With the help of newer communications technology, utilities are able to obtain detailed information on energy usage patterns and thereby optimize the distribution of energy, minimize outages, and efficiently handle peak demand. Consumers are also enabled to make choices about how they utilize their energy with more information, thereby leading to energy conservation. Technologies like smart meters offer immediate feedback, enabling users to understand their usage habits and expenditure on energy and thus influence better, more sustainable decisions. This shift towards demand-side management not only helps in reducing energy waste but also goes a long way in making the grid more secure by identifying potential weak points and taking measures before they become a threat (Copeland et al., 2019). Additionally, the use of Smart Grids creates the potential for future expansion of incorporation of future technologies, such as the Internet of Things (IoT). This integration provides the ability to deploy numerous intelligent devices, ranging from sensors to intelligent actuators, that are accountable for observing, analyzing, and controlling the grid. Support for numerous network functionalities along the energy generation, transmission, distribution, and consumption stages harmonizes with the IoT integration to eventually transform the operation environment. All this promotes a great level of automation, enabling potential corrections to the grid to be achieved swiftly and effectively. Consequently, Smart Grids not only provide a safe system of energy governance but also lay the grounds for technology innovation in the future, further optimizing energy systems and contributing to environmental sustainability (Copeland et al., 2019) (Figure 2.7).

Figure 2.7. Smart grid benefits.

Source: https://media.joeducation.eu/joeducation.eu/wp-content/uploads/2024/06/Smart-grids-2-EN-1024x557.jpg.webp.

2.2.1. Enhanced Energy Efficiency

The concept of Enhanced Energy Efficiency is core when it comes to Smart Grid Architecture, particularly with its efforts to combine various sources of energy into a single unified system. Enhanced energy efficiency not only refers to the maximization of energy consumption but also includes the calculated utilization of renewable resources such as solar and wind power. These assets, promising as they may be, will normally be accompanied by difficulties emanating from their cyclic and volatile nature. As observed in recent studies, energy storage systems (ESS) are important in supply and demand balancing in such a situation. Applying IoT-enabled ESS technologies complements grid reliability through real-time monitoring, predictive maintenance, and optimal energy management. With the help of smart sensors and analytics, the smart grid can dynamically manage the distribution of energy, thus improving efficiency and minimizing wastage. Additionally, synergistic interconnections among energy generation, transmission, and consumption are encouraged by integration of smart grid technologies. Enhanced Energy Efficiency is also realized by leveraging high-end communication technology that enables robust data exchange between grid components. These systems are characterized by the security controls layered in a way that ensures there is strength against disruption. It is possible to explain interactions between various components of the grid and between potential attackers and the system using a game-theoretic framework. This model, as documented in recent literature, focuses on ensuring that operational procedures are secured for each layer of the grid architecture. The combination of these technologies ensures that improved energy efficiency is not a single aim but is connected with security and resilience towards the overall robustness of the smart grid system. Moreover, real-world applications of enhanced energy efficiency by smart grid architecture demonstrate how it can be utilized to mitigate the adverse effects of traditional energy systems. Such technologies as microgrids use local energy sources together with ESS and IoT technologies, bringing increased sustainability and reduced fossil fuel reliance. Case studies indicate that implementation of the smart grid model not only makes energy distribution more efficient but also reduces the operational cost as well as increasing system resilience to environmental and cyber-attacks. But it is important to address the potential constraints on such technologies, such as costs of deployment and information overload. The exploration of such challenges is important in realizing maximum gains in greater energy efficiency within the smart grid ecosystem, towards the end goal of a sustainable future (Adebiyi et al., 2025; Hauser et al., 2018).

2.2.2. Improved Reliability and Resilience

The creation of smart grid infrastructure significantly enhances electrical system reliability and resilience by integrating advanced technologies. A primary reason for this is the incorporation of Internet of Things (IoT) capability, which enables real-time monitoring and analysis of data. With solar and wind from renewable resources becoming increasingly prominent, they introduce variability that can push the limits of traditional grid stability. IoT-based energy storage systems (ESS) are a significant solution to this issue, facilitating dynamic energy management with robust demand-supply balancing. Through data analytics, machine-to-machine communication, and smart sensors, not only is the energy flow optimized, but grid stability is also improved to support a more reliable power supply that can adapt to variability in renewable sources. Thus, the integration of IoT technologies is a highly critical element to enhance the overall reliability and resilience of the grid under the smart grid framework (Ojo et al., 2025).

Furthermore, the emergence of blockchain technology has introduced a new model for smart grid operation reliability and resilience improvement. The decentralized and indelible nature of blockchain assures secure, transparent data trading among many stakeholders, i.e., power consumers and producers. With blockchain, power flow data transactions and market transactions are recorded on an unalterable ledger, greatly reducing the possibility of fraudulent activity and data tampering. In general, the decentralized structure of blockchain eliminates one point of vulnerability that is easily accessible to cyber-attacks. With the application of consensus algorithms such as Proof of Work (PoW) or Proof of Stake (PoS), integrity of data is guaranteed, thus enhancing security and reliability. Blockchain is a novel technology that renders blockchain a breakthrough instrument for attaining more reliable smart grid systems as well as preventing vulnerabilities inherent to conventional grid designs (Seshasai et al., 2025).

Besides IoT and blockchain, the availability of advanced analytics and machine learning algorithms also improves the reliability and resilience of the smart grid. These systems can analyze vast quantities of operational data to pinpoint potential failures or disruptions in advance so that utilities can make pre-emptive choices. Through predictive maintenance practices, utilities are able to schedule repairs ahead of time with insights driven by data, thereby limiting downtime as well as improving the utilization of assets. Moreover, machine learning algorithms can also improve the accuracy of load forecasting to allow grids to respond better to changing consumer demands. Therefore, smart grid architecture not only becomes more responsive but also more able to withstand the ambiguities of energy supply and demand. The union of data analytics and the infrastructural backbone of the smart grid ultimately leads to greater reliability and resilience, offering sustainable energy opportunities for the future (Ojo et al., 2025).

2.2.3. Integration of Renewable Energy Sources

The integration of renewable energy sources into smart grid planning is a paradigm shift towards addressing the energy issues of the present. Conventional power systems are based mainly on fossil fuel sources that are centrally located and contribute greatly to environmental degradation. On the contrary, smart grids allow for a two-way flow of energy and information that supports the use of multiple renewable energy technologies like solar, wind, and hydroelectric systems. This shift not only lowers carbon emissions but also improves energy security through the diversification of the energy supply. The application of smart technologies enables immediate data collection and analysis, improving the management of renewable resources. Therefore, the shift to smart grids is significant in the future of sustainable energy while simultaneously promoting the decentralized collection and redistribution of energy, significant in addressing the boosted need for the integration of renewable energy. Moreover, the success of integrating renewable energy into the grid relies heavily on advanced communication and information technologies. Smart grids utilize Internet of Things (IoT) devices—sensors, smart meters, and actuators—to enhance monitoring and control of supply and demand sides of the power system. Such infrastructure enables improved balancing between ramping renewable generation and consumer demand, minimizing waste energy, and enhancing reliability. The capability of autonomous reaction to shifting demand and supply underlines the importance of IoT in modern energy infrastructure because it ensures an acclimatized and resilient grid.

By applying these technologies, smart grids will be in better position to integrate increasing penetration of variable renewable energy sources, scheduling energy production according to patterns of use without compromising grid security and stability, as outlined in Borghetti et al. (2013) and Copeland et al. (2019). Finally, concerns regarding the integration of renewable energy sources in smart grid architecture have to be closely investigated so that their full potential could be used. Above all, regulatory and technical guidelines need to adjust in accordance with the shifting energy market dynamics created by renewable energy sources. Policymakers play a significant role in developing conducive regulatory guidelines that enable investment in smart grid devices and renewable energy projects. Furthermore, issues such as cybersecurity, grid interoperability, and system strength necessitate ongoing research and development to battle potential deterrents. Engaging stakeholders—like utilities, governments, and consumers—during the smart grid development process is indispensable to ensure public acceptability and cooperation. Such openness can accelerate innovation and advance the building of infrastructures capable of successfully integrating renewable energy sources, and thus pave the way for a low-carbon, sustainable energy future in conformity with global environmental goals (Figure 2.8).

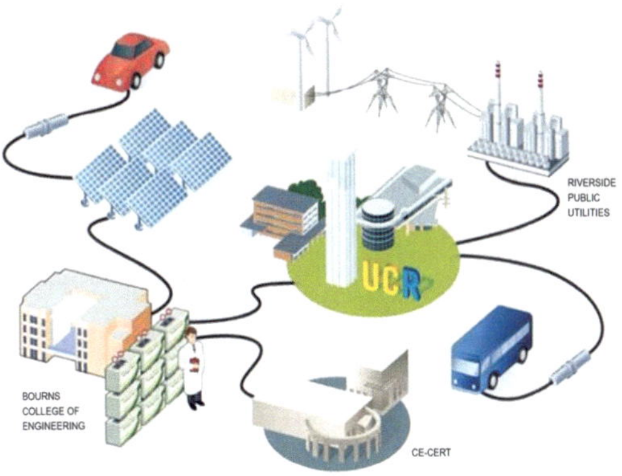

Figure 2.8. Renewable sources in smart grid.

Source: https://figures.semanticscholar.org/12e4c56fccb882f575f37aeed58ea15b2be800d0/2-Figure2-1.png.

2.3. CHALLENGES IN SMART GRID DEVELOPMENT

Construction of intelligent grids is confronted with a number of challenges that need to be addressed in order to successfully implement them. The biggest of these challenges may be integrating legacy infrastructure and new technology. As opposed to new technology, old electric grids were not designed to accommodate alternate power sources and widespread data exchange, and thus are short on reliability and flexibility. The inadequacy of existing infrastructures is capable of derailing the broader operation of smart grids, intensifying issues of demand management and energy efficiency. Moreover, aligning this inherited framework entails a colossal undertaking involving concerted effort by many stakeholders, which can be daunting in a climate of disparate interests and regulatory systems (Borghetti et al., 2013). Unless there are robust investment models and integration strategies, the dream of seamless transition to smart grid technology is beyond reach. A second challenge, of technological in nature, has to do with security and resilience of the underlying information and communication technologies (ICT). Smart grids, being highly dependent on data transmission, make the security of such streams of data paramount. The increased connectivity of the network that characterizes smart grids exposes them to cyber-attacks, which can have catastrophic consequences on power reliability as well as national security. Maintaining the heterogeneity of the devices within the smart grids that are capable of communicating with one another does present some technical challenges. The need for standardized protocols and sound infrastructures to enable such interoperability cannot be overstressed, as these are the building blocks in guarding against potential failures and optimizing energy management systems (Peffer et al., 2010). A multi-faceted policy then needs to be employed, incorporating technological innovation with sound security policies.

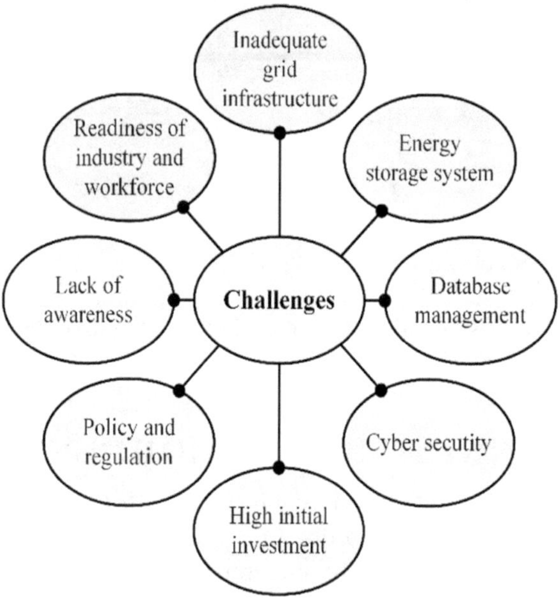

Figure 2.9. Major challenges associated with the application of a smart grid in Nepal.

Source: https://www.researchgate.net/publication/358471313/figure/fig15/AS:11431281360745880@1744096952527/Major-challenges-associated-with-the-application-of-a-smart-grid-in-Nepal.png.

Regulatory and policy guidelines also are significant challenges in the development of smart grids. The transition toward more advanced electricity markets demands an examination of existing rules that may be rigid and sluggish to adopt the latest technological advances. Policymakers, regulators, and utilities must work together in creating a conducive atmosphere for innovation while safeguarding consumers and ensuring equitable distribution of energy. This is normally an exercise of navigating complex political and economic landscapes where embedded interests can conflict. The establishment of effective and responsive regulatory structures plays a crucial role to facilitate the development of investment needed for smart grid innovation and ensure the implementation of these technologies to be cost-effective and profitable to all stakeholders. In meeting such challenges, concerted efforts are needed to enhance technology-policymaking-infrastructure synergy in smart grid development (Borghetti et al., 2013) (Figure 2.9).

2.3.1. Cybersecurity Risks

In the rapidly evolving context of smart grid architecture, cybersecurity threats are unfolding major challenges that must be addressed to provide the security and functionality of such complex systems. The transition to smart grids from traditional power grids introduces several vulnerabilities, particularly with the addition of Internet of Things (IoT) devices and real-time data communication. Because smart grids allow for two-way communication between the utilities and consumers, the number of potential entry points for cyberattacks increases exponentially. Issues such as unauthorized access and data breach may produce heavy outcomes, for instance, power supply disruptions, customer data compromised, and even the threat of public safety. Strong cybersecurity adoption is therefore essential to minimize the risk with the guarantee of service deliverability and consumer confidence. One of the inherent components of providing security for smart grid infrastructures is the application of advanced cyber defense mechanisms tailored to counter specific threats inherent in the technology. Cybersecurity measures have to evolve with the incredible advancements in smart grid infrastructures that are typically founded on Wireless Sensor Networks (WSNs) to ensure real-time monitoring and management. Recent studies emphasize the significance of applying techniques such as traffic analysis and machine learning algorithms in ensuring the effective security of WSNs. With predictive models, the operators of smart grids can enhance their cyberattack resistance and detect anomalies ahead of time that could indicate an ongoing attack. These future-oriented approaches eventually result in a safer, more dependable system of energy delivery and safeguarding critical infrastructure against sophisticated security threats, as highlighted in current research on intelligent cities (Samonte et al., 2024). Moreover, the integration of large-scale cybersecurity models is required to sufficiently cope with the sophisticated complexity of smart grid architecture and ensure its long-term viability. Research indicates that the use of multi-layered security products can help effectively counter cyber-attacks, especially with the evolution of smart grids. The use of sophisticated algorithms, such as those made for intrusion detection systems, is likely to enhance the security standing of smart grid networks. For instance, the use of machine learning techniques like the Random Forest model has proven effective in identifying various kinds of attacks with high precision and recall. The above advancements owe their existence to the life-supporting embrace between technology and cybersecurity, calling for unabating innovativeness in safeguarding smart grids against smarter cyberattacks (Zhukabayeva et al., 2024). With growing complexity of such systems, enhancing cybersecurity will be a crucial task to make modern energy infrastructures reliable and secure (Figure 2.10).

Figure 2.10. Cybersecurity risk.

*Source:*https://smartxtech.co/wp-content/uploads/2023/05/Cyber-security-%E2%80%93-29-05-23.png.

2.3.2. High Initial Costs

Transition to a smart grid configuration is denoted with high initial costs, and these create significant challenges for the global adoption. Smart grid technology implementation involves a range of different expenditures, such as investment in advanced infrastructure, the deployment of sophisticated technology, and significant employee training. For most utilities, the initial amount of capital that must be committed to replace aged infrastructure using smart technology is a formidable financial barrier. Utilities may be dissuaded from investing by these expenditures, especially where existing power infrastructure is functional, if old-fashioned. Thus, the initial costs must be addressed tactfully in a way that stimulates utilities to progress towards the adaptive and efficient capabilities of the smart grid without damaging near-term service reliability. Secondly, the technical complexity of developing a decentralized monitoring architecture also adds to smart grid high initial costs. The necessity for creative solutions, such as decentralized voltage monitoring systems, calls for massive research and development investment. The integration of smart and intelligent cooperative entities in existing grids not only requires sophisticated technology but also the restructuring of operating paradigms. As observed in the smart grid complexities forums, it is exacerbated by the demands on scalable control mechanisms that can effectively manage distributed generation sources (Støa et al., 2011). Such technology innovation with increased operation efficiency comes at a cost that might have an effect on stakeholder decision against embracing the smart grid system. Beyond the technology and infrastructure expenses upfront, market conditions and regulatory challenges can exacerbate the costs of smart grid projects. Policy regimes in place now lag behind technology advancement and cause uncertainty about return on investment for utilities contemplating smart grid options (Borghetti et al., 2013). Additionally, market forces, including competitive pressures and integration pressures for renewable resources, demand further investment in smart grid technology. As companies grapple with these demands, the temptation of high front-end costs can overwhelm future returns, and a paradox is created that inhibits innovation. To fully realize the advantages proposed by smart grid designs, every stakeholder must move together to offset up-front investments and develop supportive regulatory environments that have a beneficial impact on investment in new energy solutions.

2.3.3. Regulatory and Policy Barriers

Developing the smart grid is significantly hindered by several policy and regulatory hurdles that complicate the installation of new technologies. The traditional electricity grid is not only outdated but also not able to handle the integration of renewable energy sources and increased energy efficiency requirements, a demonstration of the urgent need for an overall policy makeover. This emphasizes coming up with a clear definition of what constitutes a smart grid, an aspect where stakeholders do not see eye to eye. In the absence of a shared perception, regulation mechanisms needed to enable effective supply and demand management can collapse to foster the efficient use of crucial factors such as two-way communication infrastructure and safe handling of data processes needed to enable the successful implementation of smart grid technologies (Axon et al., 2014). Public acceptability and associated risks to privacy, data security, and cybersecurity also present significant barriers to smart grid programs. Concerns about the use and spread of individuals' personal information can stimulate resistance by citizens, and it is therefore critical that regulators address such concerns with care. Furthermore, the competitive nature of the energy market can complicate matters as various stakeholders like utility companies, technology suppliers, and consumers hold opposing interests and concerns. This intricate interaction between actors necessitates a robust policy framework that considers the socio-technical dimensions of smart grid technology, lest their breakdown would ruin the progress towards carbon reduction targets and other essential energy goals (Borghetti et al., 2013). Against the above challenges, regulatory regimes must evolve in an effort to hasten transitions towards smart grid architectures. Policymakers must emphasize the necessity for cooperation between business players and the government to create interlinked systems of regulation that promote innovation while ensuring consumers' interests are protected. Such systems must not only facilitate simplification in the approval of new technology but also provide incentives for business players to invest in smart grid systems. Through fostering an experimentation and innovation culture, stakeholders can collectively get past regulation hurdles and maximize the smart grid potential to drive the future electricity supply chain towards advanced efficiency and sustainability for generations to come (Axon et al., 2014).

2.4. SMART GRID TECHNOLOGIES

Innovation in electricity delivery systems has created a requirement for technologies that react to the intricacies of modern energy requirements and environmental concerns. The traditional grids are not adequate to counter the intricacies of alternative energy sources and increasing power demand. The incorporation of renewable power technologies, such as solar, wind, and hydroelectric power, requires a more advanced infrastructure to counter variability in energy output. Smart grid technologies come to the forefront as a solution, leveraging Information and Communication Technologies (ICT) flexible and reliable enough to transform the existing grid into a responsive, efficient one. According to Borghetti et al. (2013), the smart grid enables the integration of distributed energy resources while optimizing the overall management and maximization of electricity supply. One significant characteristic of smart grid infrastructure is its reliance on sophisticated communication networks that facilitate real-time information exchange between utility providers and consumers. Real-time reporting facilitates closer surveillance over patterns of energy use and efficient planning for demand response. With the addition of advanced metering infrastructure (AMI) to the smart grid system, utilities can optimize load balancing, make precise predictions of energy demand, and reduce operational costs. Moreover, monitoring capabilities provide consumers with detailed information regarding their energy use, and this instills enhanced decision-making for energy consumption. Harmonization of technology and communication within the smart grid not only increases efficiency but is also a key aspect of meeting regulatory requirements and energy-saving regulations as observed in Borghetti et al. (2013). Security is a priority concern in designing smart grid technologies because of the increased connectivity and reliance upon digital infrastructure. With interconnectivity among various technologies and sharing data, these systems are attractive options for cyber-attacks, which could be threatening the reliability of the power supply. As such, robust cybersecurity measures must be incorporated into every aspect of smart grid architecture in order to protect sensitive infrastructure. This includes sophisticated encryption techniques and sound risk management processes to secure against threats. Further, an ongoing interrelation between stakeholders, including regulatory bodies and technology developers, is essential to develop security frameworks that become updated with upcoming risks. Addressing these security issues is essential for realizing public trust and the highest possible potential of smart grid systems (Figure 2.11).

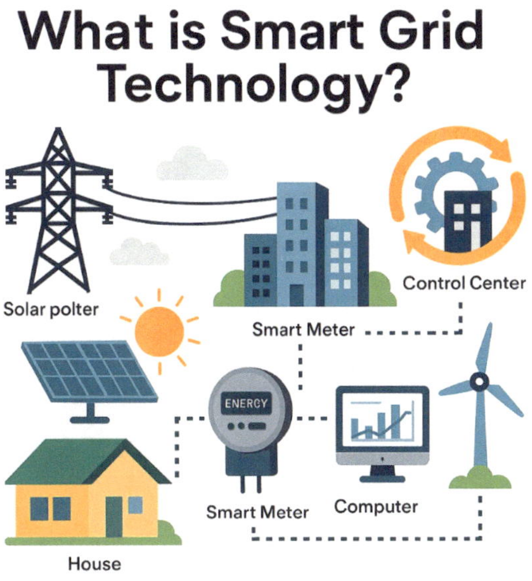

Figure 2.11. What is smart grid technology?

*Source:*https://iotdunia.com/wp-content/uploads/2019/04/what-is-smart-grid-technology-featured.webp.

2.4.1. Demand Response Technologies (DRT)

The concept of DRT lies at the heart of smart grid design, basically transforming the demand-supply interface of energy. DRT is a collection of means by which electricity consumers adjust their consumption based on real-time electricity prices or grid needs. By inviting users to switch off or reschedule their electricity demand from peak hours, demand response enhances the stability of the grid, optimizes resource utilization, and reduces the requirement for additional power generation. Demand response is particularly significant because conventional structures of electricity networks are unable to adapt to increasing energy demand as well as the variability that comes with renewable sources of energy. As discussed, the existing legacy electricity grid infrastructure has not been adequate in terms of dealing with issues of today such as alternative sources of energy (Borghetti et al., 2013). This inadequacy has necessitated a technology shift towards smart grids where DRT becomes key. Application of Demand Response Technologies has a very strong correlation with technology development in communications and information systems, specifically through embracing the Internet of Things (IoT). Smart grids utilize various IoT devices, such as smart sensors and smart meters, to enable real-time data collection and customer-to-utility communication.

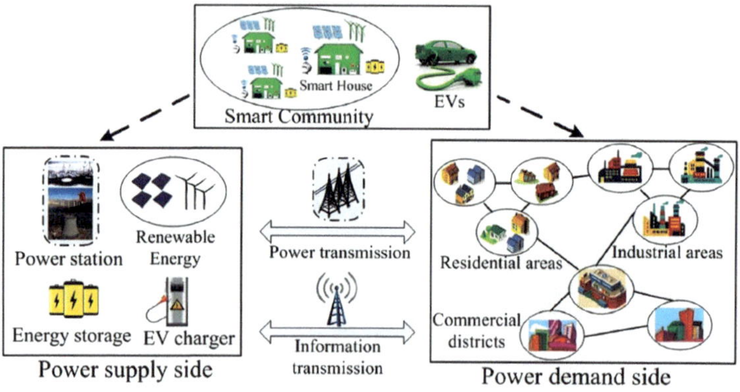

Figure 2.12. Demand response in the smart grid.

Source: https://www.researchgate.net/profile/Sabita-Maharjan-2/publication/298902712/figure/fig2/AS:347841848463361@1459943367431/Demand-response-in-the-smart-grid.png.

Two-way communication allows customers to make intelligent decisions regarding their energy consumption while making it easier for utilities to manage demand. As stated, smart grids enable bi-directional energy transmission between the consumers and service providers, including power generation, transmission, distribution and utilization systems (Copeland et al., 2019). With IoT capabilities, DRT can offer increased consumer participation whilst offering utility operators with means of providing dynamic pricing models, which also increase demand flexibility. Regulation surrounding Demand Response Technologies plays a crucial role in making them effective and embraced. It is the responsibility of policymakers to develop policies that promote energy efficiency and the use of DRT in smart grid systems. Policies can incentivize consumers to use demand response programs of utilities while ensuring consumer protection and equitable access to energy savings. While developing smart grids, regulatory policies need to prioritize incorporating renewable sources of energy and maximizing the use of demand-side management. Emphasizing these regulatory dimensions not only bridges the implementation barriers but also forms a viable energy system compatible with emerging new energy policies and technological trends. Through these structured initiatives, Demand Response Technologies emerge as a pillar of sophisticated smart grid design (Figure 2.12).

2.4.2. Energy Storage Solutions

Energy storage systems are an essential part of optimizing the functionality and performance of smart grid systems. With increasing dependence on renewable energy sources for power generation, the intermittent nature of solar and wind energy implies that robust storage systems are needed. Storage options such as battery storage, pumped hydro storage, and thermal energy storage level the supply-demand mismatch in the grid. For instance, during peak generation hours, surplus energy can be saved and fed back into the grid during peak demands. This capability not only optimizes resource usage but is also beneficial for grid reliability and hence a key component of future energy infrastructure. Equally, enhanced energy storage technology, enabled by big data, enables more efficient management of these resources and hence reduces utilities' and consumers' operating costs (Gharehpetian et al., 2019). The integration of electric vehicles (EVs) in the energy storage segment presents additional opportunity for smart grid upgradation. As the demand for EVs is picking up, not only can their batteries serve as the power source for transportation, but the batteries can also be utilized as mobile energy storage systems. A bidirectional flow of energy is thus possible, where EVs can return power to the grid when needed or recharge energy when there is surplus energy. The ability of the EVs to contribute to improved grid stability and participate in demand response systems is proof of the new role of energy storage in sustainable transport alternatives. According to European Union policy debate, which cites Aymen et al. (2016), the stimulation of such technological advancement is required in order to provide ambitious environmental targets at the same time as reduced greenhouse emissions. Besides, application of energy storage systems is vital to enhance the dependability of smart grids on sudden outages. Natural disasters or unexpected peaks in the demand for energy can test the performance of grids, and energy storage systems serve as a cushioning measure to combat them. With energy storage in favorable times, smart grids can continuously supply customers even when traditional supplies of energy are interrupted. The strategic implementation of the solutions is also supported by the enhanced diffusion of IoT technology and big data analysis that allows real-time monitoring and management of energy resources. Such synergy not only improves grid security but also brings about a more sustainable energy ecosystem, which is favorable to global targets for cleaner consumption of energy and increased reliance on renewable sources (Gharehpetian et al., 2019) (Figure 2.13).

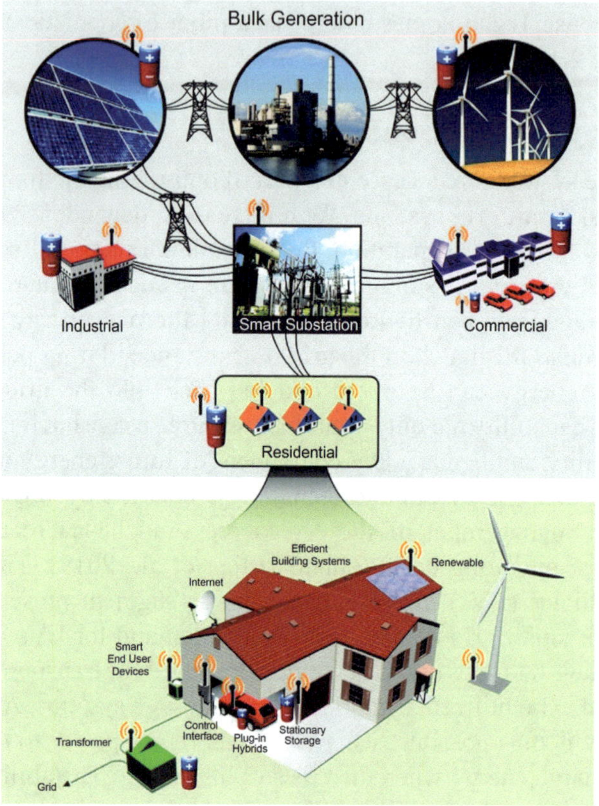

Figure 2.13. Energy storage system integrated into a smart grid with intermittent renewable energy sources.

*Source:*https://www.researchgate.net/profile/Ulderico-Ulissi/publication/327138893/figure/fig1/AS:662055103254528@1534857645014/Energy-storage-system-integrated-into-a-smart-grid-with-intermittent-renewable-energy.png.

2.4.3. Distributed Generation Systems

With new energy needs being influenced, distributed generation systems have become the cornerstone of smart grid infrastructure. Distributed generation systems, such as renewable resources of solar, wind, and bioenergy, enable local generation of electricity, thereby eliminating problems associated with conventional energy networks. Distributed generation, as opposed to centralized power generation, allows for power production close to consumption points, minimizing transmission losses and generally making it a more efficient method. The shift not only enhances sustainability but also conforms to global attempts to reduce carbon emissions. As technology advances and consumer involvement increases, distributed generation systems become an integral part of the energy supply system, ensuring resilience and flexibility in the energy network. The availability of distributed generation systems within smart grids allows better communication and control of energy sources.

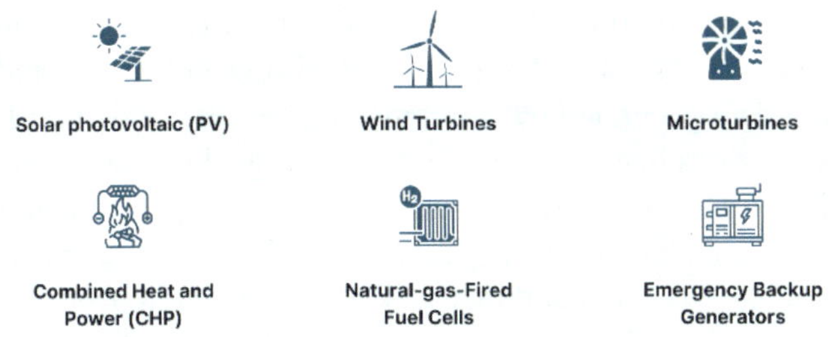

Figure 2.14. Types of distributed generation (DG) systems.

Source: https://cdn.prod.website-files.com/67f01372a54a8deb099415fa/6824f64c5efac87a7f47ef8c_Types_of_DG_Systems.png.

The coordination between the various parts of a grid, such as generation, transmission, and distribution, is highly dependent upon real-time data exchange through Information and Communication Technologies (ICT). All such technologies are essential in an effort to maximize energy utilization and ensure the stable operation of smart grids. As offered in recent research, coordination of such technologies bridges gaps that presently exist in grid responsiveness and effectiveness, ultimately leading to a more reliable energy system (Borghetti et al., 2013). System operators can hence utilize such communications to effectively balance supply and demand and offer transparent integration of renewable energy sources into the grid while keeping it efficient and secure. In addition, management and optimization of energy flow in distributed generation systems is especially difficult and promising. Interoperability between technologies in a system is crucial to obtain the maximum power of smart grids. Research indicates that overcoming such communication barriers is paramount to designing secure and reliable networks in distributed generation systems (Chin et al., 2011). By implementing standard protocols and developing synergistic activities throughout regions, stakeholders are able to enhance the scalability and dependability of distributed generation systems. The integrated approach not only enhances energy management and regulatory adherence, but also innovation, ultimately resulting in a more sustainable and robust energy future (Figure 2.14).

2.5. ROLE OF DATA ANALYTICS IN SMART GRIDS

Analytics has an important part to play in smart grids for achieving the highest possible operational efficiency and realizing maximum benefit from this new class of energy infrastructure. Smart grids facilitate two-way communication between energy consumers and suppliers, enabling collecting and analyzing vast amounts of data generated by smart meters, sensors, and other IoT devices. Data analysis provides insights to facilitate better grid management, leading to increased reliability, energy efficiency, and decision-making. As the traditional power grid transforms into a smart grid system, it becomes increasingly evident that the use of data-based solutions has value in solving issues of energy wastage and demand escalation from consumers. This paradigm shift is aligned with the necessity for comprehensive monitoring and analytics capabilities that are central to maintaining maximum operational efficiency of the entire energy network, effectively paving the way for further innovations in smart grid technology (Copeland et al., 2019). Data analytics becomes increasingly significant in predictive maintenance and fault detection for smart grids as well. Through the examination of historical data and sensor readings in real-time, utilities can recognize patterns that anticipate failures or inefficiencies. Predictive models can even predict equipment failures prior to their occurrence, and this allows timely interventions that minimize downtime and the cost of repairs. The proactive intervention not only guarantees service continuity but also helps optimize resource allocation since utilities can focus maintenance activities on the most critical areas identified through data-driven risk analysis. Moreover, the integration of machine learning methods into data analytics allows for continuous improvement in prediction ability, thereby enhancing the security and dependability of the smart grid operation. This emphasis on data analytics not only improves the schedule optimization of maintenance but also strengthens the robustness of the smart grid system in addressing current and future energy concerns (Farahbakhsh et al., 2017). Apart from improving efficiency, analytics is also crucial in enhancing customer engagement and demand response programs for smart grids. Based on consumer data, utility firms can craft customized energy management systems, encouraging consumers to manage their usage patterns according to fluctuating energy prices. Not only does this save consumers money, but it also stabilizes the grid during peak demand periods. Besides, advanced prediction algorithms can predict consumption patterns, and this assists the utilities in optimizing generation and distribution systems better. With the ability to leverage big data, energy providers as well as consumers can actively participate in energy conservation, which is a sign of a significant cultural attitude shift towards more judicious use of energy. Lastly, integration of data analytics in smart grid infrastructure not only promotes operational efficiency but also fosters a more inclusive and sustainable energy landscape (Copeland et al., 2019).

2.5.1. Real-time Monitoring and Management

Real-time management and monitoring are central components of smart grid infrastructure, significantly enhancing the power system's operational efficiency and reliability. Through the use of smart meters, sensors, and communications technologies, utilities can track electricity consumption patterns and infrastructure health in real-time. Real-time monitoring enables utilities to detect unusual patterns such as faulty equipment, energy theft, or off-pattern consumption spikes that may compromise grid stability. Application of advanced data analytics techniques, as highlighted by the growing literature, enables utilities to distinguish normal patterns of consumption from exceptions, giving them the freedom to take corrective actions on time. This pre-emptive action not only discourages interruptions but also enhances customer service quality since it ensures a more stable supply of electricity, which is crucial in the current energy-based economy (Palety et al., 2025). The high-tech nature of real-time monitoring systems is also established through the synergies of Artificial Intelligence (AI) and machine learning algorithms. These technologies allow smart grids to predict electricity demand more accurately, facilitating optimized energy distribution. AI can analyze enormous volumes of data and detect trends and anomalies in energy consumption, and as a result, grid operators can modulate energy allocation dynamically. The integration of these technologies provides the ability for real-time automated response to conditions, resulting in enhanced efficiency and cost reduction in operations. Secondly, this predictive capability assists in the functionality of effectively managing renewable energy resources that inherently are variable and require agile balancing in the grid (Kumar, 2025). Accordingly, real-time management systems play a critical role to deal with the challenges that these resources provide while ensuring grid resiliency. In addition, it is not easy to overcome the challenges that come with real-time monitoring and management systems. It is essential to deal with data privacy, system interoperability, and ensuring strong cybersecurity measures are in place in order to harness the complete potential of smart grid architecture. Since energy consumption data is personal, it is crucial to protect this data from illegal access. Innovative privacy protection mechanisms and advanced security encryption methods are necessary to support consumer trust while allowing intelligent analytics that drive smart grid innovations. Furthermore, the interoperability of various systems and platforms allows efficient communication and data exchange, promoting a single grid management approach. Overall, effective integration of real-time monitoring and management in smart grid structure holds immense potential for energy efficiency, sustainability, and reliability in the future energy era (Kumar, 2025).

2.5.2. Predictive Maintenance

Integration of predictive maintenance in smart grid structure is a paradigm shift in attaining efficiency in energy management. Predictive maintenance allows identification of equipment faults before they cause major failures using data analytics and IoT technologies. In the context of smart grids, where renewable sources such as solar and wind can lead to system instability, predictive maintenance enables utilities to achieve their highest levels of efficiency and reliability from their infrastructure. The method utilizes historical performance data, real-time monitoring, and machine learning algorithms to forecast future maintenance needs, thereby attaching itself to the broader concept of grid resilience. As highlighted in recent researches, the adoption of predictive maintenance methodologies can result in significant cost savings on operation expenditure and downtime, thereby allowing utilities to maintain an uninterrupted power supply that can meet fluctuating demand (Ojo et al., 2025). Additionally, predictive maintenance systems adoption is founded on the energy storage system (ESS) architecture and functionality that powers smart grids. The converged integration of hardware and software components—e.g., sensors, data analytics software platforms, and communications networks—facilitates real-time monitoring and measurement of the condition of assets. This architecture not only supports the short-term operational needs of the grid but also helps address long-term structural problems of eroding infrastructure and the intermittency of renewable sources. Research has indicated that having case-specific predictive maintenance models fit for various components—ranging from transformers to distributed energy resources—has the potential to dramatically enhance the overall functionality and dependability of the grid (Dr. Bidgoli et al., 2025). Utility companies are thus able to have real-time actionable insights that ensure interventions are data-driven rather than reactive in nature. With the era of smart grids evolving further every minute, the application of predictive maintenance will become increasingly vital day by day. Artificial intelligence and machine learning advancements have opened up new frontiers for enhancing prediction algorithms, resulting in improved accuracy and performance in the prediction of equipment failure. For instance, employing hybrid models that combine temporal and spatial data analysis can provide utilities with an end-to-end view of grid dynamics and allow intervention at the appropriate time to mitigate risks and enhance performance. Also, the discussion of cybersecurity cannot be ruled out, since predictive maintenance systems increasingly rely on networked sensors and IoT devices vulnerable to cyber-attacks. It preserves the reliability and safety of such systems and assists in maintaining stakeholders' confidence in smart grid technologies, ultimately achieving a resilient and sustainable energy future by employing best practices in predictive maintenance (Dr. Bidgoli et al., 2025; Ojo et al., 2025) (Figure 2.15).

Figure 2.15. How data analytics and predictive maintenance are boosting efficiency in smart grids.

Source: https://ik.imagekit.io/clouglobal/img/wp-content/uploads/2023/04/How-Data-Analytics-and-Predictive-Maintenance-are-Boosting-Efficiency-in-Smart-Grids-Symbol-image-credit-CLOU.jpg.

2.5.3. Consumer Behavior Analysis

Analysis of consumer behavior is most crucial in designing a successful Smart Grid Architecture since it will notify stakeholders of how consumers behave toward energy consumption and management systems. Consumer preferences and behavior shape the uptake of smart grid technology, designed to enhance energy efficiency and sustainability. By taking into account key consumer behavior traits, such as new technology adoption and response to price dynamic strategies, developers can fine-tune their products according to users' needs. This individualized adaptation not only increases the level of users' satisfaction but also encourages wider acceptance of smart grid solutions. Consumer behavior study must also consider the external variables of environmental awareness and socio-economic status that are very influential in affecting energy consumption patterns and technology take-up. Apart from the individual preferences, the integration of the data-based decision-making capacity and automation capabilities with the Internet of Things (IoT) also makes Smart Grids more responsive to consumer needs. Various applications of IoT enhance users' ability to automate the use of energy, increase efficiency, and enhance convenience in their daily lives. Installation of the technology is also hindered by the aspect of interoperability as well as security. According to Hong-Dai et al. (2019), the incorporation of blockchain technology into IoT can address the challenges and make the user experience safer. Providing consumers with a seamless and intuitive interface for the control of energy generates active participation and investment in smart grid technology. Consequently, uncovering how the consumer truly interacts with these technologies can drive improved design and implementation strategies in order to promote sustainable energy habits. The privacy and security concerns of IoT applications are also critical in influencing consumer behavior through the Smart Grid Architecture framework. Consumers are always hesitant to adopt new technologies and infrastructure if privacy loopholes and potential dangers are perceived. Effective consumer behavior analysis has to thus incorporate these security concerns so that the architecture of the smart grid is not only designed for automation but also for trust by the consumers. As per Hassija et al. (2019), the implementation of such technologies as blockchain, fog computing, and machine learning will enhance security and drive consumers' trust in energy management systems. Through the solution of these problems and ensuring robust security mechanisms, developers are able to market their smart grid products better and entice consumers, leading to enhanced adoption of innovative energy practices.

2.6. CASE STUDIES OF SMART GRID IMPLEMENTATION

Across the globe, smart grid technology adoption has gained speed, leading to unprecedented innovation in energy management and distribution. One such example comes in the form of a model that works to maximize energy consumption management in Smart Home environments. The model uses cooperative intelligent objects in the context of Smart Meters, allowing devices to collaborate and efficiently manage energy consumption. By allowing appliances within a condominium or neighborhood to share data, the system not only deals with electric power supply but also incorporates renewable sources like solar panels. The paradigm of communication implemented with M-Bus for data collection is a classic example of new technology integration in smart grids, illustrating that such infrastructures are capable of maintaining their operational lifespan while maximizing energy efficiency, especially in cases where controlled devices have an irregular usage pattern (Atzori et al., 2014). Another pivotal case study is that of the UC Berkeley Building-to-Grid testbed, which illustrates the all-encompassing approach undertaken with smart grid deployment. A multidisciplinary team, in the space of nearly a year, accepted the challenge of optimizing energy flows within Cory Hall. The process began with mapping the building's load tree, and then extensive research was done on the variety of metering components necessary for adequate monitoring. While navigating sets of design, procurement, and testing, the team effectively installed a network of metering components, connecting advanced data storage and visualization software. This ongoing process not only satisfied the immediate needs of the building but also laid the groundwork for other critical discussions with nearby utilities, facilitating the establishment of energy information communication connections required in integrating the buildings into the overall electric grid (Peffer et al., 2010). The lessons learned through these case studies illustrate the different methods adopted to enhance energy management and grid performance. Through the assistance of smart meter technologies and sophisticated communication platforms, as represented in the Smart Home model, or through collaborative building projects like that which UC Berkeley is pursuing, stakeholders are able to satisfy both current energy requirements as well as future sustainability goals. Every case highlights the importance of incorporating renewable energy sources and supporting real-time communication among household entities and utilities, ultimately demonstrating the feasibility of pilot projects as a foundation for wide-scale application of smart grids. This combination of various components is a built-in characteristic of Smart Grid Architecture, which results in a more efficient and resilient energy paradigm as societies look towards the modernization with increasing energy efficiency (Atzori et al., 2014; Peffer et al., 2010).

2.6.1. Smart Grid in the United States

Smart Grid development in the United States is a turning point in energy infrastructure, fundamentally solving the problems of conventional electric systems. This initiative combines innovative technologies into electrical distribution, providing a more efficient, trustworthy, and sustainable energy system. The Smart Grid enables power and consumer interactive communications in real-time, thereby fostering adaptive energy management. This interactivity allows the customers to get the most out of their power while allowing utilities to enhance the grid stability. All these are important in today's times when the renewable energies are on the increase, proving that a Smart Grid is capable of accommodating the variability of energy consumption and generation. By designing a decentralized system that can react quickly to demand and disturbances, the Smart Grid facilitates overall objectives of energy security and resiliency for the United States. Despite the challenges in deploying Smart Grid technologies, solving problems regarding voltage management is among the most critical challenges. Traditional hierarchical monitoring schemes are likely to fail to meet the requirements of modern electric systems due to their escalating complexity. In order to manage this complexity effectively, new solutions need to be implemented, such as the proposal of a non-hierarchical decentralized voltage monitoring architecture. This architecture makes use of smart and cooperative agents that utilize traditional sensors with coupled oscillators to provide an overall view of grid status. Voltages can be monitored and controlled by utilities more effectively through this architecture, ultimately resulting in more reliable and efficient distribution of power. As explained in studies, to overcome these challenges is to ensure the development of robust energy networks that are capable of adapting and thriving within the rapidly changing context of energy production and consumption (Støa et al., 2011). Interconnection of Smart Grid systems aligns with policy agendas at the national level aimed towards achieving sustainability and energy independence. Exercises across various states have brought into focus the need for technological investment and strategic planning towards these ends, with priority being accorded to improved energy efficiency, reduced emissions, and improved renewable energy. Efforts in reports prepared for the California Energy Commission, for instance, point to how intelligent technologies are being utilized to spur grid modernization. As a result, the Smart Grid not only enhances business operations but also makes possible a collaborative strategy among energy actors through encouraging public-private sector partnership. Such joint investment is crucial towards ensuring a green energy climate, placing the United States firmly at the fore of the global energy revolution (Burstein et al., 2011).

2.6.2. Smart Grid Initiatives in Europe

The energy supply and management paradigm in Europe is being dramatically altered, driven mostly by smart grid initiatives aimed at enhancing the efficiency, reliability, and sustainability of the power grid. The initiatives encourage the utilization of cutting-edge technologies, such as the Internet of Things (IoT), which enhances seamless communication among devices and optimizes trends of energy use. Smart metering and real-time data analysis are the enablers that make it possible for consumers as well as utilities and give them unprecedented levels of operational visibility and control. As cities transform into smart urban spaces, these initiatives lay the basis for energy consumption to be sustainable, reducing carbon emissions while fulfilling the escalating requirements for electricity from emerging technology and online services (Shafique et al., 2020). In addressing the challenges of moving towards more intelligent energy grids, European nations are investing in key infrastructural elements enabling enhanced integration of renewable energy as well as grid resilience. One aspect of these investments is the development of digital twin (DT) technology, which integrates physical and virtual environments to monitor and optimize grid activity in real time. Such integration not only involves the optimization of operating efficiency but also the prediction of probable failures and maintenance needs by advanced data analysis. European smart grid initiatives are progressively adopting this double-pronged approach, which is part and parcel of achieving the aspirations of Industry 4.0 and working towards a cleaner energy future. As various sectors harmonize their operations with such blueprints, the arrival of smart grids marks an era set for a makeover by innovation and plenty of economic opportunities (Tao et al., 2019). Besides, harmonization and collaboration among European member states play a vital role in ensuring the success of smart grid initiatives. The European Union promotes cross-border activities actively, calling for interoperability and cooperation on best practices among member states. This co-operative context allows countries to leverage each other's experience while addressing common problems, such as mixing varied energy sources and optimizing peak loads. With new smart grid technologies on the horizon, regulatory policy focus and funding structures become more critical by the day. By installing a robust regulatory system, Europe not only encourages private investment but also propels the transition towards a more networked and sustainable energy system, creating synergies that benefit consumers as well as the ecosystem.

2.6.3. Emerging Smart Grid Projects in Asia

The proliferation of smart grid initiatives in Asia is indicative of the regions sensitivity to the immediate need for sustainable energy solutions. India, China, and Japan are some of the nations that are driving the change because of the twin challenge of climate change and urbanization. The smart grid architectures being implemented focus on enhancing efficiency and reliability of energy as well as integrating renewable sources of energy. For instance, Indias initiative under the National Smart Grid Mission aims to establish a smart grid of electricity that not only optimizes energy consumption better but also facilitates the integration of renewable sources, thus addressing the large percentage of greenhouse gas emissions in fossil fuel-powered electricity generation (Kabeyi et al., 2022). This multi-faceted approach aims to create strong energy grids that are adaptable to the changing trends of energy consumption and production, thus encouraging a further dedication towards sustainability along with green responsibility. Moreover, all the countries in Asia are now beginning to realize the importance of embedding new technologies in their smart grids. Information and communication technology (ICT) integration has proved to be the foremost catalyst, supporting real-time monitoring along with control of energy resources. South Korea and Singapore, among others, are leveraging smart technologies to make grids more efficient and safer to operate. As more high-tech systems become integrated, data privacy issues as well as exposure of the grid to vulnerabilities also grow. Existing security architectures fall short in most cases, prompting the use of decentralized options like blockchain technology for assured enhanced security and cyberattack protection (Bodkhe et al., 2020). This new technology is crucial in solving the problem of scalability and ensuring high-performance functionality in smart grid systems, thereby building higher stakeholder trust in leveraging these emerging technologies. With continued development in these smart grid programs, collaboration among governments, the private sector, and universities becomes the driving force for their progress. Collaborations facilitate the sharing of information and resources, allowing for faster development and deployment of new solutions. In addition, policy guidelines should pave the way for the transition by articulating appropriate regulations and incentives for the adoption of smart grid technologies. The cooperation should also be guided by principles of sustainability in analyzing the environmental, economic, and social dimensions of energy innovation. The physical outcomes of such synergetic endeavors can lead to the creation of a sustainable grid infrastructure that is eco-friendly and compliant with the international agreements for sustainable development, as outlined in agreements such as the Paris Agreement (Kabeyi et al., 2022). This comprehensive method is necessary to ensure that future smart grid projects in Asia meet the existing energy needs but also guarantee the move towards a cleaner future.

SUMMARY

In summarizing the Smart Grid Architecture controversy, it is obvious that a dynamic strategy is critical to optimize residential energy usage. The traditional perspectives have been focused on the technical operations of Smart Home Energy Management Systems (SHEMS), i.e., load control and energy savings. However, as indicated in recent studies, there exists a massive knowledge gap when it comes to user behavior and interaction in SHEMS. McIlvennie et al.

(2020) stresses that user-centric solutions should also be included alongside techno-centric approaches. The type of all-encompassing vision as outlined is required as it can create a higher level of bond between consumers and technology that in itself can produce more sustainable household energy consumption patterns as well as overall energy grid patterns. Further, Smart Grid networks demand stringent cybersecurity measures to protect confidential data and provide seamless operation. As systems become increasingly interconnected with state-of-the-art communication gateways, the risk of cyber-attacks looms larger. Research depicted in Marksteiner et al. (2017) describes that rigorous security is required in guarding the communication among electric storage devices and the grid. In the absence of protection, trust in Smart Grid technologies vanishes, hindering their adoption. Hence, an integrated strategy towards addressing cybersecurity, alongside energy management technology development, is warranted to bolster resilience for SHEMs. Finally, the way to an efficient Smart Grid is not merely technological but societally embedded. The future research must change direction towards understanding the complexities of user engagement with SHEMS and the subsequent implications for demand-side management programs. Results of interdisciplinary studies can drive innovation and maximize current systems to expand their reach. By inviting a broader discussion involving the interest of stakeholders—from policymakers to consumers—researchers and professionals can collaborate to create processes for enhancing energy efficiency and maximizing Smart Grid functionality. In such a dynamic environment, being responsive and adaptive to user demand and technological advancements is the only way to achieve the high promises of Smart Grid Architecture.

In Smart Grid Architecture, the incorporation of EVs without causing any disruption is an essential move toward green energy usage. The proposed AI-fortified smart grid architecture addresses the imperative intersection of EV infrastructure and cybersecurity by recognizing the necessity to deal with cyber-physical threats. With electric vehicles at the forefront of reducing greenhouse gas emissions, the architecture concerns itself with the need to develop efficient and secure solutions for charging. By incorporating AI-powered models and anomaly detection technology, the platform not only enhances grid resilience but also optimizes energy utilization with innovative vehicle-to-grid (V2G) technologies. Such measures are, in the end, critical in enabling rising demand for charging infrastructure in addressing latent vulnerabilities within this rapidly evolving sector (Sharma et al., 2025). Another key component of the Smart Grid Architecture is the integration of renewable energy sources (RESs) and the formation of energy communities (ECs). The development of a Smart Grid Architecture Model (SGAM)-compliant software platform provides an easy method for integrating heterogeneous energy assets in these communities.

This type of platform is particularly suited to upgrade energy dispatching, self-consumption, and legacy infrastructure compatibility. Case studies such as Polígono Industrial Las Cabezas in Spain and the IMP R&D campus in Serbia are used to closely analyze the problems and solutions encountered during implementation. What can be learned from these experiences captures the complexity of device integration, user take-up, and regulatory configurations, and ultimately illustrates the role of bespoke solutions in developing effective energy community policies (Janev et al., 2025). Secondly, the architecture showcases comprehensive understanding of how technical and operational challenges in smart grid installations are reconciled. Through the factors of production forecasting and user engagement concerns, the architecture seeks to facilitate energy efficiency and decarbonization ease. The findings of the aforementioned case studies highlight efficient methods while also shedding light on challenges encountered, and as such provide useful lessons for subsequent projects. As the energy scenario continues to change, incorporating smart grid infrastructures is a critical necessity to foster energy sustainability and resilience. Combined, all the foregoing points create a global commitment to innovation in smart grid architecture towards establishing a secure and sustainable energy environment in the midst of future technological advancements (Janev et al., 2025).

REFERENCES

1. Adebiyi, O. E., Ezekiel, S. E., Habeeb, A. O., Ojo, M., et al. (2025). IoT-enabled energy storage systems for renewable energy grid integration. https://core.ac.uk/download/657008560.pdf (accessed on 30th September 2025).
2. Ai, A., Beck, A., Belkacemi, R., Bellifemine, F., Brown, N., Catterson, V., et al. (2012). Practical applications of multi-agent systems in electric power systems. https://core.ac.uk/download/9033895.pdf (accessed on 30th September 2025).
3. Atzori, L., Cherchi, R., Colistra, G., & Pilloni, V. (2014). Energy consumption management in smart homes: An M-Bus communication system. https://core.ac.uk/download/pdf/54602568.pdf (accessed on 30th September 2025).
4. Axon, C., Balta-Ozkan, N., Cipcigan, L., Connor, P., et al. (2014). Scenarios for the development of smart grids in the UK: Literature review. https://core.ac.uk/download/155776124.pdf (accessed on 30th September 2025).
5. Bidgoli, A. J., & Al Yousef, M. (2025). AI-driven framework to optimize smart grid operations, enhance energy efficiency, and facilitate seamless integration using hybrid (LSTM-CNN) models. *Journal of Information Systems Engineering and Management*. https://www.semanticscholar.org/paper/9ab0b5e2606db83784a9fe8e978788a940c2b444 (accessed on 30th September 2025).
6. Bodkhe, U., Tanwar, S., Parekh, K., Khanpara, P., Tyagi, S., Kumar, N., & Alazab, M. (2020). Blockchain for Industry 4.0: A comprehensive review. *IEEE Access, 8*, 79764–79800. https://doi.org/10.1109/access.2020.2988579 (accessed on 30th September 2025).
7. Bonepalli, R. (2025). Sustainable smart grid architecture: Carbon footprint reduction with Oracle Utilities Software. *International Journal of Scientific Research in Computer Science, Engineering and Information Technology*. https://www.semanticscholar.org/paper/b5cdfd8b2221df9eeed59e12a07a2b60026a73df (accessed on 30th September 2025).
8. Borghetti, G., Menga, G., Procaccianti, G., Kok, K., Ardito, L., Morisio, M., Mo, D., et al. (2013). Smart grid technologies in Europe: An overview. https://core.ac.uk/download/pdf/11431923.pdf (accessed on 30th September 2025).
9. Burstein, A. J., Mulligan, D. K., & Wang, L. (2011). Privacy in the smart grid: An information flow analysis. https://core.ac.uk/download/288432753.pdf (accessed on 30th September 2025).

10. Chaouachi, A., Covrig, C. F., & Fulli, G. (2016). Framework for electric vehicles and photovoltaic synergies. https://core.ac.uk/download/38632047.pdf (accessed on 30th September 2025).
11. Chin, W. H., Efthymiou, C., Fan, Z., Gormus, A., et al. (2011). Smart grid communications: Overview of research challenges, solutions, and standardization activities. http://arxiv.org/abs/1112.3516 (accessed on 30th September 2025).
12. Copeland, R., Crespi, N., Rehmani, M. H., Saleem, Y., et al. (2019). Internet of Things-aided smart grid: Technologies, architectures, applications, prototypes, and future research directions. http://arxiv.org/abs/1704.08977 (accessed on 30th September 2025).
13. Dai, H. N., Zheng, Z., & Zhang, Y. (2019). Blockchain for Internet of Things: A survey. *IEEE Internet of Things Journal, 6*, 8076–8094. https://doi.org/10.1109/jiot.2019.2920987 (accessed on 30th September 2025).
14. Develder, C., Haerick, W., & Mets, K. (2010). A simulator for the control network of smart grid architectures. https://core.ac.uk/download/55823207.pdf (accessed on 30th September 2025).
15. Farahbakhsh, R., Farahzadia, A., Rezazadeh, J., Shams, Z., et al. (2017). Middleware technologies for cloud of things–A survey. http://arxiv.org/abs/1705.00387 (accessed on 30th September 2025).
16. Gharehpetian, G. B., Marzband, M., Moradi, J., Nafisi, H., et al. (2019). Attributes of big data analytics for data-driven decision making in cyber-physical power systems. https://core.ac.uk/download/305121394.pdf (accessed on 30th September 2025).
17. Hassija, V., Chamola, V., Saxena, V., Jain, D., Goyal, P., & Sikdar, B. (2019). A survey on IoT security: Application areas, security threats, and solution architectures. *IEEE Access, 7*, 82721–82743. https://doi.org/10.1109/access.2019.2924045 (accessed on 30th September 2025).
18. Hauser, C. H., Lorenz, D. H., Fudenberg, D., Santacana, E., Kuipers, F., Wu, F. F., Xue, G., et al. (2018). Multi-layer cyber-physical security and resilience for smart grid. http://arxiv.org/abs/1810.00282 (accessed on 30th September 2025).
19. Janev, V., Berbakov, L., Tomašević, N., Martin-Borja Sotoca, J., & Lujan, S. (2025). Validating the Smart Grid Architecture Model for sustainable energy community implementation: Challenges, solutions, and lessons learned. *Energies*. https://www.semanticscholar.org/paper/da89eebce4040da97a9231d6f1c1412690c10e54 (accessed on 30th September 2025).

20. Kabeyi, M. J. B., & Olanrewaju, O. A. (2022). Sustainable energy transition for renewable and low carbon grid electricity generation and supply. *Frontiers in Energy Research, 9*. https://doi.org/10.3389/fenrg.2021.743114 (accessed on 30th September 2025).

21. Kaif, A. D., Shahjahan, K., Alam, S. K., Guo, D., Chen, S., Islam, L. F. I. S. M., Muyeen, F. I., et al. (2025). Blockchain-integrated cyber-physical smart meter design and implementation for secured energy trading in virtual power plants. *IEEE Transactions on Automation Science and Engineering, 22*, 15083–15093. https://www.semanticscholar.org/paper/706e1248258e38a51e4d5a1ba55dcb7ae666de54 (accessed on 30th September 2025).

22. Kumar, K. S. S. (2025). Distributed intelligence for smart grid management: Architectures, applications, and future. *World Journal of Advanced Engineering Technology and Sciences*. https://www.semanticscholar.org/paper/a24db53571a4ae0460a85bf194d9265894eb0786 (accessed on 30th September 2025).

23. Marksteiner, S., & Vallant, H. (2017). Towards a secure smart grid storage communications gateway. http://arxiv.org/abs/1710.03017 (accessed on 30th September 2025).

24. McIlvennie, C., Pritoni, M., & Sanguinetti, A. (2020). Of impacts, agents, and functions: An interdisciplinary meta-review of smart home energy management systems research. https://core.ac.uk/download/323062198.pdf (accessed on 30th September 2025).

25. Ojo, O. T., Habeeb, A. O., Adebiyi, O. E., Olanrewaju, S. A., Olabode, O. J., & Ezekiel, S. E. (2025). IoT-enabled energy storage systems for renewable energy grid integration. *Path of Science*. https://www.semanticscholar.org/paper/dd1e35630310eb32a22530a52c129e7f4084c77d (accessed on 30th September 2025).

26. Palety, B. N., Mahalakshmi, C., & Reddy, P. N. (2025). Anomaly detection in residential electricity consumption with GAN-CNN for enabling smart grid security and efficiency monitoring. *Journal of Advances in Information Technology*. https://www.semanticscholar.org/paper/e1bef08de8519c35f2da3edbaa6ea252ab400d96 (accessed on 30th September 2025).

27. Peffer, T. (2010). UC Berkeley's Cory Hall: Evaluation of challenges and potential applications of building-to-grid implementation. https://core.ac.uk/download/288432727.pdf (accessed on 30th September 2025).

28. Samonte, M., Panganiban, K. D. H., Laqueo, L. M. D., Lazaro, J. R. M., & Ogaya, L. A. (2024). A critical analysis on cybersecurity of Internet-

of-Things in system integration and architecture of smart cities. *2024 8th International Conference on Smart Grid and Smart Cities (ICSGSC)*. https://www.semanticscholar.org/paper/763ef29dade03f8ee9d2df2a399bae43435c0793 (accessed on 30th September 2025).

29. Seshasai, B., Koley, E., & Ghosh, S. (2025). Blockchain for secure and decentralized power system operation in smart grid systems. *2025 Fourth International Conference on Power, Control and Computing Technologies (ICPC2T)*. https://www.semanticscholar.org/paper/6d83644b356f439472bff9c18ee18918df5c7416 (accessed on 30th September 2025).
30. Shafique, K., Khawaja, B. A., Sabir, F., Qazi, S., & Mustaqim, M. (2020). Internet of Things (IoT) for next-generation smart systems: A review of current challenges, future trends and prospects for emerging 5G-IoT scenarios. *IEEE Access, 8*, 23022–23040. https://doi.org/10.1109/access.2020.2970118 (accessed on 30th September 2025).
31. Sharma, A., Rani, S., & Shabaz, M. (2025). Artificial intelligence-augmented smart grid architecture for cyber intrusion detection and mitigation in electric vehicle charging infrastructure. *Scientific Reports, 15*. https://www.semanticscholar.org/paper/48b1a55c4ab39b83e4c59f9a5f85bab28abc8dff (accessed on 30th September 2025).
32. Støa, P., Vaccaro, A., Zeineldin, H. H., Zobaa, A. F., et al. (2011). The role of smart sensor networks for voltage monitoring in smart grids. https://core.ac.uk/download/30340002.pdf (accessed on 30th September 2025).
33. Tao, F., Zhang, H., Liu, A., & Nee, A. Y. C. (2019). Digital twin in industry: State-of-the-art. *IEEE Transactions on Industrial Informatics, 15*, 2405–2415. https://doi.org/10.1109/tii.2018.2873186 (accessed on 30th September 2025).
34. Zhukabayeva, T., Pervez, A., Mardenov, Y., Othman, M., Karabayev, N., & Ahmad, Z. (2024). A traffic analysis and node categorization-aware machine learning-integrated framework for cybersecurity intrusion detection and prevention of WSNs in smart grids. *IEEE Access, 12*, 91715–91733. https://www.semanticscholar.org/paper/d7e76ae1d22a635304ae0df8c87246f62b461f83 (accessed on 30th September 2025).

CHAPTER 3

SMART METERING AND ADVANCED METERING INFRASTRUCTURE (AMI)

Contents

INTRODUCTION

As the pace of embracing green energy solutions picks up speed, the demand for sophisticated metering technologies becomes self-evident. The transition from traditional metering infrastructure to Smart Metering and Advanced Metering Infrastructure (AMI) is a breakthrough in the regulation of energy consumption, generation, and supply. The technologies enable real-time data collection, enabling utilities and consumers to monitor energy consumption with higher accuracy. Apart from providing detailed data on consumption patterns, smart meters also enable the integration of renewable energy sources, which are essential in developing a stronger and more sustainable energy grid. This integration reflects growing awareness of the need for the modernization of energy infrastructures in meeting contemporary challenges, from unsteadfast energy consumption to regulatory measures in pursuit of energy efficiency and management.

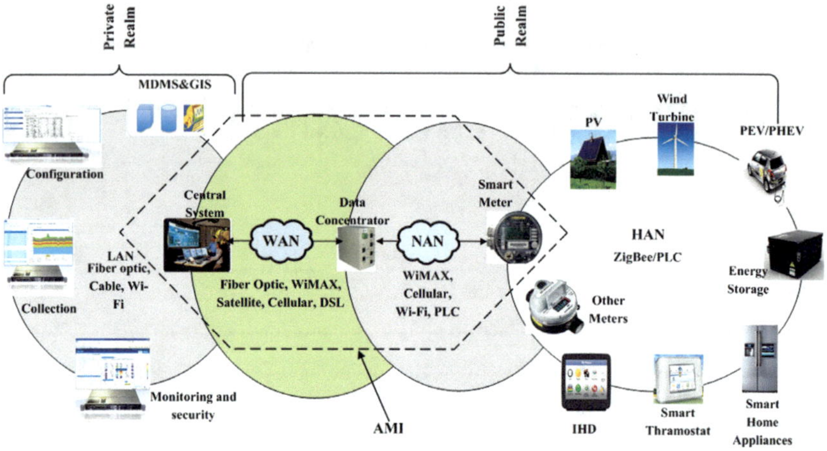

Figure 3.1. Advanced metering infrastructure (AMI) overview architecture.

Source: https://cdnintech.com/media/chapter/50727/1512345123/media/fig3.png.

The implementation of Smart Metering and AMI also heavily depends on the evolution of Information and Communication Technologies (ICT). As stated in the research, the present electricity network infrastructure has been proven to be inadequate to manage modern needs due to the introduction of alternative power sources and energy conservation measures. ICT has become reliable and adaptable enough to enable the implementation of a smart grid concept that enhances efficiency and security of operations within the energy sector. This transition not only optimizes real-time use of energy but also achieves significant concerns of safety and reliability in energy management systems. Consequently, their integration is a strategic move to strengthen the energy grid, allowing for more efficient communication between generation, transmission, and distribution assets (Borghetti et al., 2013). Furthermore, Smart Metering and AMI development needs harmonized standardization efforts in order to facilitate interoperability between various communication technologies and protocols. Designing an all-embracing framework for smart grid communication poses a number of challenges and opportunities. The principal areas of security, privacy, and standardization must be thoroughly analyzed and coordinated across different technologies to facilitate an integrated energy management system. As noted, future-proofing the smart metering networks is of prime importance; thus, ongoing research is essential in addressing these problems while employing existing networking technologies (Chin et al., 2011). In the end, the successful implementation of Smart Metering and AMI can not only propel overall grid management but also promote a greener future of energy that supports global initiatives towards green energy policies (Figure 3.1).

Definition of Smart Metering

Smart metering is a significant development in energy management technology that provides a two-way communication channel between consumers and utility companies. These advanced meters are capable of measuring real-time electricity usage, which not only provides for more accurate billing but also facilitates dynamic pricing plans. In contrast to traditional metering technologies, with their restricted information and data, smart meters can empower customers to manage their energy consumption and thereby take part in more sustainable practices. Through the integration of smart metering within the overall architecture of AMI, utilities are better positioned to achieve sophisticated operational efficiencies that are essential in addressing modern energy issues. Smart metering is a key driver in the evolution of a smart grid, which leverages ICT in maximizing electricity distribution and usage ultimately towards the construction of an even more robust and responsive energy system (Borghetti et al., 2013). Smart metering is central to the transformation towards distributed energy resources and renewable sources of energy that are fed into the grid. The flexible technology gives the consumer the power of necessary information on their energy consumption behaviors in order to make an informed decision on how to utilize their energy and, in the process, lower their carbon footprint. Besides, utilities can benefit from this system in the aspects of greater grid reliability, real-time detection of outages, and improved demand management. Smart metering infrastructure also enhances demand-response program integration, which will prompt customers to reduce or shift their use of electricity during peak times. As these systems expand, data generated is becoming increasingly essential to utilities and policymakers in shaping future energy policies that are in sync with decarbonization targets (Burstein et al., 2011). Furthermore, the policy environment for smart metering is evolving to support its installation and deployment. Governments and regulators are increasingly aware of the need for policies facilitating the roll-out of smart meters as part of broader energy modernization measures. Such policies have a critical role to play in setting standards for metering technology, data privacy, security, and consumer interaction. In an environment that encourages cooperation between all the parties involved, including the utilities, tech companies, and consumers, the entire potential of smart metering can be realized. As the energy sector transforms through new technology and operational frameworks, smart metering is a central element that will shape the future of energy availability and management for both the utility and consumer (Borghetti et al., 2013) (Figure 3.2).

Figure 3.2. Smart metering.

Source: https://www.telit.com/wp-content/uploads/2021/11/Telit_Rapid_Prototyping_Smart_Grid_Smart-Metering.jpg.

Overview of Advanced Metering Infrastructure (AMI)

The AMI represents a new frontier in the administration of energy delivery and consumption. Through connecting smart meters to communication technology, AMI offers real-time data acquisition and management, significantly enhancing the efficiency of utility delivery infrastructures. Two-way communication between utilities and customers is one of the key features of AMI, paving the way for more productive demand response. The integration not only advantage utilities by allowing them to control supply more effectively but also allows consumers to guide their decisions regarding how they use energy. As part of the change, the unification of various grid elements across generation, transmission, and distribution becomes more harmonized, a requirement for the success of smart grids in the future (Chin et al., 2011). Furthermore, the utilization of AMI facilitates the integration of renewable energy sources into the grid based on rising environmental standards. Infrastructure supports a distributed energy scheme that allows the integration of various green technologies that promote efficiency and sustainability in utilizing energy. Utilities can use data to optimize energy distribution and consequently reduce wastage by utilizing AMI. Yet, though the developments promise much, the implementation of AMI is complicated by a variety of cybersecurity problems. The electronic nature of smart metering opens systems up to a range of potential threats that could undermine data integrity and customer privacy. Addressing these challenges must be comprehensive in scope, taking into account all the vulnerabilities that such interdependent systems may be exposed to Ghazi et al. (2018). With the evolution of the energy topography with AMI, standardized and secure initiatives are imperative. The utility players and regulatory authorities must come together and evolve as well as implement robust protocols that can withstand the changing cybersecurity threats such infrastructure presents. The devices, systems, and applications must be interoperable to ensure complete dependability and user confidence in AMI. Furthermore, a proactive cybersecurity strategy, encompassing new technologies, will be needed to establish defenses against sophisticated assaults on critical infrastructure. Along with this rapidly changing environment, a continuous influx of research and advancement in communications technologies will play a pivotal role in reducing such security threats, thereby enhancing the efficiency and security of smart grids and the associated metering infrastructures (Chin et al., 2011) (Figure 3.3).

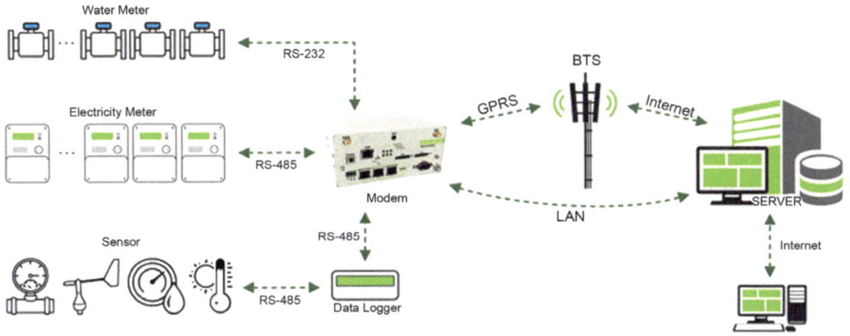

Figure 3.3. How advanced metering infrastructure (AMI) works.

Source: https://www.german-metering.com/wp-content/uploads/advanced-metering-infrastructure-ami.jpg.

Importance of Smart Metering in Modern Utilities

Smart metering has become a hallmark of modern utility management, breaking down the inefficiencies of traditional metering systems. Smart meters allow for the collection of granular use data that can be accessed in near real-time, providing both customers and utilities with actionable intelligence. These advancements promote effective energy management and planning for resources, which are essential at a time that is characterized by fluctuating energy demands and the infusion of intermittent energy supplies. With data-driven approaches, utilities are in a position to optimize their business operations, resulting in better grid efficiency and reliability. The integration of smart metering points to the necessity of an adaptive and responsive energy infrastructure that addresses contemporary energy consumption problems while, concurrently, fostering sustainable approaches in the sector (Chin et al., 2011). Besides the improvement in performance, smart metering plays an important role in customer involvement and satisfaction. With accurate consumption data and simple interfaces, consumers gain more knowledge about the way they consume energy, triggering more responsible behavior. For instance, smart meters enable customers to monitor their electricity consumption in real time, which enables them to identify when they consume the most and make intelligent adjustments in order to reduce their overall bills. Alongside, the extensive availability of such information facilitates a variety of demand-side management programs that can lead to cost savings for consumers and utility companies both. Essentially, the value of smart metering extends beyond mere technological innovation; it promotes a more collaborative partnership among utilities and consumers, as well as energy efficiency schemes (Borghetti et al., 2013). In addition, smart metering opens the way to more sophisticated management of the grid with the help of AMI. Smart meter infrastructure allows two-way communication between the smart meter and the utility, facilitating not only real-time observation but also rapid response to grid disturbances. In this regard, smart meters are one part of the smart grid solution required to solve the increasingly complicated issues of electricity distribution in modern economies. The technological innovation in information and communication technology has greatly facilitated the credibility of smart grid operations to provide adequate space for alternative sources of energy without compromising security and privacy as primary necessities. Thus, smart metering is not that much of a technology stride; it is a shift from conventional wisdom in the direction of smart energy systems that are significant for sustainable growth (Chin et al., 2011).

3.1. HISTORICAL CONTEXT

The development history of smart metering and Advanced Metering Infrastructure (AMI) is inextricably linked with the evolution of the energy distribution grid throughout history. The first system was a predominantly uni-directional one in which electricity would flow from central power plants to users with no instantaneous feedback loop. With increasing energy demands and realization of the complexities of power generation and transmission, it became apparent that a more sophisticated system was needed. The evolution toward Smart Grids (SGs) brought with it a watershed shift in this paradigm, with the potential for bidirectional flow of energy as well as enhanced communication among service providers and consumers. This evolution was accompanied not only by the redesigning of physical infrastructures but also by the adoption of new technologies, such as the Internet of Things (IoT), to facilitate real-time monitoring and control, providing the foundation for the integration of smart metering solutions that characterize modern energy systems (Copeland et al., 2019).

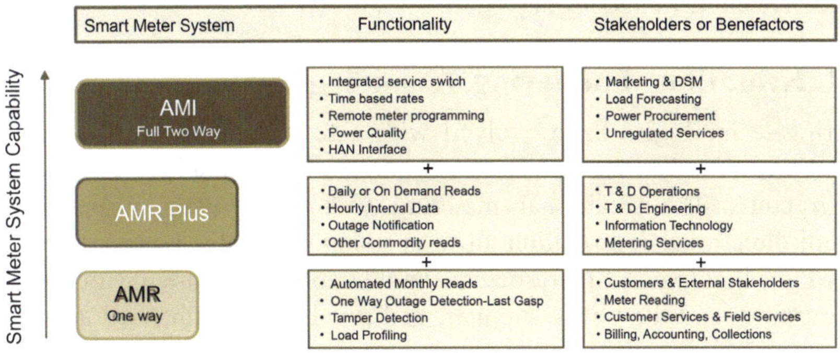

Figure 3.4. Evolution of smart metering.

Source: https://www.mdpi.com/applsci/applsci-06-00068/article_deploy/html/images/applsci-06-00068-g001.png.

Expansion of renewable energy sources, particularly wind power, has dominated the installation of AMI. As countries such as South Africa began introducing wind farms into their National Grids, issues with the volatility and unpredictability of wind power became a concern. Issues that required new solutions to maintain the security and reliability of power grids. As a response to these problems, utilities worldwide began exploring more into Demand Side Management (DSM) and Demand Response (DR) programs. These programs allowed for greater management of energy consumption and generation during peak times and unexpected fluctuations. Nevertheless, the successful implementation of DSM and DR programs was dependent on advanced technologies, including smart meters and advanced communication systems, underscoring the historical significance of AMI for managing and optimizing the distribution of energy in a greener power environment (Dayabhai et al., 2015). The historical context of smart metering and AMI is the coincidence of technological advancement and the need for improved measures of energy management. Having been demonstrated in early days through adoption, the focus then shifted towards enhancing grid resilience and integrating renewable sources of energy. Electric utilities and policymakers were aware that AMI not only facilitated real-time data collection but also empowered the consumer by driving a participatory approach to energy consumption. The shift to decentralized forms of energy generation and consumption offered greater flexibility and efficiency, culminating in sustainable energy consumption practices. In addition, with on-going developments in the energy infrastructure, lessons learned from early deployments of AMI will be instrumental in directing future development so that smart metering remains at the vanguard of addressing both new and prospective energy challenges (Figure 3.4).

3.1.1. Evolution of Metering Technologies

Metering technologies have evolved with a revolutionary transformation in the last two decades due to advancements in telecommunications and digital infrastructure. The traditional metering technology with manual reading and uni-directional information flow is being gradually replaced by Smart Metering technologies. Smart meters enable two-way communication between the consumer and the utility supplier, and this leads to enhanced accuracy in data transmission and better energy management. Worldwide, this shift has been key to addressing the implications of increasing energy demand, energy wastage, and the requirement for increased reliability and security, informing the inherent objectives of Smart Grids that continue to reshape the metering paradigm. When these new technologies come to be utilized by the utilities, they also address the complexities of integrating various devices in generation, transmission, and consumption, ultimately paving the way for an integrated and efficient energy system. Internet of Things (IoT) integration within Smart Metering technologies is central to the shift of paradigm from legacy metering solutions to Advanced Metering Infrastructure (AMI). IoT devices like sensors and actuators have an inescapable function in tracking and managing energy use at various locations in the power supply chain. This bi-directional energy flow not only allows utilities to optimize energy consumption and grid operation but also allows consumers to receive real-time data to make effective decisions on their energy consumption patterns. Through greater connectivity and automation between the elements of the grid, IoT applications highlight the potential for greater operational efficiency and cost savings. Metering technologies, especially, are not mere measurement devices anymore; they are part of a sophisticated network serving as key components to support energy management in Smart Grids, following the need for innovation in the industry. While the promising advancements, the development of metering technologies also has some challenges, primarily interoperability, security, and standardization. As various manufacturers create devices for Smart Grids and AMI, the requirement for standardized communication formats becomes even more vital to enable seamless interconnection of heterogeneous technologies. Research has also indicated the necessity of concerted standardization efforts, specifically in Europe, to attempt to standardize communication protocols that can be implemented on a wide range of devices used throughout the entire energy supply chain. Solutions to these communications-related problems are critical to creating future-proof smart grid networks that not only enhance efficiency and resilience but also support new models of consumer engagement and energy management strategies. Generally speaking, metering technology development must struggle with these complexities while remaining consistent with the preeminent goal of building a smarter and more sensitive energy grid, whereby customers benefit from enhanced services and utilities achieve maximum operating capacity (Chin et al., 2011; Copeland et al., 2019).

3.1.2. Transition from Analog to Digital Meters

Analog to digital meters migration is the giant step in metering technology necessary to develop Smart Metering and Advanced Metering Infrastructure (AMI). Analog meters, the traditional standard, utilize mechanical components to measure energy consumption, leaning towards causing errors and limited data collection capability. Digital meters, in contrast, use advanced electronic systems which allow for accurate measurement of energy consumption with real-time data. This shift not only enhances the accuracy of measurements but also enables two-way communication between the utility companies and consumers, enabling real-time feedback for consumption patterns. As mentioned in existing literature, the integration of digital metering technologies forms the basis of smart grids, which are able to optimize energy handling and delivery (Axon et al., 2014). The benefits of digital meters extend beyond the mere gain in precision; they are also key to promoting energy conservation and efficiency. Digital meters provide accurate data regarding consumption habits on the consumer's side, allowing for better energy management choices. Consumers are able to adjust their consumption patterns in line with instantaneous feedback, thereby preventing wastage and energy expense. Besides, digital technology makes possible the utilization of time-of-use pricing, where customers can opt to pay less off-peak. This not only empowers the customer but also stabilizes the grid as a whole by demand-side management. Moreover, effective rollout of digital metering campaigns, as noted in European Commission reports, shows growing awareness and deployment of these technologies in various energy markets (CATALIN-FELIX et al., 2013). In addition to improving accuracy and efficiency, digital meter transition poses significant issues of data protection and user privacy. Owing to the vast amounts of information generated by digital meters, concerns about the security of personal data and risk of unauthorized access to information on utility consumption become a priority. As regulators and utility companies look to incorporate these new technologies into the grid, they must implement robust cybersecurity policies and regulatory programs to protect customer data while supporting innovation. The task lies in balancing the benefits of smart meter technologies with the need for increased protections because successful smart grid programs depend as much on public acceptance and trust as they do on technical advancement (Axon et al., 2014).

3.1.3. Regulatory Changes Influencing Advanced Metering Infrastructure (AMI) Adoption

Regulatory reform is one of the most important drivers in laying the groundwork for AMI deployment. Regulatory bodies and governments establish the parameters under which utilities roll out smart metering systems, and the key objective is enhanced business effectiveness and consumer engagement. With the requirement for energy on the rise and the shift to alternative sources of energy gaining momentum, effective Demand Side Management (DSM) is an essential requirement. Application of policies promoting DSM programs is an important function in mitigating the intermittency of renewable energy resources, such as wind power, and ensuring the stability of power systems. Utilities are thus encouraged to implement AMI as they are regulated for investment in technology when necessary to facilitate better collection of data and energy management. Such an investment environment reflects that of changing regulatory priorities with a goal of harmonizing energy sustainability and economic viability, eventually enhancing the performance and acceptability of AMI systems across various regions. Second, global cooperation has begun to standardize regulations around AMI adoption. The initiatives around Smart Grids, particularly those emanating out of EU and US collaborative models, underscore the necessity of an integrated approach to measurement techniques. In this respect, laws develop to embrace shared experience and learned outcomes of different jurisdictions to promote a harmonious regulatory environment. These international debates work towards harmonizing terminology and methods for Smart Grid technology to allow countries to frame their policies according to the best practices for implementing AMI. As highlighted in various pilot projects in the US and Europe, effective regulation tailored to local conditions can significantly aid in the building and integration of the AMI infrastructure, providing a solid foundation for end-to-end energy management systems. By harmonizing regulatory approaches, stakeholders will be able to enable improved adoption of AMI and overall efficiency in energy consumption. Finally, regulatory changes also address consumer data confidentiality and security concerns regarding the rollout of AMI. While smart meters collect and report extensive usage information, regulatory frameworks must ensure the privacy of such sensitive information while promoting efficient use of the data for energy management. Balancing exposure of consumer data for utility optimization with personal privacy can drive greater public acceptance and trust to roll out AMI technologies. As electricity utilities navigate this evolving landscape, consumer-oriented regulation that ensures data integrity and consumer protection must be in place to facilitate broader adoption of smart metering technologies. By defining standards and guidelines for data openness, regulators can pave the way for AMI technology and business process innovation, towards an even more sustainable and consumer-friendly energy future. The ongoing debate and evolution of these guidelines will be crucial in shaping the direction of AMI adoption at different levels and contexts.

3.2. COMPONENTS OF ADVANCED METERING INFRASTRUCTURE (AMI)

There are three fundamental building blocks that underpin AMI: smart meters, networks, and data management systems. Smart meters are the edge technology that is replacing legacy meters and enabling bidirectional consumer-utility communication. Smart devices not only record usage but also communicate in real-time, enhancing billing accuracy and consumption profiling. Furthermore, the deployment of smart meters is a must for the utility company in order to monitor the variations in demand and manage energy distribution optimally. This transformation is crucial in the optimal utilization of energy and aligns with the final aim of creating an efficient and responsive energy network, as highlighted in the latest literature (Chin et al., 2011). Enabling the smart meters, the communication networks are the bonding fiber of AMI to provide immediate data exchange. These grids utilize a variety of communications technology, such as radio frequency, cellular, and power line carrier, to offer secure and effective data streams in order to address the requirements of modern energy systems. Through solid communications links, utilities can execute near real-time monitoring of the grid, which lowers operating inefficiency by far. Further, the integration of these communication technologies is not a seamless process since interoperability and security aspects must be duly taken into account to ensure the protection of valuable consumer data and ensure reliability in energy management. As the literature observes, addressing these problems is key to unlocking the full potential of smart metering and the broader smart grid infrastructure (Copeland et al., 2019). Data management systems are the final but necessary component of AMI, offering the capability to store, process, and utilize the vast amounts of data collected through smart meters. Data management systems convert usage data into actionable intelligence so that energy management habits are made efficient. Advanced analytics enable utilities to anticipate demand behavior, manage energy distribution, and engage consumers with energy-saving programs designed specifically for them. In addition, the addition of Internet of Things (IoT) technologies to data management systems enhances inter-device connectivity further enhancing operations within the smart grid system. The mutual operation of these components not only promotes efficiency but also aids in the accomplishment of the general objective of sustainability, evidence of the paradigm-shifting impact of AMI in modern energy systems (Chin et al., 2011; Copeland et al., 2019) (Figure 3.5).

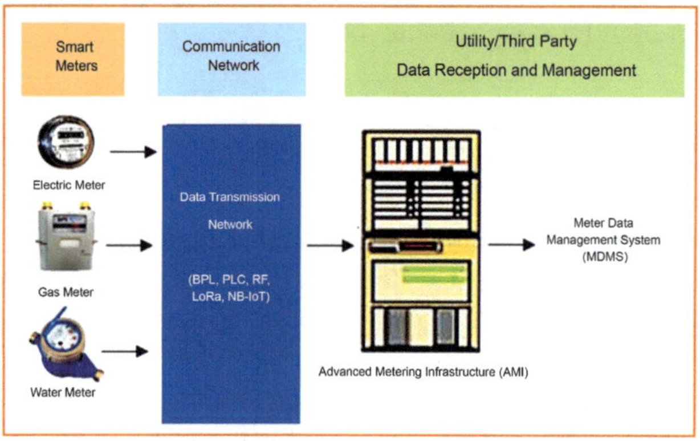

Figure 3.5. The components of advanced meter infrastructure (AMI).

Source: https://www.researchgate.net/profile/Muhammad-Imam-Nashiruddin-2/publication/339976572/figure/fig1/AS:908133459308546@1593527297460/The-Components-of-Advanced-Meter-Infrastructure3.jpg.

3.2.1. Smart Meters and Their Functionality

The advent of smart meters is a milestone in energy management systems, making possible a revolutionary way of energy consumption and communication for the grid. Manual reading was traditional metering, which typically meant time lags in achieving data and lower consumer engagement with their own electricity consumption. Conversely, smart meters enable real-time data, which supports consumers to track their electricity usage down to an hourly basis. Furthermore, support for advanced metering infrastructure (AMI) enables two-way communication between the consumer and utility, enhancing grid efficiency. As captured in findings on optimal energy consumption in smart grids, the near-real-time communication capability with the introduction of smart meters is the cornerstone of energy management systems designed for the future (Chin et al., 2011). This capability extends well beyond mere monitoring; it is a foundational element in the move towards renewable energy integration and dynamic load management. Coupled with the Internet of Things (IoT), smart meters enable the overall ethos of smart grids, significantly boosting operating efficiency. Smart grids are a highly advanced energy system where bi-directional energy exchange between providers and consumers is of prime significance. This development relies on devices that are inter-networked, such as smart meters, which monitor and control the flow of energy from consumption to generation. Therefore, smart meters not only record use but also enable automatic feedback to dynamic energy requirements, maintaining grid stability. This two-way system addresses most of the problems associated with traditional power systems, such as energy loss and reliability. Smart meters are a strategic IoT device deployment hub through the ability of complete monitoring and control networks that facilitate dynamic energy management (Copeland et al., 2019). The technology for smart meters has extensive implications in the areas of security and privacy with new challenges that must be addressed as part of the implementation of AMI. With more and more devices being networked together, the volume of data being generated raises issues regarding consumer data protection as well as grid integrity. The inclusion of smart meters in the energy framework necessitates a personal understanding of security measures to ensure that sensitive data is not exposed to unauthorized individuals. Additionally, the push towards communication standard harmonization, as in the context of smart grid interoperability, emphasizes the importance of stakeholders embracing concerted efforts to advocate for safe and efficient energy systems (Chin et al., 2011). Thus, while smart meters promise heightened functionality and efficiency, their deployment relies on robust frameworks for data security and interoperability on smart metering infrastructures (Figure 3.6).

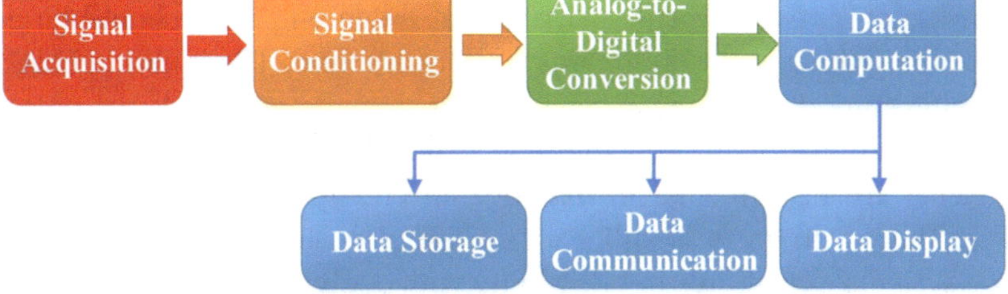

Figure 3.6. Smart meter functional block diagram.

Source: https://www.researchgate.net/publication/333654345/figure/fig1/AS:1143128 1244610831@1715965795914/Smart-meter-functional-block-diagram.png.

3.2.2. Communication Networks Used in Advanced Metering Infrastructure (AMI)

Communication networks used in AMI are important in facilitating the efficient management of energy consumption data. The networks enable two-way communication between smart meters and utility firms, a feature required for real-time monitoring and analysis. By gathering data on energy consumption behavior, the utilities are able to streamline their operations and thus provide better resource utilization and billing accuracy to the consumers. However, with the use of internet-based communications in SMNs, there are serious security implications involved. Besides that, with more devices connected in the network, cyber-attacks become increasingly feasible, necessitating robust security controls to avoid data leaks and ensure user privacy. Abdullah et al. (2015) explains that the infrastructure of such communication networks should have defensive countermeasures as a part of its structure to enhance security and operational dependability. The communication architectures in AMI involve different forms of networks, such as Home Area Networks (HANs), Neighborhood Area Networks (NANs), and Wide Area Networks (WANs). Each of these forms of networks is structured to have a specific role in the infrastructure that supports the overall operational functionality and efficiency of smart metering systems. For instance, HANs facilitate communication between smart devices within a home, while NANs connect different homes to utility companies. WANs offer greater geographic coverage. Recent advancements in Low-Power Wide-Area Networks (LPWANs) introduced new solutions for AMI communication with the ability to expand network capabilities while reducing power consumption. They also ensure improved efficiency and enable compatibility of disparate applications within the Internet of Things (IoT) scenario. Palit et al. (2021) highlights the paradigmatic significance of these multi-layered networks in supporting the dynamic nature of modern energy systems. In solving the challenges and opportunities in AMI communication networks, it is apparent that much emphasis should be given to ascertaining the current limitations of infrastructure. The building blocks of AMI architecture such as data management systems and analysis tools play an important role in load modeling and demand-response applications. Such components are instrumental in maximizing energy efficiency and reducing wastage. But as progress is still being made, it is necessary to compare the existing systems' scalability and flexibility with evolving technology trends and consumer needs. Innovations such as state-of-the-art communication protocols and data encryption mechanisms are crucial in order to meet the evolving world of smart meter functionality and security. As researchers continue to explore the optimization of such networks, possibilities for further integrations and upgrading in smart meter technology are limitless, paving the way for smarter energy solutions in the future (Figure 3.7).

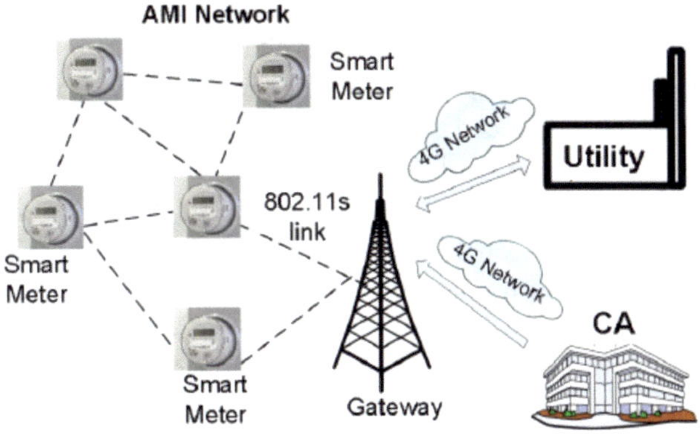

Figure 3.7. The considered advanced metering infrastructure (AMI) communication network.

Source: https://www.researchgate.net/profile/Kalid-Rabeh/publication/275893041/figure/fig1/AS:294507062022145@1447227363677/The-considered-AMI-communication-network.png.

3.2.3. Data Management Systems and Software

Evolution of smart metering and advanced metering infrastructure (AMI) has had a catastrophic impact on the collection, management, and analysis of energy data. At the heart of AMI is advanced data management systems and programs that enable near-real-time communication among different grid elements. These systems assist in efficient monitoring and optimization of electricity use on generation, transmission, and distribution grids. As detailed in existing research, interoperability among these systems has many challenges that require a cooperative approach to standardization across technologies and protocols (Chin et al., 2011). This partnership is critical to enhancing privacy and security and making sure data management systems can adapt to future innovation in smart grid technology. Data management systems within the AMI system perform a variety of critical functions, including data collection, storage, and analysis. They aggregate large amounts of data from deployed smart meters distributed throughout residential and commercial landscapes, allowing utilities to derive actionable insights into energy consumption behavior. For instance, the UC Berkeley study is an example of a systematic approach to building a comprehensive data framework, showing the integration of various sensing technologies as well as visualization tools for facilitating energy management (Peffer et al., 2010). These works are used to highlight the importance of building robust infrastructure that can facilitate proper communication between the end-users and the grid, which helps push energy conservation initiatives and build a more reliable grid. Secondly, the effective implementation of data management systems relies on the ability to harness big data analytics. With advanced software tools, utilities can analyze consumer behavior, forecast demand, and design accurate energy-saving programs. The analytical tools help energy companies make rational decisions, thereby boosting operational efficiency as well as customer satisfaction. With the development of technologies, greater emphasis will be placed on reinforcing data security and user privacy without compromising the efficiency of energy flow management. The progress in standardization activities as presented in scholarly literature shows that frequent engagement between industry stakeholders and researchers is needed in addressing the evolving needs to be fulfilled by smart metering systems in the future (Chin et al., 2011) (Figure 3.8).

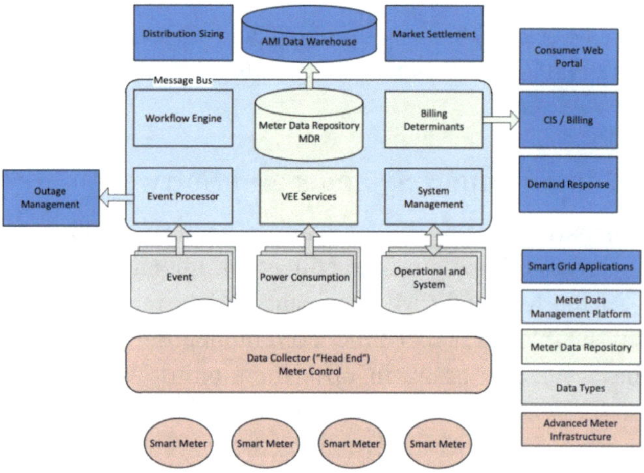

Figure 3.8. Master data management system infrastructure.

Source: https://research.aimultiple.com/wp-content/uploads/2024/01/Master-data-management-system-infrastructure.png.webp.

3.3. BENEFITS OF SMART METERING

The implementation of smart metering through Advanced Metering Infrastructure (AMI) presents numerous benefits that serve both the energy management of utilities and consumers. Two-way communication between consumers and energy suppliers is facilitated with the help of smart meters, allowing dynamic interaction in terms of energy usage and price. This transition towards bi-directional information exchange, as presented in recent literature, helps maximize energy use and promote energy-saving behavior on the consumers' side. With real-time information available, customers are now capable of making informed decisions about their energy use, which means higher efficiency and long-term cost savings. Additionally, the technology assists utilities to balance demand and supply, thus increasing the overall reliability of the power grid as production is sized to meet future needs as suggested in Duran et al. (2008). In addition, smart metering systems significantly increase increased operational efficiency for utilities.

Figure 3.9. The benefits of smart meters.

Source: https://www.innovationinbusiness.com/wp-content/uploads/2024/01/Smart-Meters.jpg.

By automating the process of data gathering, utilities are able to significantly reduce the cost of meter reading and billing errors. Incorporating smart grids with IoT devices facilitates real-time monitoring and regulation of energy distribution, resulting in improved reliability and reduced loss of energy. Utilities can install smart meters to detect outages in real time and respond more effectively to disruptions in services, thus cutting down on consumers' downtime. This change towards a stronger energy infrastructure is underscored by the revolutionary nature of smart metering as is treated in the context of future developments presented in Copeland et al. (2019). Utilities can, in return, manage resources more optimally, which results in better service quality as well as better customer relationships. Lastly, the green benefits of smart metering cannot be overlooked. By inducing energy-efficient action, smart metering has a prominent role to play in reducing greenhouse gas emissions and promoting the utilization of renewable resources. As the user maximizes their utilization according to real-time feedback provided by smart meters, energy demand can be better aligned to sustainable action. This reduction in peak-hour energy consumption not only reduces the operation time of fossil fuel-based power plants but also supports the integration of cleaner energy sources. The gradual evolution towards a Smart Grid, as outlined in Duran et al. (2008), necessitates new solutions for growing energy demand while ensuring reliability. Therefore, the benefits of smart metering extend beyond individual cost reduction to more encourage overall sustainability and environmental stewardship in the energy sector (Figure 3.9).

3.3.1. Enhanced Energy Efficiency

The advent of Smart Metering and Advanced Metering Infrastructure (AMI) has initiated a significant paradigm shift towards enhanced energy efficiency in contemporary power systems. With bi-directional communication via smart meters in place, consumers now have the ability to view their energy consumption patterns in a different manner. This transparency allows users to make smart decisions regarding their usage, allowing them to better control usage at peak hours. Heightened energy awareness, in conjunction with up-to-date information, not only helps to reduce individual energy costs but also becomes a quite significant means of alleviating stress on the overall power grid. As a result, utilities optimize distribution and generation capacity, which leads to reduced energy wastage and improved energy reliability. Smart metering function is therefore pivotal in developing energy awareness culture for both consumers and utilities, ultimately driving system changes in energy efficiency. In addition, greater energy efficiency through AMI extends beyond consumer behavior and touches on the very infrastructure of energy supply and distribution. IoT-enabled smart grids offer real-time management and monitoring of the flow of energy, significantly improving the coordination between generation, transmission, and consumption. The integrated framework offers near-real-time communication among the elements with scope for adaptive change to changing energy needs and circumstances. As highlighted in research, such systems address key issues of lost energy, peak demand, and grid stability by applying a networked approach to energy management (Copeland et al., 2019). In addition to enhancing efficiency in the consumption of energy, such optimization also encourages more environmentally friendly consumption of energy through lower environmental footprints and efficient incorporation of renewable sources into the grid. Also, the complementarities of smart metering technology and communication innovations are accompanied by numerous challenges and prospects for improved energy efficiency in smart grids. The capability to install interoperable and secure standards of communication is paramount in guaranteeing the successful deployment of intelligent smart grid systems. Issues of this sort, such as privacy and the need for robust security controls, represent obstacles that must be addressed in order to achieve consumer confidence in these technologies (Chin et al., 2011). By focusing on inclusive communication strategies, the energy sector can achieve peak efficiency while simultaneously sustaining heterogeneous energy resources and augmenting overall system resilience. In brief, in crossing the hurdles, the convergence of smart metering, advanced communication technologies, and smart grid management systems can transform energy efficiency into a reality, making the energy industry more sustainable as well as fiscally prudent (Figure 3.10).

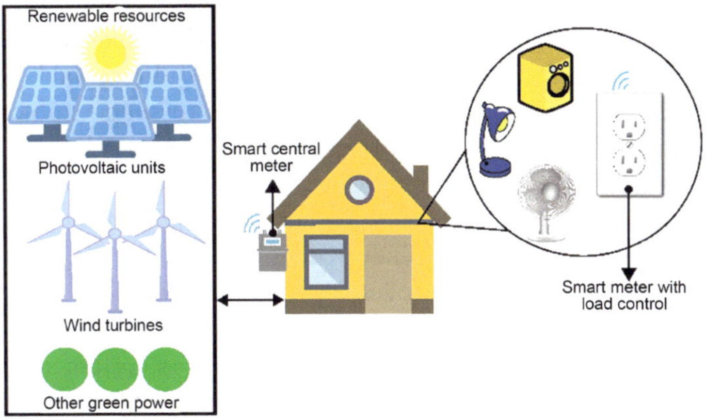

Figure 3.10. Environment of usage of the smart meter with load control.

Source: https://www.mdpi.com/sensors/sensors-22-07536/article_deploy/html/images/sensors-22-07536-g001.png.

3.3.2. Improved Customer Engagement and Awareness

The use of smart metering and advanced metering infrastructure (AMI) has transformed customer awareness and engagement in the energy sector significantly. By enabling real-time data access on consumption, consumers are enabled to monitor their habits better. This helps them make informed decisions on energy usage, lowering costs and contributing to greener lifestyle choices. In the context of the Smart Meter Implementation Program (SMIP) in the UK, the program aims to fit smart meters in every household based on an assumption that when consumers become more knowledgeable about energy, this will lead to energy savings ranging from 5%–15% (Hielscher et al., 2017). The worth of savings not only propels proactive action in optimizing energy but also engages consumers in broader discussions around energy efficiency and sustainability.

Moreover, the socio-technical barriers that go into the utilization of AMI render it pertinent to consider whether both technological competence and consumer mindsets should be addressed. Successful adoption not only needs to resolve technicalities, such as perfect data transmission and user-friendly interfaces, but also incorporate consideration of the social forces that may affect consumer adoption. The SMIP experience reflects that overcoming consumer resistance and ambivalence is critical, particularly in the context of data privacy issues and technology complexity perceptions of smart meters (Hielscher et al., 2017). By engaging in a debate on these matters, energy utilities can enhance customer participation as well as stimulate a transparency and trust culture, thereby facilitating an easier transition to smart energy solutions. Lastly, the contribution of smart grids and AMI to the low-carbon energy future should not be overemphasized. Smart grids are made to blend renewable sources of energy as well as stream electricity supplies so as to support global climate objectives. Regardless, the success of these efforts largely relies on consumer participation and acceptance by the public. Consumer outreach programs that educate consumers on the benefits and functions of smart meters can facilitate comprehension and acceptability of new technology. By directing technological advances at consumer needs and reducing concerns regarding safety, privacy, and cost, energy utilities can raise customer awareness and promote uptake of smarter energy habits. This approach not only increases customer engagement but also contributes towards the bigger objective of having sustainable energy infrastructure (Figure 3.11).

Figure 3.11. An intelligent way to reengage customers with smart meters.

*Source:*https://escalent.co/wp-content/uploads/2021/06/Engaging-Through-Smart-Meters.jpg.

3.3.3. Real-time Data Collection and Analysis

The innovation of smart metering technologies and Advanced Metering Infrastructure (AMI) has transformed the landscape of energy data management through real-time data collection and analysis. This new technology enables utilities to harvest a real-time flow of data regarding energy consumption patterns, thus billing more accurately and operating more efficiently. With the deployment of sophisticated sensors and communications technologies-based smart meters, utilities can gather data in real-time rather than relying on periodic readings. This real-time supports dynamic pricing systems so that customers can reschedule their usage to off-peak periods, ultimately leading to energy consumption optimization and reduced stress on the grid.

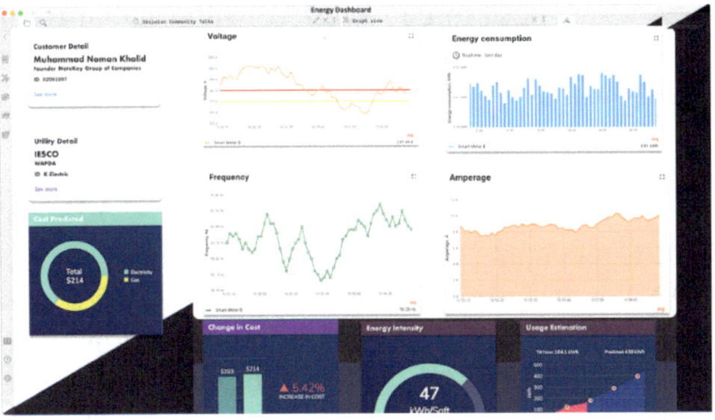

Figure 3.12. Smart meter data analytics.

*Source:*https://miro.medium.com/v2/resize:fit:1100/format:webp/1*_my7RUHYnZvFwe-c01yWRA.png.

Additionally, real-time information allows utilities to monitor system behavior constantly, thereby being able to proactively predict and react to outages and service problems, according to smart grid communications challenges literature (Chin et al., 2011). In addition to optimizing operational efficiencies, real-time data capture and analytics enable consumer interaction and energy management function. With detailed insight into their energy consumption behaviors presented to them, consumers are more informed and proactive in how they use energy. With such openness, residential houses and commercial buildings are able to participate in demand response programs actively, where they can adjust their usage according to real-time price signals or the requirements of the grid. Through end-user empowerment, AMI not only fosters energy efficiency but also shapes behavioral changes that can lead to real cuts in overall energy consumption. In a time where discussion of energy efficiency and sustainability is on the rise, real-time data is a valuable tool to promote consumers and businesses toward sustainable energy practices (Burstein et al., 2011). Alongside, the implications of real-time data extend beyond utility management and on-the-fly user engagement; they also constitute a vital piece in the development of smarter, tougher grid infrastructures. Integrated communications platforms allow for seamless exchange of information between various components of the energy system, from generation to transmission and usage. Interconnectedness is vital in promoting interoperability in smart grid systems for facilitating seamless coordination among various stakeholders. Periodic synchronization of data enables utilities to refine energy distribution, streamline grid operation, and undertake immediate maintenance actions. Effortless communication in such a fashion is beset with numerous challenges in the guise of security threats and data privacy, which need to be navigated with caution for the maximum utilization of real-time data in smart metering and AMI systems (Chin et al., 2011) (Figure 3.12).

3.4. CHALLENGES IN IMPLEMENTATION

Implementation of migrating to smart metering and advanced metering infrastructure (AMI) is not an easy process, as it involves a myriad of challenges. One of the key challenges is the present electricity grid infrastructure, which in most instances is outdated and not efficient enough to support advanced technologies. Legacy grid is one-way energy flow-based and not flexible to take care of alternative energy sources and dynamic electricity demand. As indicated in the observation that the aging power grid has been inadequate to manage challenges at hand, a wide-overhaul of technology as well as management strategies is required for successful reorganization into a smart grid (Borghetti et al., 2013). Without addressing these root issues, the implementation of smart metering solutions may face significant hurdles, from energy distribution wastage to being unable to harness renewable resources effectively. In addition to infrastructural problems, technological complexity is another key barrier to smart metering and AMI system adoption. The operation of a smart grid relies heavily on near-real-time communication between various segments of the network, i.e., generation, transmission, distribution, and loads of consumers. Interoperability of these elements relies on addressing a series of communication-related challenges. As indicated, network security and privacy are critical issues that should be taken seriously when installing smart metering systems (Chin et al., 2011). Poor control of such communication challenges can result in increased vulnerability to cyberattacks, which may deter consumer trust and discourage the extensive use of smart metering technologies. Moreover, the regulatory framework can significantly influence the challenges faced with the installation of smart metering and AMI. Development of a robust regulatory environment is critical in defining the right standards for interoperability and data privacy and collaboration among entities. However, the existing regulations do not necessarily adapt to the innovations brought by smart grid technologies. This is likely to cause inconsistency and confusion in the regulations, further creating barriers for utility providers wanting to deploy smart metering solutions effectively. To surmount the above challenges, the stakeholders must engage in proactive communication to make sure regulations evolve as technology advances. Through such collaboration, the stakeholders may establish a framework of innovation while addressing crucial industry challenges, ultimately enhancing the success of deploying smart metering (Figure 3.13).

Figure 3.13. Challenges in smart metering deployment for utilities.

Source: https://cdn.prod.website-files.com/67139b4944f3d6b890cda082/6888a50f2133a4169f836359_67989bc56f51bf4f70b19159_imresizer-1738054457397.jpeg.

3.4.1. High Initial Costs and Investment

The transition to Smart Metering and Advanced Metering Infrastructure (AMI) is always accompanied by the problem of huge initial investment and plenty of investments. It is the cost of money that arises from several reasons including the procurement of better metering devices, installation charges, and the need to embark on comprehensive grid upgrade. Utilities and governments must make considerable sums not only for the meters themselves but also for the ancillary communication infrastructure that enables data to be transmitted and the system to be controlled. The upfront financial investment can be a dampener for stakeholders, especially in societies already running tight financial tightropes. But proponents assert that the long-term benefits of enhanced energy efficiency and reduced operational costs ultimately could pay for themselves. Also, integrating Smart Metering systems in existing networks requires significant financial investments in training the workforce and installing cybersecurity measures. A capable workforce to handle new technologies becomes a necessary requirement, for which investment in training programs and upkeep is needed. Where traditional metering systems are already in place, the cost incurred in system swapping can be heavily daunting. Despite these challenges, the growing need for new energy infrastructure, supplemented by increasing consumer demand for holistic energy management systems, necessitates strategic financial planning and investment frameworks. The alignment can help bridge the transition and amortize the perceived risks of up-front costs. Finally, the roll-out of Smart Metering is influenced by various extraneous factors which ease or exacerbate up-front cost burdens. As the electricity sector matures, innovative models for financing and government subsidies are being contemplated to spur investments in AMI. Research indicates that successful adoption can potentially realize extremely substantial long-term savings by saving energy, improving demand response, and making distributed generation choices available. The realization of these benefits depends on the dedication of public organizations and private investors to invest wisely in AMI and Smart Metering technologies, promoting a paradigm shift towards effective and sustainable energy usage, as referred to in varied studies (Burstein et al., 2011; Pita et al., 2016).

3.4.2. Data Privacy and Security Concerns

As Smart Metering and Advanced Metering Infrastructure (AMI) technologies spread, security and privacy concerns about data become critical challenges that must be addressed. These systems gather enormous amounts of data, including intimate details about consumer energy consumption patterns. The same qualities that make AMI features attractive to consumers make it difficult to secure this information, making it a prime target for cyber-attacks. As noted in studies, the deployment of advanced functionalities in AMI constitutes significant cybersecurity threats that demand robust security systems to protect consumer privacy and keep sensitive information away from unauthorized access and breaches (Hansen et al., 2019). Secure system use is not just critical for individual privacy but also for building public confidence in smart grid technologies, which are themselves critical to the successful integration of renewable energy options. Additionally, the vulnerabilities embedded in AMI systems create real risks to consumers and utility companies. Smart meters, communication networks, and data management systems all constitute various attack surfaces that can be exploited by malicious parties. This reality emphasizes the importance of developing advanced security measures, particularly in Key Management Systems (KMS), which have been discovered to be a significant area of research in addressing security threats (Conti et al., 2018). Lacking robust

KMS strategies, energy data may be manipulated by adversaries, service delivery may be derailed, and the reliability of smart grid infrastructures undermined as a whole. Counteracting these security threats protects not just consumers but also strengthens the building blocks necessary for a smooth paradigm shift toward smarter, cleaner energy systems. With such issues in mind, the implementation of a broad security framework is critical to the future of AMI. Such a system must have a multi-faceted approach that integrates technological fixes with policy guidelines that define handling data and privacy controls. As seen in the AMI vulnerability example, it is necessary to integrate stringent security controls and targeted countermeasures aimed at mitigating constantly evolving cyber-attacks (Hansen et al., 2019). Furthermore, studies on secure key management techniques and ongoing security policy assessments must be prioritized. These initiatives will not only safeguard consumer data but enhance overall security of the power grid as well. With the culture of security-conscious innovation, stakeholders are able to ensure that smart metering technologies can benefit without taking a toll on data privacy and security (Figure 3.14).

Figure 3.14. Data privacy and security concerns with smart meters.

Source: https://media.licdn.com/dms/image/v2/D5612AQEXfnce7M_Nug/article-cover_image-shrink_720_1280/article-cover_image-shrink_720_1280/0/1723525512497?e=2147483647&v=beta&t=nSOqt3Hf-r4XssHqTn3oN04SIuaPmgEG-U8EnUHxqIp8.

3.4.3. Resistance from Consumers and Stakeholders

Roll-out of Smart Metering and Advanced Metering Infrastructure (AMI) has been severely obstructed by customer and stakeholder resistance, a large part of which is founded on concerns regarding privacy, data security, and so-called energy use control. Customers regard smart meters as intruding technologies that invade their personal lives, with there being worries regarding the harvesting of their information and its abuse. This resistance is further exacerbated by low levels of understanding about the advantages of smart metering systems, which are not properly communicated by electricity providers. As it has been noted, most smart metering initiatives globally are met with consumer resistance owing to these issues, highlighting the need to synchronize the aims of AMI with consumer demands to counter resistance and ensure market acceptance (Yesudas et al., 2020). Moreover, stakeholder engagement is crucial when fighting smart metering program resistance. Locally derived knowledge and neighborhood issues are highly effective in shaping the public perception and acceptance of smart metering programs. Utilities must take the initiative to engage with various stakeholder groups such as advocacy groups to learn and collaborate in overcoming concerns. Successful stakeholder involvement in discussions may lead to increased awareness of consumer needs and issues surrounding smart metering technologies. By aggregating these needs and responding to concerns voiced by consumers, stakeholders can ensure functionalities and designs of smart meters are in harmony with the needs and values of different consumer segments (Burstein et al., 2011). Such cooperation is not only required for successful introduction of smart metering systems, but also to promote trust in users. Finally, overcoming resistance from consumers and stakeholders involves telling a good story that generates the material benefits of smart metering technologies. The utilities need to enlighten the consumers on how smart meters will help in maximizing energy efficiency, cost savings, and efficient management of energy consumption. If consumers see the opportunity of lower energy bills and improved sustainability performance, resistance can be mitigated. Furthermore, training programs to demystify the technology and outline its safety aspects can lower fears regarding data privacy and security. Through the use of effective communication strategies combined with sound stakeholder engagement, utilities can create a climate which not only discourages protest but also facilitates consumers in embracing the transition to smart metering and AMI technologies (Yesudas et al., 2020).

3.5. CASE STUDIES OF ADVANCED METERING INFRASTRUCTURE (AMI) IMPLEMENTATION

The roll-out of AMI has been starkly illustrated in a number of case studies, all exhibiting both the strengths and weaknesses of smart metering technology. An excellent example in the US shows how a utility integrated AMI to enhance operations effectiveness as well as customer service. By using the smart meters and other sophisticated communication technologies, the utility obtained near real-time information gathering and analysis. This not only optimized the supply of energy but also empowered customers by equipping them with information regarding their energy consumption patterns. This case also underlined the imperative need for robust security measures to protect sensitive data transmitted over the grid, a challenge echoed by studies affirming the importance of interoperability in smart grid systems (Chin et al., 2011). Another typical example can be witnessed in European cities, where AMI has been implemented as a part of overall sustainability projects. Those schemes were utilized to reduce wastage of energy, enhance reliability, and move towards utilising renewable forms of energy. Utilities in those cases employed smart meters not as a substitute for ongoing infrastructure but as a component of a Smart Grid infrastructure. With bi-directional communication via advanced technology, utilities are at a better position to monitor and regulate energy flow. Importantly, the integration of Internet of Things (IoT) devices has transformed the potential for real-time energy management while simultaneously having the potential to offer greater consumer engagement. Still, the applications are challenged in terms of standardized communication protocols for effective operation on varied technologies (Copeland et al., 2019). AMI implementation case studies not only bring into focus progress but also shed light on the imperative for coordinated efforts in standardization that is crucial to unlock maximum potential of such technologies. For example, some nations have initiated collaborative initiatives between multiple stakeholders towards developing harmonized protocols offering long-term performance and security. This is particularly applicable in the case of technological advancements, wherein existing network facilities are required to evolve in order to maintain pace with the demands of modern energy management. In this regard, the experience of such implementations points toward the necessity of continued research and development of communication technologies for smart grid systems. Ongoing efforts by utility firms to address these challenges simultaneously indicate that there is dedication towards shaping energy management competence for the future (Chin et al., 2011).

3.5.1. Successful Advanced Metering Infrastructure (AMI) Projects in Urban Areas

Successful implementation of AMI in cities has evidenced immense potential to enhance energy management and sustainability. As urban centers grapple with increasing energy consumption and the demand for stable power supply distribution, AMI systems enable two-way communication among utilities and consumers. This connectivity not only promotes efficiency in energy usage through informed consumption but also allows the resilience of the electrical grid to operate. Urban environments benefit from AMIs ability to integrate green energy sources and smart grid technologies, hence paving the way for innovative energy solutions more specifically tailored to the unique needs of high-density urban environments. Specifically, these systems can sense inefficiencies and reliability problems, ultimately reducing utilities cost operation and encouraging higher customer satisfaction. Riding the wave of advancing communication technology has been critical to the success of AMI initiatives in cities. With the application of real-time data analytics, utilities are capable of predicting usage patterns and managing peak demand.

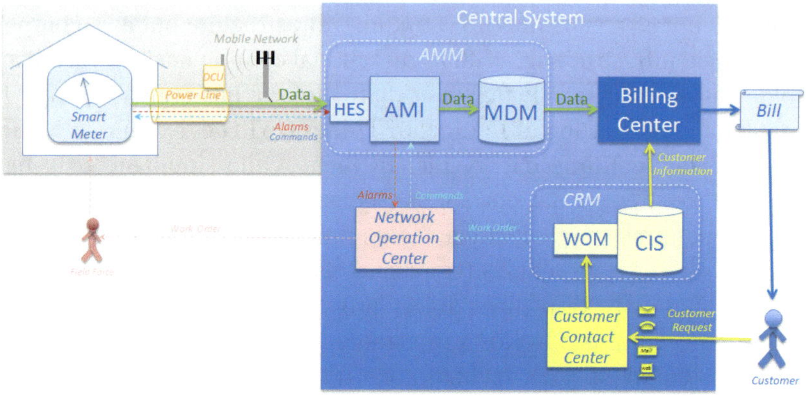

Figure 3.15. Successful advanced metering infrastructure (AMI) projects.

Source: https://media.licdn.com/dms/image/v2/D5612AQEIiKnh3jxdkQ/article-cover_image-shrink_600_2000/article-cover_image-shrink_600_2000/0/1705600272563?e=2147483647&v=beta&t=t5ZjgkLlocGqf8EgU8KoyeQPXlI0C30nfUApI4PbdnQ.

Through the integration of Internet of Things (IoT) devices such as smart meters and sensors, city grids are capable of enjoying enhanced monitoring and control over every component of the grid, from generation to consumption. This shift, characterized by the shift from traditional power grids to Smart Grids (SGs), addresses not only usual issues such as wastage of energy and security but also fosters concerted efforts towards standardization by stakeholders. Such concerted efforts, particularly in the European continent, aim to harmonize communications protocols to ensure future-proof operation and interconnectivity in many urban environments. Such advancements are also essential to deal with interoperability and grid resiliency issues, as noted by recent literature Chin et al. (2011) and Copeland et al. (2019). Urban AMI initiatives also honor the principles of social equity by correcting disparities in the availability and cost of energy. By comparing consumers with real usage data in real-time, AMI allows them to make intelligent choices about their energy usage patterns, thus saving them money. Moreover, the large data collected by smart meters also highlights the energy usage patterns of various demographic segments, thus allowing targeted programs to assist low-income groups with greater ease. The findings identified through AMI systems can shape policy and program design for energy efficiency and conservation to serve larger sustainability goals. Through urban transformation, the ongoing injection of smart metering technologies will play an increasingly prominent role in shaping the economic and environmental landscape of cities. Thus, the advantages of such successful AMI projects are not only operationally oriented but also establish a culture for sustainable energy consumption to the benefit of all urban residents (Figure 3.15).

3.5.2. Lessons Learned from Failed Implementations

Study of lessons learned from unsuccessful smart metering implementations is valuable in that it presents significant lessons on challenges related to the installation of Advanced Metering Infrastructure (AMI). An examination of past failure reveals that inadequate stakeholder interaction has been a pointer towards project failure. This includes inadequate communication between the utilities, regulators, and consumers, which can lead to conflicting expectations and resistance from consumers. Past projects in Europe, for instance, were resisted due to inadequate disclosure of the benefits of smart metering technology, resulting in lower adoption than anticipated. These participatory gaps only underscore the necessity of stakeholder analysis and plans for stakeholder engagement. Through embedding feedback loops in planning stages, utilities are able to build a supportive and informed user base, hence ensuring higher possibilities of successful implementations, as exemplified by the successful cataloging campaigns undertaken by the European Commission detailed in CATALIN-FELIX et al. (2013). Also, misalignment of technology and infrastructural needs underestimation of the infrastructural needs of smart metering systems have been detrimental in previous implementations. In the majority of the utilities, AMI projects were undertaken with an unduly optimistic vision of the current digital infrastructures capability without factoring the necessity for upgrades and interoperability with existing systems. The realization that robust data communication and management structures are integral components of successful smart grid evolution has gained prominence with the increasing passage of time. According to Ancillotti et al. (2013), a sophisticated understanding of communication systems is crucial for facile integration. The lessons learned from the failure underscore that technological development planning is poorly executed in a timely manner, leading to considerable economic losses as well as service disruptions. Hence, making sure that there is solid infrastructure set up before proceeding towards implementation has the potential to tackle the issues caused by technology adoption. In addition to stakeholder interactions and technology issues, economic factors are a major factor in the success of smart metering programs. Instances of combining budget constraints with excessive technological aspiration are a common trap. The majority of the utilities initially underestimated the long-term cost that went into staff training, back-end systems, and dealing with large volumes of data analytics. Particularly, it is critical that the human resource be equipped with the skills required for operational efficiency. Specifically, as noted through careful assessments of smart grid initiatives, effectiveness of operations is often derailed through inadequate investment in learning and development that leads to the general decline of projects. Therefore, it is essential for utilities to keep investments in funds towards technology and human capital investments as well to have a diversified approach in utilizing AMI successfully as propounded by Ancillotti et al. (2013).

3.5.3. Comparative Analysis of Different Regions

Smart Metering and Advanced Metering Infrastructure (AMI) rollout varies immensely from region to region, depending on local policies, technological advancement, and socio-economic considerations. In North America, for instance, intense focus on grid modernization and consumer engagement has led to large-scale implementation of AMI technology. Here, the utilities have precedence in installing smart meters that enable two-way communication and real-time data analytics in order to be capable of managing energy distribution optimally. As mentioned, the transformation of traditional power grids into Smart Grids (SGs) resolves long-standing problems such as wastage of energy and issues of reliability, enhancing capacity of service providers to accommodate rising energy demands (Copeland et al., 2019). This effort focuses on the North American regions' dedication to efficiency and sustainability with energy consumption. Europe, on the other hand, has a mixed landscape for AMI uptake, where different countries utilize differing methods with strict regulatory systems and environmental goals. Nations like Scandinavia lead high levels of penetration for renewable sources through sophisticated implementations of smart grid infrastructure. Germany and other nations utilize smart metering to assist their Energiewende policy, a nationwide ambitious energy transformation program aimed at maximizing energy efficiency and minimizing greenhouse gas emissions. European utilities have also embraced ICT-based technologies to extend monitoring and control intelligence across the power grid. This focus on the integration of a wide variety of renewable energy sources is a vision that is indicative of the region's objective of creating more robust energy systems, where reliability and environmental compatibility coexist (Adebisi et al., 2015). Developing countries are, however, at the initial stage of their AMI developments, which often experience unique challenges such as poor infrastructure, regulatory frameworks, and finance constraints. Sub-Saharan nations, for example, have massive challenges in implementing smart metering due to unstable power supplies and immature technological infrastructures. Nonetheless, there has been growing recognition of the potential of AMI, and pilot projects and international cooperation are aimed at improving energy availability and reliability. These regions are experimenting with low-cost solutions that utilize mobile technologies, which are able to circumvent traditional infrastructure challenges. Moreover, proper use of IoT devices in AMI systems can provide such regions with revolutionary opportunities to improve energy management, thereby facilitating economic development and meeting increased energy demands (Copeland et al., 2019).

3.6. FUTURE TRENDS IN SMART METERING

As technology advances, the future of smart metering is sure to see revolutionary changes that will significantly improve energy management and consumer awareness. Through the integration of the Advanced Metering Infrastructure (AMI), smart meters are improved in their ability to provide real-time data that makes customers better energy monitors. Such enhanced detail of data not only promotes energy efficiency but also makes it possible to design personalized energy solutions based on the individual consumption patterns. As an illustration, with the capability of consumers to see detailed statistics about their energy consumption, they are able to adjust their behavior, which may lead to reductions in energy expenditure and costs on the environment. Future technological innovations in smart metering will necessarily emphasize user involvement and data, going further in promoting energy efficiency and sustainability. The analytical capability embedded in next-generation smart metering solutions has the ability to transform the analysis and management of energy usage patterns. With greater data available, utilities can utilize next-generation algorithms and artificial intelligence to identify abnormal usage behavior in rapid succession. For example, the utilization of machine learning algorithms can facilitate the identification of deviations from standard usage patterns, allowing for early notifications to consumers and utilities. Such preventive measures can come in handy to help prevent energy wastage and address root cause health concerns, as has been outlined in recent research into behavioral trends on the basis of smart meter data. Using systems that detect subtle fluctuations in energy consumption, like 95.96% accuracy for monitoring sleep disorder in high-risk populations (Chalmers et al., 2025), the smart metering possibilities extend far beyond mere energy consumption to wide-ranging public health benefits. In addition, the coupling of smart metering technologies with other system integration within the smart grid will enhance overall efficiency and reliability of energy distribution. Successful deployment of a strong AMI allows the different stakeholders, including regulatory bodies, consumers, and utility companies, to share relevant energy data that supports shared decision-making processes. Trends indicate that as the smart grid keeps growing, so will the functions of smart meters to share information not just on consumption, but also the state of the grid. Thus, real-time analytics and data sharing will help utilities optimize load management more effectively and combine renewable sources. Detection of anomalous energy consumption with stunning accuracy — as much as 99.45% as is shown in studies employing advanced classification techniques (Chalmers et al., 2025) — indicates the potential of smart metering systems in creating a sustainable and adaptive energy future.

3.6.1. Integration with Smart Grids

The interfacing of smart grid and smart metering systems represents a revolutionary change in energy handling and distribution. As existing electric infrastructures are strained to cope with the modernity of current energy demands, the introduction of smart grids offers a solution that is fueled by emerging technologies. At its root is the ability of smart grids to enable real-time communication between grid components, from generation and transmission to points of use. This dynamic interaction is needed not only to achieve maximum energy dispatch but also to achieve maximum grid robustness and agility in reacting to demand and supply changes. The challenge of achieving such interconnectivity is multi-angled, involving technical, or operational issues but also necessitating security and privacy frameworks, which are absolutely needed in order to establish consumer trust and streamlined functionality in smart grid programs (Chin et al., 2011). Moreover, the incorporation of information and communication technologies (ICT) has played a crucial role in the transition to smart grids. The traditional electricity infrastructures—whose design is inflexible and unable to integrate renewable energy sources—have limited functionality, which has made it important to rethink the architecture of electricity distribution systems. ICT provides the reliability and flexibility necessary for the deployment of progressive energy solutions, thereby driving the development of interconnected and optimized smart grid systems. Through the application of smart metering infrastructure (AMI), utilities can collect detailed usage information that enables the application of intelligent demand-side management strategies. Proper application of ICT not only results in enhanced efficiency in operation but also plays a vital part in meeting regulatory policies, particularly those in pursuit of energy saving and sustainable conduct (Borghetti et al., 2013). Finally, interoperability is the driving principle behind grid smartening and AMI integration. To achieve the optimum extent of utilization of the potential of smart metering systems, there has to be standardization of communication protocols so that equipment of varying manufacturers and capacities can interact with and work with one another easily. This aspect becomes all the more necessary in handling the diverse technology environments that are present within geographical areas or countries. As stated, coordinated efforts in Europe are aimed at creating harmonized standards for these developments to allow these breakthroughs to ensure that the developed smart grid technology gains broader acceptance and operational functionality. By creating a supportive regulatory environment along with technological breakthroughs, the transition to widespread smart grids can help bring an energy future that is more sustainable, empowering consumers and enhancing system efficiencies (Chin et al., 2011).

3.6.2. Role of Internet of Things (IoT) in Enhancing Advanced Metering Infrastructure (AMI)

Combining the IoT with AMI significantly enhances the data analysis and gathering capacity, and it has become a most important part of modern power systems. Monitoring in real-time is enabled by IoT through the installation of a large base of smart meters and sensors that gather, report, and analyze usage data. This information exchange enables utilities to move away from traditional methods to more dynamic management systems, which are crucial in optimizing energy distribution and reducing wastage. Secondly, the two-way data exchange made possible by IoT devices offers greater sensitivity to fluctuations in energy demand, critical towards enhancing the overall efficiency and reliability of the electricity grid. By solving energy wastage and growing demand problems, IoT becomes a central pillar of the objectives of Smart Grids, ultimately achieving a more sustainable energy infrastructure (Copeland et al., 2019). Moreover, the application of IoT in AMI also enables the integration of renewable energy sources to the power grid. As the grid grows more sophisticated with the integration of distributed energy resources such as solar and wind generators, smooth interoperability among heterogeneous devices is crucial. IoT technologies play a key role in managing the complexity by enabling heterogeneous devices to effectively communicate with each other. Interoperability is usually facilitated by middleware architecture that unifies communication protocols so that it becomes easy to manage the challenge of integrating heterogeneous hardware. These advancements lead to better energy management policies that, apart from ensuring maximum efficiency in energy delivery, also enable the capacity of the grid to address variable inputs from renewables (Kammen et al., 2018). Furthermore, consumer engagement and empowerment are enabled by IoT in AMI through increased visibility of energy usage and costs. IoT-supported smart meters allow consumers to monitor real-time energy consumption via mobile apps or website portals. Such unmediated access to information on energy use patterns allows for a more informed consumer, and thus, encourages behavioral changes towards energy efficiency. Utilities, in return, can subsequently utilize available data to craft programs rewarding consumers for low usage during peak-demand times. As the consumer becomes actively involved in energy management, there arises the potential for mass energy savings to the advantage of both the consumer and the grid. Therefore, the role of IoT in complementing AMI is not solely technological innovation but changes the utility-consumer relationship dynamics for a collaborative model of energy management.

3.6.3. Potential for Renewable Energy Integration

The transition to Smart Grids (SGs) is a great future opportunity for renewable energy source integration. By using advanced metering infrastructure (AMI), SGs facilitate bidirectional information exchange between the consumer and the energy provider, enabling more responsive energy management systems. Such transition is paramount because traditional power grids are marred with inefficiency because of a one-way transfer of information and resources. With near-real-time communication, SGs are able to optimize the process of energy distribution more effectively so that variable renewable sources like wind and solar can be integrated into the system. This integration is important in order to meet rising energy demands as well as enhance reliability and sustainability within energy systems. By improved efficiency, AMI can potentially eliminate wastage of energy and design an energy system that is more responsive, setting the stage for a reduced carbon footprint and a shining, cleaner future (Chin et al., 2011). One of the greatest challenges that the integration of renewable energy presents is that of variability and intermittency in supply. Smart metering bridges this gap by enabling real-time measurement and analysis of the patterns of energy consumption and generation. This way, utilities can provide more precise projections of energy needs and schedule load distribution accordingly. For instance, during high renewable generation hours, smart meters can stimulate consumers to shift their use, which optimizes energy flow. In addition, as there are continuous advancements in communication technologies, SGs can encompass sophisticated prediction models that monitor consumption patterns to improve the decision-making of energy consumers and producers. The ultimate outcome is an improved ability to deal with renewable energy fluctuations, ultimately resulting in a flexible and stable power grid to accommodate the world's transition to cleaner sources of energy (Copeland et al., 2019). The incorporation of renewables into AMI and Smart Grids, in addition to its promise at the technological capability level, also promises societal and economic benefits. In enabling more efficient use of energy, AMI encourages energy conservation practices from the users themselves, consequently resulting in reduced bills and encouraging greater consumer engagement in energy management. Decentralized power generation, enabled by AMI, brings consumers into the role of active participants in the energy market, with the potential to produce new revenue through a system like net metering. More local power generation and use of renewable sources can enhance energy security by reducing fossil fuel dependence and reducing power transmission losses due to long-distance transport. Thus, the application of renewable energy through advanced metering infrastructure not only improves the operation of the grids but also propels broader societal change toward sustainability and economic resilience (Chin et al., 2011).

SUMMARY

Overall, the implementation of Smart Metering and AMI is a significant step in the management of energy and monitoring of consumption. These technologies offer a comprehensive platform for utilities to collect and analyze data in real-time, enhancing their operational efficiency and customer engagement. As attested to by the extensive database of smart grid and smart metering programs documented by the European Commission, the diversity of deployments throughout the region attests to an overall dedication towards energizing modernization. All such efforts not only contribute towards better resource management, but also make it possible to include renewable resources in the grid, and hence it's a huge leap towards sustainability (CATALIN-FELIX et al., 2013). Also, the concept of Internet of Agent introduces new ideas towards decentralization of energy management at residential scales. Through the treatment of interlinked appliances as separate agents capable of negotiating energy consumption, this method focuses on a paradigm shift in household energy consumption management. Some of the advantages include optimized energy utilization and enhanced load balancing at peak times, hence reducing the cost of electricity for consumers (Bui et al., 2018). Such innovations, therefore, mark an essential milestone towards a smart and efficient energy system. Lastly, the continued innovation of smart metering and AMI technologies will necessarily shape the future energy management landscape. It is up to policymakers, utility companies, and customers to embrace and accommodate these advances so that the benefits of improved data analysis and real-time data are maximized. As the global energy mix grows more sustainability-, grid-reliability-, and consumer-engagement-challenged, smart metering and AMI will be at the center of navigating the transition to smarter, tougher energy systems. Cumulative experience from ongoing projects and emerging technologies holds a lot for the new energy world of the future (Bui et al., 2018; CATALIN-FELIX et al., 2013).

This shift from traditional power grids to SGs is a paradigm shift in the management and distribution of energy. The process is fundamentally rooted in transcending the limitations of existing infrastructures, which are uni-directional data flow and lead to inefficiencies and wastage of energy. By adopting bi-directional flows of energy, SGs facilitate more dynamic interaction between service providers and consumers in the generation, transmission, distribution, and consumption of power. The convergence of AMI, encompassing smart meters, plays a vital role in this process because it facilitates the acquisition of real-time data and communication. As these devices are installed on a big scale across various industries, connectivity and automation become inevitable in order to monitor and monitor energy usage in an efficient manner. Thus, smart metering is more of an addition but a foundation stone of the Smart Grid program as a whole, which vows better efficiency in operations and reliability (Copeland et al., 2019). Apart from operational betterment, employment of smart metering technologies leads to more environmentally friendly practice. The smart grid employs information technology to optimize energy delivery in ways that promote the usage of renewable energy sources, optimizing efficiently to deal with increasing environmental concerns. By presenting exhaustively the patterns of energy usage to consumers, smart metering helps in fostering conservational interventions on an educated level. Consumers and companies, as a result, make more conscious choices regarding their use, leading to less energy waste and less carbon footprint. Furthermore, the expanded communication capability of AMI allows utility companies to optimize their supply based on real-time data regarding variations in demand, ensuring higher energy efficiency. In line with this, the smart grid not only solves issues of short-term operating but also positions itself within large-scale environmental agendas to make possible a sustainable future of energy (Ghazi et al., 2018). The move to smart metering and AMI is not without challenge, however, with one of the main challenges being cybersecurity. Application of two-way communication channels exposes the grid to cyber-attacks that may compromise the integrity of consumer information and energy supply. There are weaknesses in most areas that can be hacked from various angles and hence an in-depth study of potential attacks targeting confidentiality, integrity, availability, and overall accountability of the system is necessary. While multiple security issues have been explored in earlier literature, no single approach exists that consolidates various countermeasures under a single framework that can protect the entire infrastructure through it. As such, existing research is essential to develop robust cybersecurity initiatives that can adapt based on emerging threats so that the advantages of smart metering and advanced metering infrastructure are not succeeded by unintended vulnerabilities (Ghazi et al., 2018).

REFERENCES

1. Abdullah, M. D. H., Ahmad, Z., Mohamed, Z., Mohamad Afendee, M. H., et al. (2015). Attacks, vulnerabilities and security requirements in smart metering networks. https://core.ac.uk/download/153823362.pdf (accessed on 30th September 2025).
2. Adebisi, B., & Ikpehai, A. (2015). 6LoPLC for smart grid applications. https://core.ac.uk/download/189171263.pdf (accessed on 30th September 2025).
3. Ancillotti, E., Bruno, R., & Conti, M. (2013). The role of communication systems in smart grids: Architectures, technical solutions and research challenges. https://core.ac.uk/download/pdf/37831683.pdf (accessed on 30th September 2025).
4. Angulo Pita, I., De la Vega Moreno, D., Hernández, L., Uribe-Pérez, N., et al. (2016). State of the art and trends review of smart metering in electricity grids. https://core.ac.uk/download/188819846.pdf (accessed on 30th September 2025).
5. Axon, C., Balta-Ozkan, N., Cipcigan, L., Connor, P., et al. (2014). Scenarios for the development of smart grids in the UK: Literature review. https://core.ac.uk/download/155776124.pdf (accessed on 30th September 2025).
6. Borghetti, G., Menga, G., Procaccianti, G., Kok, K., Ardito, L., Morisio, M., Mo, D., et al. (2013). Smart grid technologies in Europe: An overview. https://core.ac.uk/download/pdf/11431923.pdf (accessed on 30th September 2025).
7. Bossart, S., & Giordano, V. (2012). Assessing smart grid benefits and impacts: EU and U.S. initiatives. https://core.ac.uk/download/38627650.pdf (accessed on 30th September 2025).
8. Bui, K. H. N., Camacho, D., & Jung, J. J. (2018). Consensual negotiation-based decision making for connected appliances in smart home management systems. https://repositorio.uam.es/bitstream/handle/10486/687652/consensual_hoai_Sensors_2018.pdf?sequence=2&isAllowed=y (accessed on 30th September 2025).
9. Burstein, A. J., Mulligan, D. K., & Wang, L. (2011). Privacy in the smart grid: An information flow analysis. https://core.ac.uk/download/288432753.pdf (accessed on 30th September 2025).
10. Chalmers, C., Fergus, P., Hurst, W., MacKay, M., et al. (2025). A smart health monitoring technology. https://core.ac.uk/download/42478735.pdf (accessed on 30th September 2025).

11. Chalmers, C., Fergus, P., Hurst, W., MacKay, M., et al. (2025). Profiling users in the smart grid. https://core.ac.uk/download/42477065.pdf (accessed on 30th September 2025).
12. Chin, W. H., Efthymiou, C., Fan, Z., Gormus, A., et al. (2011). Smart grid communications: Overview of research challenges, solutions, and standardization activities. http://arxiv.org/abs/1112.3516 (accessed on 30th September 2025).
13. Conti, M., & Ghosal, A. (2018). Key management systems for smart grid advanced metering infrastructure: A survey. http://arxiv.org/abs/1806.00121 (accessed on 30th September 2025).
14. Copeland, R., Crespi, N., Rehmani, M. H., Saleem, Y., et al. (2019). Internet of Things-aided smart grid: Technologies, architectures, applications, prototypes, and future research directions. http://arxiv.org/abs/1704.08977 (accessed on 30th September 2025).
15. Covrig, C. F., Fulli, G., & Olariaga-Guardiola, M. (2013). Insights from the inventory of smart grid projects in Europe: 2012 update. https://core.ac.uk/download/38630060.pdf (accessed on 30th September 2025).
16. Dayabhai, S. (2015). Investigation into the impact of wind power generation on demand side management (DSM) practices. https://core.ac.uk/download/39676257.pdf (accessed on 30th September 2025).
17. Duran, L., Ebrahim, T., Faulhaber, K., Serface, C., et al. (2008). CleanTX analysis on the smart grid. https://core.ac.uk/download/211336585.pdf (accessed on 30th September 2025).
18. El Ghazi, H., El Ghazi, H., Kaabouch, N., Mrabet, H., et al. (2018). Cyber-security in smart grid: Survey and challenges. http://arxiv.org/abs/1809.02609 (accessed on 30th September 2025).
19. Hansen, A., Barnicoat, G., Pepermans, G., Leiva, J., Shuaib, K., Nabeel, M., Schultz, P. W., et al. (2019). Smart meter security: Vulnerabilities, threat impacts, and countermeasures. https://core.ac.uk/download/480764249.pdf (accessed on 30th September 2025).
20. Hielscher, S., Jenkins, K., Kivimaa, P., Sovacool, B. K., et al. (2017). Vulnerability and resistance in the United Kingdom's smart meter transition. https://core.ac.uk/download/84310213.pdf (accessed on 30th September 2025).
21. Kammen, D. M., & Rodriguez-Molina, J. (2018). Middleware architectures for the smart grid: A survey on the state-of-the-art, taxonomy and main open issues. https://core.ac.uk/download/323310963.pdf (accessed on 30th September 2025).

22. Palit, R., & Paul, B. (2021). Smart metering in infrastructure-less communication environments and applicability of LoRa technology. https://core.ac.uk/download/401618974.pdf (accessed on 30th September 2025).
23. Peffer, T. (2010). UC Berkeley's Cory Hall: Evaluation of challenges and potential applications of building-to-grid implementation. https://core.ac.uk/download/288432727.pdf (accessed on 30th September 2025).
24. Yesudas, R., & Clarke, R. (2020). Measures to improve public acceptance of smart metering system identifying requirements for residential consumers. https://core.ac.uk/download/357380873.pdf (accessed on 30th September 2025).

CHAPTER 4

COMMUNICATION TECHNOLOGIES IN SMART GRIDS

Contents

INTRODUCTION

The transition towards sustainable energy systems calls for the incorporation of advanced communication technologies into smart grid networks. Smart grids represent a paradigm shift in electric power distribution that facilitates more effective interaction between energy producers and consumers. The focal point of this revolution is the integration of near-real-time data exchange capabilities, which support timely and efficient critical decision-making. As Axon et al.

(2014) observes, however, there continues to be a building of consensus in the fundamental definitions and components of smart grids. Such ambiguity is an expression of the complexity in creating effective communication structures, which must cater for technology integration and operational definitions that are subject to alteration over time.

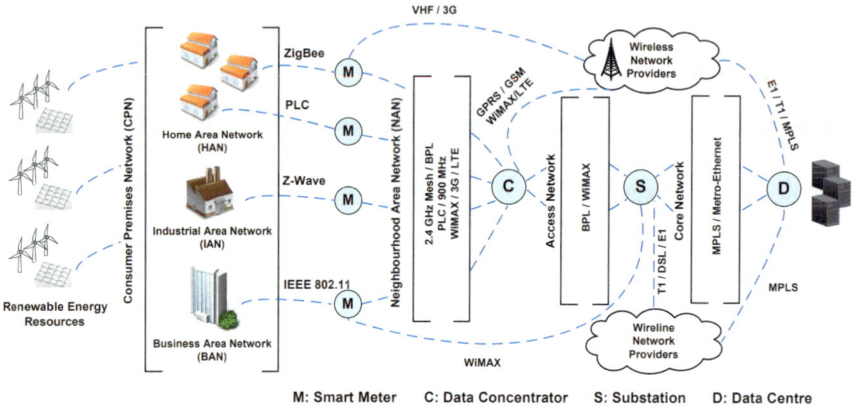

Figure 4.1. Communication technologies in smart grid.

*Source:*https://www.researchgate.net/profile/Mitja-Kolenc/publication/284159647/figure/fig1/AS:669498948452365@1536632396874/Communication-technologies-in-Smart-Grid-22.png.

Effective communication technologies in smart grids are pivotal to realizing maximum energy usage and enhancing the general resilience of electrical infrastructure. The interaction among different components of the grid—between generation and use—requires frictionless interoperable systems that can be immune to the dynamic forces of energy requirements. As explained in Chin et al. (2011), some among the issues encountered are establishing robust standards for communication protocols and ensuring data protection in smart grid systems. Resolving these challenges is critical because the deployment of smart grids will have many stakeholders involved, meaning there must be a coordinated effort towards ensuring functionality and user confidence across the system. Other than technical considerations, the deployment of smart grid technology entails significant policy and regulatory considerations which society will need to navigate. Smart grid deployment has broad political and financial backing, fueled by potential carbon-emission savings and higher economic efficiency. In order to actually achieve these advantages, however, it is essential that public acceptance and the sociotechnical circumstances surrounding these innovations are considered. As Axon et al. (2014) illustrated, efficient deployment of communication technologies relies not just on technological advancements but also on addressing individuals' issues regarding data security and privacy. Therefore, gaining a proper understanding of these various aspects will unveil an easier agenda for further innovation in smart grid communication technologies (Figure 4.1).

Definition of Smart Grids

Smart grid is a revolutionary step in power distribution technologies, characterized by the integration of information and communication technologies along with traditional energy infrastructure. Smart grids leverage advanced ICTs to provide real-time information exchange among utilities, consumers, and various sources of energy. The revolution facilitates better monitoring, management, and optimization of electricity grids with a view to delivering greater reliability and responsiveness in electricity supply. With the application of smart grid technologies, utilities can more precisely predict energy demand, integrate renewable energy sources, and better identify and rectify power outages. In the end, the smart grid paradigm enhances the efficiency of energy distribution and consumption in general while supporting initiatives for sustainability and reducing greenhouse gas emissions. The fusion of cyber-physical systems (CPS) and the Internet of Things (IoT) is at the root of what smart grids are all about. CPS consists of engineered systems capable of working together with physical and computational systems, enabling interaction with the world in an enhanced manner. In smart grids, this fusion provides devices such as smart meters, sensors, and controllable loads with the capacity to communicate and dynamically respond to real-time market prices and operating conditions. With the advent of the latest technologies in autonomous systems like driver-less cars and automated aviation, the fundaments of CPS are rapidly changing. The smart grid is therefore a prototypical case of how technologies can be applied to bring operational performance to unprecedented heights, ushering in smarter, more integrated energy systems (Marwedel, 2021). Application of edge computing (EC) in intelligent grids is one of the main innovations that continues to augment the role of ICT solutions in this field. This emerging model of computing allows for data processing and analytics at the vicinity of the location of information, thus lowering delays and raising responsiveness in electricity management. As outlined in EC and smart grid synergies research, interoperation of the two technologies facilitates a more effective energy distribution and utilization process. Through gathering large data from smart grid-enabled devices, utilities are able to achieve maximum operational efficiency, enhance demand response levels, and offer support for distributed energy resource management. This synergistic relationship not only enhances the performance of smart grids but underscores the importance of sustained research and innovation in edge computing applications in this new-wave energy landscape (Feng et al., 2020) (Figure 4.2).

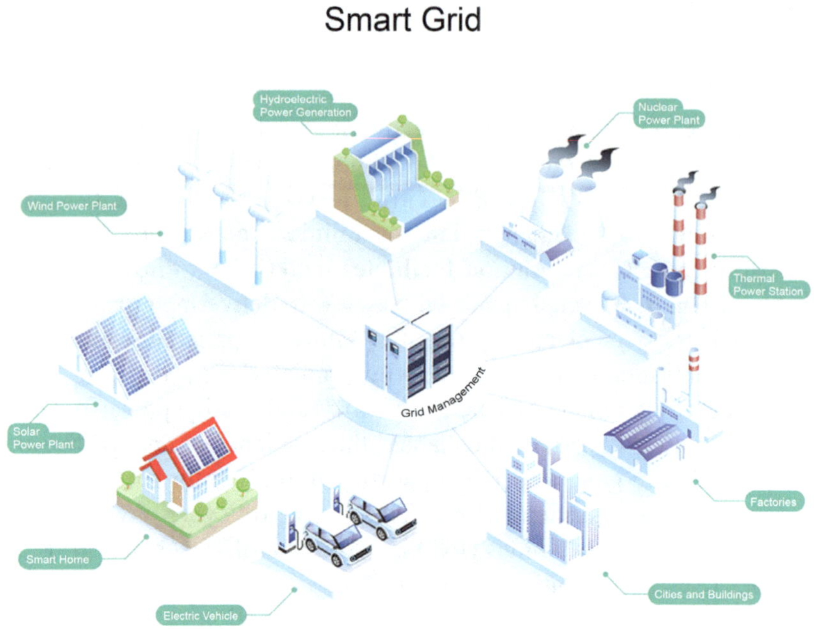

Figure 4.2. Definition of smart grids.

Source: https://codibly.com/wp-content/uploads/2022/12/What-are-smart-grids_5-1024x863.png.

Importance of Communication Technologies

The evolution of communications technologies has come to serve as a backbone of next-gen electric grids, giving the push towards smarter home and business infrastructure. With the fundamental shift towards low-carbon energy systems, smart grids have gained prominence as an answer to managing the increased complexity brought in by distributed energy resources efficiently. Communications technology facilitates bidirectional communication between consumers and producers, which is essential for attaining maximum efficiency and reliability of energy transmission. Not only can these technologies facilitate the real-time transfer of data, but also maximize the management of supply and demand variability. Further, as smart grids develop, the integration of high-end communication technologies can reduce energy loss tremendously and optimize grid stability. The prominent role that communication technologies play in facilitating smart grids cannot be overemphasized, as they provide the foundation for sustainable energy solutions in today's highly interconnected world. Privacy and data security are now at the forefront during the installation of smart grid infrastructure, with the need for secure communication networks protecting sensitive information. The traditional electrical infrastructure was beset with a host of problems, ranging from vulnerability to attacks and inefficiencies in distributing resources. Smart grids leverage advanced Information and Communication Technologies (ICT) that provide a flexible, reliable platform capable of managing existing energy needs. Through the provision of secure communication channels, actors can ensure the integrity and confidentiality of data captured by smart meters and other devices. In addition, efficient communication systems enable cooperation between various players like consumers, utility companies, and regulators. Achieving the acceptability of such technologies is based on lessening the possible risks involved in data safety and cybersecurity, which are crucial to the building of consumer trust and providing a global acceptance of smart grid projects (Axon et al., 2014). The deployment of smart grids effectively also relies not solely on technology but on dynamic policy and regulatory interaction with these communication networks. Policymakers must consider the socio-technical factors influencing the uptake of smart grid technologies, as the very complex environment involves enormous interactions among stakeholders, public institutions, private companies, and consumers. Incentives for smart grid initiatives from governments have been crucial in providing funding and resources for innovation (Borghetti et al., 2013). Further, the establishment of regulatory frameworks ensures that communication technologies in smart grids do not diverge from specific safety and operational standards, allowing for effective integration of various energy sources. As these frameworks continue to evolve, they will significantly shape the future of energy systems. Therefore, an understanding of the part played by communications technologies in these regulatory pathways is central to a successful transition to smart grids.

4.1. OVERVIEW OF SMART GRID TECHNOLOGY

Smart grid technology advancement is a revolutionary transformation in electrical power system management. One-way communication and centralized control grids are being dislodged by smart grids that emphasize two-way data interaction and decentralized decision-making. This transformation is primarily initiated by the requirement for more efficiency and sustainability in the use of energy. Information and communication technologies are central to this transformation because they enable the complementarity of renewable energy sources, facilitate demand-response programs, and support advanced metering infrastructures. By facilitating various elements of the grid to communicate and act in concert with one another, smart grid technology has the potential to optimize energy distribution and enhance system robustness, in reaction to infrastructure degradation and increasing power demands (Ancillotti et al., 2013). Fundamental to the functioning of smart grids is the communication infrastructure that facilitates interference-free interaction among devices. Advanced metering infrastructure (AMI), for instance, delivers the framework for data gathering and coordination between smart meters, data aggregators, and control centers.

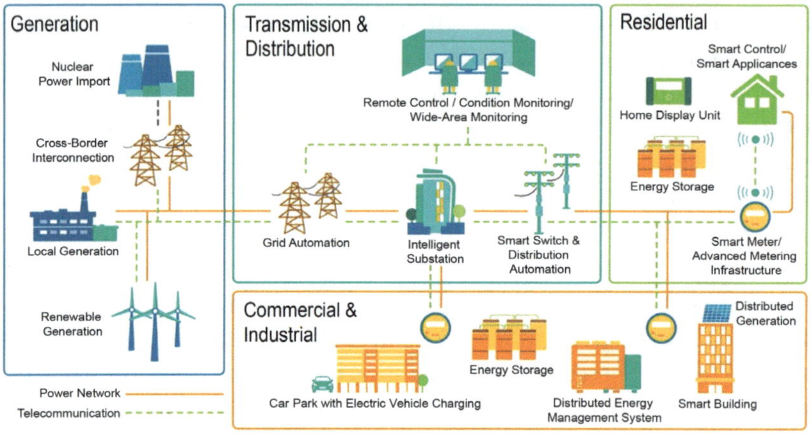

Figure 4.3. Overview of smart grid technology.

*Source:*https://www.elprocus.com/wp-content/uploads/2017/02/grid-overview.png.

The hierarchical structure of smart grids enables the inclusion of many small-scale networks, allowing for the scale required to span extremely large geographic areas (Ancillotti et al., 2013). This multi-layer approach not only optimizes the energy transmission efficiency but also the flexibility of the systems towards variable energy needs. Therefore, the effective engineering of network solutions is key to ensuring that the communication paths are robust enough to resist the dynamic nature of modern energy systems and adapt challenges such as bandwidth uncertainty and reliability (Ancillotti et al., 2013). Further, the communication issues in smart grids are two-pronged in nature and include reliability, timing, and quality of service (QoS) issues. Because smart grids interconnect a vast range of devices, being efficient with the communication as well as being resilient is among the highest priorities. A number of protocols have been developed to counter the limitations of traditional communication channels, including the Internet Protocol version 6 (IPv6) Adaptation for Low Power and Lossy Networks, specifically to counter the specific needs of AMI applications. Research has established that though these protocols enhance average network performance, possible vulnerabilities are present, particularly concerning packet loss and path dependability (Ancillotti et al., 2013). Resolution of such issues is paramount to realizing the full potential of smart grid technologies, enabling them to efficiently meet present and future energy demands (Figure 4.3).

4.1.1. Components of Smart Grids

The components of smart grids represent a significant leap in energy distribution and control, reconciling advanced technologies to enhance overall efficiency. Smart meters form the center of this system, which enables real-time consumption transmission of consumers to suppliers, hence enabling informed energy consumption and prompt responses to variations in demand. The meters also enable two-way communication, hence supporting more variable price structures and reduced maintenance outages. The application of sustainable energy sources such as wind and sunlight also supports the operation of smart grids to a larger extent by enabling decentralized energy generation. This resilience not only helps maintain the grids through the utilization of intermittent resources but also makes the grids less susceptible to failures and supply disruptions, a great reason why these components are essential to modern energy grids (Biswas et al., 2025). Besides smart meters, there are other advanced communication technologies that are integral components of smart grids. These include sensors, actuators, and communications networks that operate together to monitor system operation and create transparency of operations. These technologies allow for real-time transmission of information throughout the grid, allowing utilities to effectively manage resources, predict maintenance needs, and balance supply with demand. Surprisingly, the incorporation of cyber-physical systems elevates the cooperation between the physical and virtual worlds in the grid to unprecedented levels, enabling far-reaching functionalities such as demand response and self-healing outages. While these technologies significantly contribute to the reliability and efficiency of the grid, they add complexity, particularly in the domain of cybersecurity, necessitating robust protection mechanisms to repel new vulnerabilities (Mutlaq et al., 2025). Successful implementation of smart grids entails addressing compelling issues based on security, efficiency, and integration of several energy sources. The most imperative is cybersecurity, as the increasing reliance on communication technologies exposes the grid to vulnerability. For instance, with the smart grids governing vast networks of interconnected devices, susceptibility to cyber-attacks becomes more likely, undermining sensitive information and infrastructural integrity. To combat these threats, innovative security solutions like blockchain and novel cryptographic algorithms are being developed to enhance data security without compromising the performance of the system. Furthermore, an integration of artificial intelligence with data analysis in smart grid equipment can further optimize energy management. With these technologies, utilities can optimize grid performance, offer reliable delivery service, and easily manage the complexity in existing energy systems, indicating the multidimensional nature of smart grid components (Biswas et al., 2025) (Figure 4.4).

Figure 4.4. Understanding smart grid components.

Source: https://www.elprocus.com/wp-content/uploads/2017/02/smart-meter.png.

4.1.2. Benefits of Smart Grids

The introduction of smart grids, fueled by advanced communication technologies, offers deep benefits that promote the sustainability and reliability of energy systems. Smart grids apply the use of digital communication to improve energy management in a manner that allows efficient electricity distribution. Smart grids allow real-time monitoring and control of energy consumption, subsequently reducing carbon emissions and the cost of operations. Smart grids thus contribute towards environmental sustainability through efficient integration of renewable energy sources, addressing challenges of their intermittent nature. This convergence not only ensures reliable power supply but also results in energy efficiency by optimizing the use of available resources. Moreover, efficiency in smart grids for managing energy distribution plays a pivotal role in pursuing cleaner energy technologies, eventually enabling international collective efforts towards sustainable development (Biswas et al., 2025). Secondly, the smart grid contributes to maximizing operational efficiency by using advanced technology such as the Internet of Things (IoT) and artificial intelligence (AI). The two technologies have enabled the creation of smart energy distribution systems that respond to real-time information. This feature is particularly crucial for management of energy storage systems that has the core role to stabilize the intermittent supply of green energy sources. Smart grids utilize IoT-enabled energy storage to facilitate dynamic management of energy, predictive maintenance, and optimized energy usage. Improved control mechanisms mean grid operators can enhance the overall stability and reduce operational expenses, thus making the energy infrastructure stronger. This operational efficiency not only increases economic viability but also addresses the growing demand for electricity in a time frame characterized by the invention of electric vehicles and smart technologies (Ojo et al., 2025). The benefits of smart grids also echo in the increased power supply reliability to customers. By integrating advanced communication technologies, smart grids provide immediate feedback regarding energy usage, which allows consumers to make informed choices regarding their energy usage. Real-time communication enables user participation and promotes energy-saving behaviors among residential and business consumers. In addition, the automated fault detection and self-healing of smart grids significantly minimize power outages, which enhances overall service reliability. As such systems become robust solutions to energy management, they also make way for microgrids, which can operate in isolation and provide electricity during broader grid outages. Through doing this, smart grids not only contribute to enhanced energy sustainability but also address inherent issues of resilience and reliability within the modern electrical landscape, and as such are a requirement to the future of energy infrastructure (Biswas et al., 2025).

4.1.3. Challenges in Implementing Smart Grids

The installation of smart grids is facing some challenges that must be addressed to make them sustainable and efficient. The biggest of these challenges is the obsolescence of the old inefficient electricity networks that are not able to cope with the complexity of modern energy requirements and renewable energy sources. As recognized in the discussion of the inadequacies in existing infrastructure, they are outdated networks that lack capability to integrate new technologies such as Information and Communication Technologies (ICT) whose functional operation is required in smart grids (Borghetti et al., 2013). By failing to create a stable basis for upgrading these infrastructures, energy utilities will be prevented from adequately dealing with increasing electricity demand and integrating distributed energy resources in an optimal way, hindering the overall progress towards a more responsive future for the energy system. Furthermore, the transformation towards smart grids needs high investment and effort from various stakeholders, including government administrations, utility companies, and customers. The expense of rolling out advanced communications technology and upgrading existing infrastructure can be deterrent to investment and limit deployment efforts. The issue is exacerbated by regulatory regimes that are not necessarily welcoming to facilitating the accelerated deployment of smart grid technology. Disparity and synchronization among regulatory authorities, utility providers, and consumers need to be established in an attempt to develop policy encouraging innovation and investment in smart grid technology. The challenge thus does not only arise from technological advancement but also from the creation of a collaborative setting to produce a culture of shared responsibility in modernizing energy plants. Moreover, cybersecurity threats pose a significant hindrance to the effective establishment of intelligent grids. The more ICT-dependent such systems become for control, monitoring, and communication, the greater chances of cyber-attacks, which could potentially compromise energy security and consumer privacy. While big data analytics are as beneficial to ensure that energy is properly distributed and efficient, it has vulnerabilities that can be hacked by hackers. As cities use these technologies to reduce greenhouse gases and optimize energy management, they will also have to prioritize the steps needed to increase communication security (Bonetto et al., 2017). The task of creating a resilient smart grid infrastructure thus has technological as well as economic considerations but also robust cybersecurity that protects the integrity and reliability of the infrastructure in general.

4.2. ROLE OF COMMUNICATION TECHNOLOGIES

The application of communication technologies to smart grids revolutionizes the energy management and supply environment. With real-time data transfer, smart grids enhance operation efficiency to enable utilities to monitor and respond to electricity demand changes with unprecedented precision. Advanced communication protocols allow for the exchange of information among grid operators, customers, and distributed energy resources. This bidirectional flow of information facilitates timely decision-making and operational adjustments, eventually resulting in a more reliable power system. Moreover, with the capability to forecast patterns within traffic in power grid data, communications technology can aid the continual optimization of grid performance, as highlighted in the state-of-the-art Transformer-LSTM method presented in recent research.

This evolution signifies the central role that communication frameworks play in increasing grid smarts and alertness (Liang et al., 2025). With increasing integration of renewable energy sources into the modern power grid, the role of communication technologies is even more prominent. The smart grid employs sophisticated communication platforms to synchronize the variable generation of renewable energy with consumer demand, thereby improving grid stability and efficiency. This dynamic communication calls for advanced communication systems capable of handling large volumes of data at high rates and effectively. Communication technologies, including internet-of-things (IoT) devices, play an important role in facilitating the operations of home energy management systems and smart metering, both of which together allow consumers to engage interactively with their consumption.

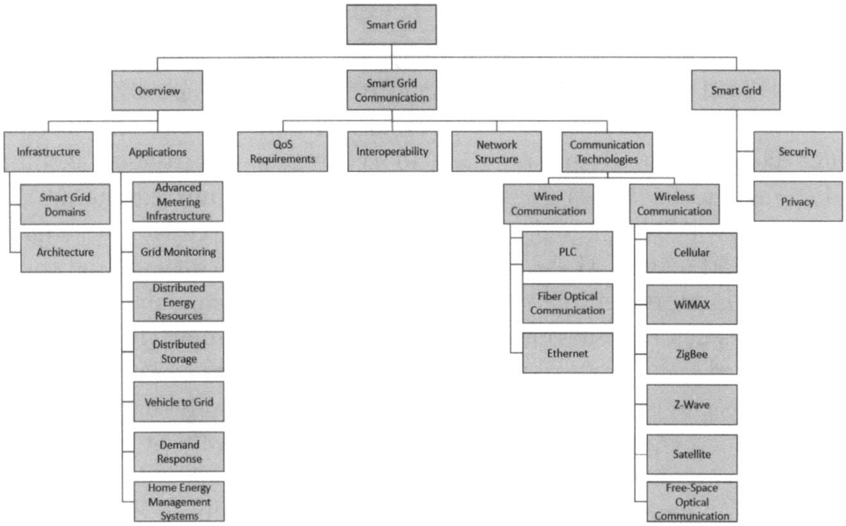

Figure 4.5. The structure of communication technologies for smart grid.

Source: https://www.mdpi.com/sensors/sensors-21-08087/article_deploy/html/images/sensors-21-08087-g001.png.

The development of such systems not only enhances customer awareness and empowerment but also promotes a more sustainable rate of energy consumption. With latest evidence indicating that further investment in smart grid technologies continues to reinforce the essential link between communication infrastructures and the utilization of renewable sources of energy (Al-Shetwi et al., 2024), the constraints involved with the incorporation of communication technologies into smart grids cannot be overlooked. Despite technological advancements that have significantly improved grid efficiency, challenges such as cybersecurity attacks and data privacy still exist. Communication systems must be designed not only to gather and transmit information but also shield against potential loopholes. With smart grids being increasingly interdependent on communication networks, security becomes increasingly important to protect sensitive data. In addition, addressing interoperability issues among heterogeneous communication standards is crucial to having various aspects of the smart grid work in harmony with each other. In order to achieve maximum utilization of the potential of communication technology in smart grids, it is necessary to overcome these challenges in a strategic manner, thereby enabling the establishment of effective, efficient, and secure energy systems that can adapt as the energy landscape changes (Al-Shetwi et al., 2024) (Figure 4.5).

4.2.1. Types of Communication Technologies Used

Smart grids' communication technologies facilitate energy management and operational optimization. At the core of all these systems lies a robust communication infrastructure that allows for real-time data exchange between various components, from producers to consumers and storage facilities. Leading these technologies are next-generation metering infrastructure (AMI) technologies, which use digital communication networks to collect consumption data from smart meters. These networks enable utilities to manage and control energy usage in an efficient manner, identify outages in advance, and enhance grid reliability. Besides, the implementation of the Internet of Things (IoT) facilitates a decentralized system where various sensors and devices communicate with one another as part of the grid system, contributing immensely to a stronger and more efficient energy system, as reviewed comprehensively in recent papers citing the operational imperatives of smart grid application (Alsuwian et al., 2022). Another important communication technology utilized in smart grids is the Vehicle-to-Grid (V2G) system. V2G is a new technology, facilitating two-way communication between electric vehicles (EVs) and the electrical power grid infrastructure. It enhances grid resilience while promoting green energy culture through the provision of stored energy by EVs to the grid during peak usage times. Information sharing between EVs and servers in the grid facilitates a symbiotic process where there is maximum balancing of energy supply and consumption. Contrarily, application of communication protocols to exchange information subjects the system to cyberattacks. Counter to such threats, research proposes privacy-based authentication schemes that ensure data integrity and anonymity of users along with secure communication, thus boosting the overall cybersecurity framework in the smart grid system (Kaveh et al., 2020). In addition to AMI and V2G, other communication technologies play an important role to enhance the functionality of the smart grid. For instance, telecommunication networks such as fiber optics and wireless communication networks facilitate high-speed connectivity required for real-time processing of data as well as decision-making. Such networks facilitate integration of distributed energy resources (DERs), e.g., wind turbines and solar panels, by providing means for them to communicate with the grid appropriately, thus facilitating optimized consumption and generation behavior. Moreover, combining such communication technologies with advanced analytics and machine learning facilitates predictive modeling and demand response strategies, making the energy management system more adaptive and responsive. As energy distribution continues to advance, the said communication technologies will continue to shape efficient smart grids, managing the very inherent problems emanating from expanded energy demands and added complexity (Alsuwian et al., 2022).

4.2.2. Importance of Real-Time Data Transmission

The transformation from traditional power systems to Smart Grids (SGs) entails real-time data transmission to enhance efficiency and reliability. Unlike their predecessors, SGs facilitate bi-directional energy flow to enable not just the supply of electricity but even its generation by consumers themselves. At the center of this transformation is the connectivity of all types of Internet of Things (IoT) devices, ranging from intelligent sensors and meters that track and analyze energy consumption at every moment. The real-time capability of data transmission allows service providers to act in response to changes in demand and supply in a timely fashion, removing wastage and maximizing utilization of resources. This dynamic capability is vital to addressing the growing energy demands and challenges of today's cities, which are experiencing unprecedented changes in terms of energy consumption patterns and supply sources (Copeland et al., 2019). Real-time information transfer is also critical in ensuring the security and stability of the electrical grid. By constantly monitoring operation parameters, SGs can automatically detect aberrations such as electrical imbalances or potential cybersecurity attacks. The use of Information and Communication Technologies (ICT) enhances the ability to not just observe but control the grid in a timely manner, enabling instant corrective measures when problems do occur. This sort of responsiveness is critical in today's world, where power production more and more depends upon renewable sources that are potentially more volatile. The integration of big data analytics in these real-time systems enhances decision-making processes, where it becomes viable for stakeholders to predict the demand change and implement load balancing approaches effectively (Bonetto et al., 2017). In addition, the significance of real-time data extends beyond the aspect of operational efficiency; it produces an active consumer base. As SGs transform into a prosumer-integrated model, real-time data facilitates better decisions on energy utilization. Transparency leads to energy-saving behavior among consumers, who are able to adjust patterns of usage based on real-time data feedback regarding prices and availability of energy. Demand response programs bolstered by real-time data communication fuels active energy management, thereby supporting sustainability goals. Such initiatives increase energy efficiency as well as create a common sense of responsibility toward saving energy, which is imperative for future energy landscapes and climate goals (Bonetto et al., 2017).

4.2.3. Impact on Grid Reliability and Efficiency

The inclusion of communication technologies in smart grids significantly increases the reliability of the grid through real-time data exchange and monitoring features. Smart grids utilize advanced sensors and communication algorithms that enable two-way communication between utilities and end-users. This continuous flow of data allows for prompt detection of outages, real-time optimization of energy supply, and immediate response to changing patterns of demand. By applying automated fault detection and isolation techniques, utilities can automatically detect and repair issues in real time, minimizing downtime and generally enhancing the reliability of the electric supply. Also, the potential of smart grids to optimize the use of renewable energy contributes to a more robust energy infrastructure that minimizes the need for traditional generation facilities. Lastly, grid reliability improvement through smart technology produces greater consumer trust and encourages sustainable energy practices as noted in current examinations of smart grid implementations (Olatunde et al., 2024). The efficiency of energy usage is enhanced significantly with the deployment of smart grids, as they enable efficient management of resources and energy usage. Advanced metering infrastructure gives users information on their consumption patterns real-time, enabling them to adjust their consumption pattern accordingly. Real-time feedback not only aids in lowering energy cost, but also towards aggregate goals of energy conservation and sustainability. Studies have shown that regions that implemented smart grids can manage a huge decrease in energy consumption, thus lowering the environmental aspect that comes with the use of fossil fuels. Additionally, the grids enable the incorporation of distributed energy resources and demand response initiatives, which optimize energy flow and reduce peak demand stresses. Thus, the overall impact on consumer participation and grid efficiency becomes a critical component of modern energy management practices that spur new innovation (Olatunde et al., 2024). In addition to enhancing efficiency and reliability, communication technologies in smart grids also face various challenges that hinder their full potential. One of the greatest challenges is the complexity of meshing various technologies and protocols between various stakeholders in the electrical supply chain. This complexity often leads to interoperability issues since systems and components do not communicate well with one another. Furthermore, the upfront capital costs of implementing smart grids remain significant barriers to most utilities, and this may stifle widespread adoption. As the research highlights, it is imperative to overcome such challenges in order to unlock the full potential of smart grids, particularly in terms of energy efficiency and reliability. Sustainable innovation and collaboration among stakeholders are essential in mitigating such barriers to ensure that the implementation of smart grid systems lives up to its potential towards a more resilient and efficient energy system.

4.3. WIRELESS COMMUNICATION IN SMART GRIDS

The use of wireless communications technologies in smart grids is a significant development in power distribution and management systems. Smart grids are unique compared to the traditional grids, which are operated using wired connections. They utilize a variety of communication technologies, both wired and wireless. Wireless technology, in specific, has many advantages due to its inherent agility and flexibility in volatile conditions. Wireless technologies enable affordable machine-to-machine and human-to-human communication, facilitating the interaction among various system components such as smart meters, renewable sources of energy, and consumer devices. This innovation fuels the efficiency of energy consumption and management as well as enhances operational dependability. Therefore, wireless communication is a critical basis for future smart grid infrastructure to provide smooth operation for home area networks (HANs), neighborhood area networks (NANs), and wide area networks (WANs) (Erol-Kantarci et al., 2015). The challenges in the application of wireless communication in smart grids are interdisciplinary, including concerns regarding spectrum availability and data security. The paucity of radio frequency spectrum necessitates innovative means such as cognitive radio technology, which can dynamically utilize unused spectrum bands to avoid congestion. The technology maximizes the ability of smart grids to operate effectively by facilitating near-real-time communication between disparate components of the grid, including generation, transmission, and distribution systems. Security and privacy issues remain paramount, given increased connectivity of devices that provide entry points for compromising vulnerability. Resolving these problems through strict communication standards and procedures is critical to the assurance of reliable functioning of smart grid systems. The standards improve interoperability among different technologies in addition to maintaining data integrity being transmitted across the network (Chin et al., 2011). Moreover, the establishment of wireless sensor networks is driving ongoing growth in monitoring and regulating smart grid operations. They allow real-time monitoring of distribution plants and transmission lines, leading to improved decision-making capabilities. The operating advantage of remote-mounted sensors is particularly valuable since they can perform required operations without human intervention. Hence, the integration of wireless communication not only increases system resilience but also provides a sustainable energy future since it enables effective management of renewable sources and avoids energy loss during transmission. With research across disciplines continuing to address the challenges and prospects of wireless communication systems for smart grids, resulting innovations hold the promise to transform energy networks into intelligent responsive networks that react to the demands of a changing energy environment (Erol-Kantarci et al., 2015) (Figure 4.6).

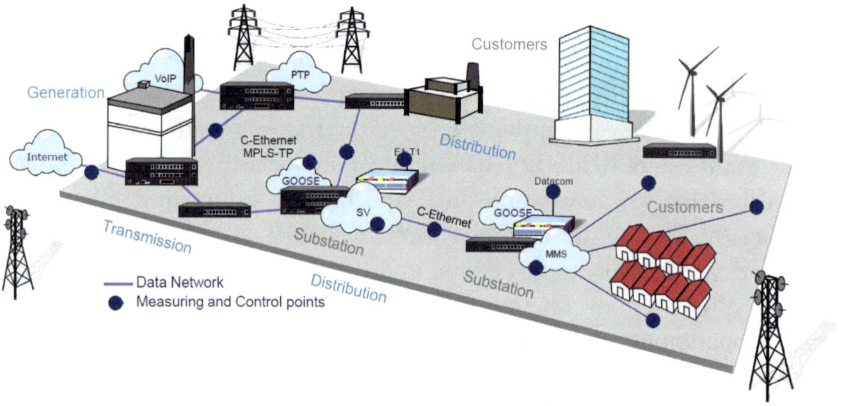

Figure 4.6. Legacy and new protocols: Ethernet / MPLS, T1/E1, GOOSE, SV, and more.

Source: https://blog.albedotelecom.com/wp-content/uploads/2023/02/PGA-9.jpg.

4.3.1. Types of Wireless Communication Technologies

Evolution in wireless communications technology is a cornerstone in the advancement of smart grid systems. Different types of wireless communications, such as cellular networks, radio frequency identification (RFID), and satellite communications, play a pivotal role in facilitating seamless and effective data exchange. Cellular networks, particularly those founded on Long-Term

Evolution (LTE) technology, provide fast communication capabilities essential to the real-time monitoring and control of electrical distribution systems. Further, RFID technology enables effective asset tracking within the grid, such as transformers and smart meters, that enhances efficiency of operations. Smart grids can maximize the flow of energy and enhance responsiveness of the energy supply system by utilizing these multidimensional wireless technologies, thus resulting in a dynamic and resilient energy grid (Copeland et al., 2019). Another valuable wireless communication technology is the inclusion of Internet of Things (IoT) devices. IoT devices integrate the sensors and actuators that are scattered all over the intelligent grid to gather and transmit vital information on energy usage and grid performance. For example, smart meters deployed at consumer sites gather real-time energy usage history and provide it back to the utility companies.

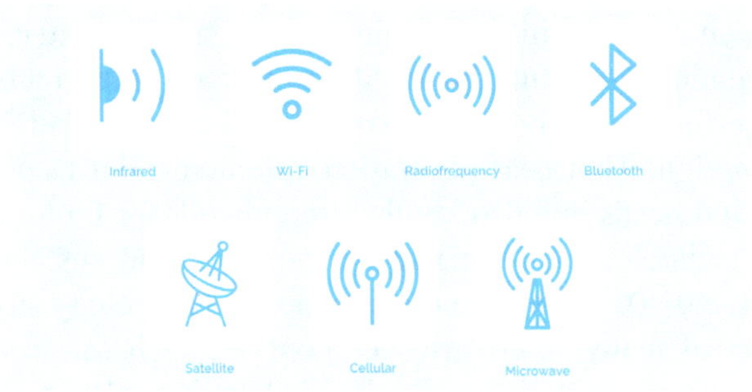

Figure 4.7. Examples of wireless communication technologies.

Source: https://tektelic.com/wp-content/uploads/34-Wireless-Communication-Technologies.svg.

This two-way conversation not only empowers customers to more effectively watch their energy usage but also aids the utilities to manage demand better (Chin et al., 2011). The robust infrastructure established through IoT connectivity is such that both conventional and non-conventional sources of power generation can be well-integrated into the grid, thus providing a platform for sustainable energy operations. In addition, the inclusion of low-power wireless technology such as Zigbee and LoRaWAN has found growing significance within smart grid scenarios. Zigbee is utilized within home area networks to facilitate smart devices communicating with one another so as to facilitate coordinated management of energy in the home. LoRaWAN, alternatively, focuses on long-distance transmission needs with low power consumption levels, thereby being adequate for large-scale sensor deployment across expansive geographical areas. These technologies facilitate smooth interoperation among various grid components, enabling monitoring and control systems to function optimally in various environments (Chin et al., 2011). Collectively, these wireless communication technologies underscore the capacity of smart grids to adapt according to energy-related issues of today and promote efficiency and reliability at various levels of the supply chain of energy (Figure 4.7).

4.3.2. Advantages of Wireless Communication

There are numerous advantages of wireless communication in smart grids, significantly enhancing operational efficiency as well as flexibility. One of the main advantages is the capacity of wireless technologies to facilitate real-time sharing of information between the components of the grid, i.e., generation, transmission, and distribution systems. Such capacity not only maximizes operations but also allows for real-time responses to changing demand and conditions, which is crucial in maintaining system stability and reliability. With wireless communications being included, utilities are able to more effectively observe grid performance and customers' usage behavior, thereby optimizing energy consumption. This next-generation communications not only support existing operation needs but also lay the groundwork for future developments in smart grid technology to enhance performance and sustainability (Erol-Kantarci et al., 2015). Wireless communication technology also allows for the deployment of many smart grid components, such as smart meters and renewable energy sources, in formerly challenging places. Because they don't require heavy physical wiring, they offer unparalleled flexibility in the expansion of grid infrastructure at little investment in hard infrastructure. This flexibility is particularly beneficial in rural or city settings where traditional wiring is logistically cumbersome or economically infeasible. Wireless sensor networks can even be installed where human access is limited, offering better monitoring and control over decentralized energy sources. These grids form the core of exploring new solutions in the area of energy management and ensuring that the grid can accommodate different approaches towards power generation in a cost-effective manner (Erol-Kantarci et al., 2015). Finally, the more critical role communication technologies need to play in supporting grid reliability and security cannot be overstated. Wireless communication supports the smart grids' reliability by promoting interoperability across different systems, which is required in the effective operation of future energy technology. As smart grids mature with more sophisticated technology, such as the Internet of Things (IoT) and machine-to-machine communication, the need for dependable wireless communication networks will be more pronounced. These not only allow efficient operations but also enhance grid security by enhancing the capability to monitor and respond quickly to incidents. Last but not least, the continuous development of wireless communications technology is a crucial factor in the successful implementation of smart grids, which renders them future-proof and capable of fulfilling increasing energy demands in an environmentally friendly way (Chin et al., 2011) (Figure 4.8).

Figure 4.8. Advantages of wireless communication.

Source: https://technoeyenet.com/wp-content/uploads/2022/08/c-blogimages-thinkstockphotos-611997072.jpg.

4.3.3. Security Concerns Associated with Wireless Technologies

The increase in wireless technologies in smart grid development has enabled tremendous energy management and efficiency gains. But with this ease comes associated extensive security challenges that need to be addressed comprehensively. As smart grids provide real-time communication across generation, transmission, distribution, and consumption units, vulnerabilities inherent within wireless networks may be exploited by malicious elements. Unauthorized access can lead to breach of data, theft of energy or even power disconnection, and thus the need for robust security measures during design and installation of these systems. Security is of essence in a bid to protect sensitive information and integrity of infrastructure given that wireless communication inherent lack of physical barriers makes it highly susceptible to all manner of cyberattacks (Chin et al., 2011). In addition, integration of WSANs in smart grids is somehow difficult with regard to security. The inherent characteristics of WSANs make them attractive to manage microgrids, yet they render the core systems more vulnerable. For instance, the wirelessly connected actuators and sensors, as much as they enhance the efficiency of operation through effective data transmission, may be exposed to interception or tampering if adequate protection is not offered. The potential for interference with command messages conveyed via these networks is a serious concern; negative alterations can lead to deadly operational malfunctions or degrade the reliability of the energy supply. Therefore, developing and deploying robust communication protocols that are secure is critical for the sustainable scalability of smart grid technologies (Curea et al., 2016). To overcome these security risks, ongoing R&D is necessary to establish strategic models that enhance the robustness of wireless technologies used in smart grids. This involves the development of standardized security standards, performing regular security audits, and rendering stakeholders sensitive to potential vulnerabilities. Industry, government, and academia need to join together to ensure that impactful strategies are designed and implemented universally. Additionally, investments in new encryption techniques, anomaly detection tools, and high-strength network equipment will provide layers of protection required to protect smart grids from new threats. As the energy management landscape continues to develop further, giving a prime importance to the security infrastructure will become imperative to realize the full potential of smart grid technologies while protecting against inherent vulnerabilities pertaining to their wireless communication networks (Chin et al., 2011) (Figure 4.9).

Figure 4.9. Wireless network security concerns.

Source: https://solutionsreview.com/wireless-network/files/2017/02/Untitled-2.jpg.

4.4. WIRED COMMUNICATION IN SMART GRIDS (SGS)

The evolution of traditional power grids into SGs is a quantum leap in energy distribution and management. This transformation remedies a wide array of issues associated with traditional power systems with uni-directional information flow, such as wastage of energy and unreliability. A critical component in this regard is application of wire-bound communication technologies, which facilitate two-way energy transfers between end-users and service providers. These systems allow for the exchange of real-time data across various segments of a system, such as power generation, transmission, and consumption. The interconnectivity of these wired network communications plays a crucial role in the effective utilization of monitoring, analysis, and control equipment on which SGs rely, ultimately aimed at improving operation efficiency and consumer engagement in energy management. The interconnecting model presents the role of communications technologies as a facilitator of the operation of Smart Grids today, making them more responsive to demand and supply dynamics. Wireline communication technologies complement Smart Grids by offering the speed and reliability of data transfer necessary for near-real-time operation. With the consumption patterns of energy growing more dynamic with aspects such as increased reliance upon renewables and electric vehicles, communication infrastructure would also be required to follow suit. The use of energy in such intelligent grids depends upon effective communication among all the elements of the grid, ranging from generation and transmission to distribution and load management.

Next-generation networking technologies like fiber optics and Ethernet play a crucial role in facilitating high-speed data transfer, enabling fast decision-making and effective energy management techniques. Such reliance on robust wired communication not only enables operational efficiencies but also allows for a high level of security and privacy within the Smart Grid infrastructure, reconciling several technological standards necessary in future-proofing (Chin et al., 2011). In addition to improving operational performance, wired communication technologies aid in the development of innovative applications in Smart Grids. The interconnection of IoT devices such as sensors, smart meters, and actuators into the wired communications infrastructure allows for complete monitoring of energy use and production. Through such devices, fine observation of energy usage flows and user behavior is realized, which can be processed to make optimizations in overall efficiency within the grid. Wired communications infrastructure supports interoperability among various grid components, so more functionality and integration can occur without problems. While more research is ongoing to address the open issues and challenges in IoT-assisted Smart Grids, the requirement for standard communication protocols is even more greatly evident. Proper communication technologies will, in due course, render SGs future-proof in fulfilling requirements so that energy management is not only efficient but also sustainable over the long term (Copeland et al., 2019).

4.4.1. Types of Wired Communication Technologies

In the smart grid world, wired communication technologies support an infrastructure which offers unhampered data flow and simple system management. The most prevalent among them is Power Line Communication (PLC) which makes use of installed power lines for data transmission and thus enjoys low infrastructure expenditure. Two-way communication between data concentrators and smart meters and real-time monitoring and control of energy consumption are supported by PLC.

Figure 4.10. Types of wired networks.

Source: https://collectionperformance.com/wp-content/uploads/2023/06/wired-networks-types.jpg.

This technology has proved particularly valuable in industrial and city centers where additional communication infrastructure installation is largely impractical. Existing research shows that with increased signal-to-noise ratios, like the situation in urban centers, PLC is more efficient and reliable than traditional wireless methods, hence making it crucial in existing applications of smart grids (Rhouma et al., 2023). Another wireless communications technology such as Fiber Optic Communication has the advantage of high-speed data transmission capacities that are priceless in the huge networks of smart grids. Fiber optics convey information using light along great distances with minimal loss and is thus suitable for both rural and urban applications where high bandwidth is required. This technology allows a broad array of smart grid functions like real-time data analytics and effective demand response strategies, hence enhancing the control over the grid. Moreover, fiber optics are immune to electromagnetic interference, enhancing the communication security and reliability throughout the grid. Due to its capability to support large-scale deployments, it is possible to integrate various distributed energy resources and advanced metering infrastructures (Tsitsimelis, 2020). Last but not least, coaxial cables are also another wired communication technology employed in smart grids where there is a requirement of high speed deployment and flexibility. Coaxial cables, capable of handling high frequency signals, provide an excellence in utilizing them in applications which also require good quality bandwidth. They can be used effectively in Neighborhood Area Networks (NANs), linking communications between smart meters and central systems in urban areas or industrial complexes. This technology is simple to implement in existing communication infrastructure, allowing for robust and efficient energy management systems. The comparison between wireless alternatives and wired alternatives implies that when there are dense networks, coaxial communications can achieve better performance and reliability (Rhouma et al., 2023) (Figure 4.10).

4.4.2. Benefits of Wired Communication Systems

For smart grids, wire-based communication systems provide numerous advantages that enhance both efficiency and reliability. Unlike wireless communication systems, wire-based communication is less susceptible to interference and external interruptions, which is necessary for the proper functioning of grid operations. Smart grids operate on reliable data transfer, and the dependability in the wired connection ensures demand and supply data is properly transferred at several layers of the network. Such dependability is essential in real-time decision-making and emergency scenarios and helps to minimize the minimum possible outages and enhance system robustness to man-made and natural barriers in grid management (Erol-Kantarci et al., 2015). Wired communication systems also support robust security controls due to their nature. Since cyberattacks on infrastructures have been increasing as threats, wired communication can offer a more secure platform less exposed to attacks typical of wireless communications. With the use of dedicated data transmission channels, utilities are better able to implement robust encryption and intrusion detection measures. These functions enhance the overall security system of smart grids, gaining the confidence of the public and encouraging user participation. This confidence is important as smart grids evolve to use renewable sources of energy and techniques of demand response management (Peffer et al., 2010). Moreover, wireline-based communications technologies integration provides scalability and interoperability for smart grid systems. With the creation of new devices and technologies, the wired infrastructure can be conveniently extended with additional devices without causing drastic changes to the existing setup. The flexibility is essential in ensuring the integration of various networks, such as Home Area Networks (HANs), Neighborhood Area Networks (NANs), and Wide Area Networks (WANs). Interoperability enables effortless communication among heterogeneous systems to ensure that all devices work harmoniously to enhance energy efficiency and minimize losses. This adaptability puts wired communication in the foreground when it comes to future development of smart grids and, by default, sustainable and more reliable energy management systems (Erol-Kantarci et al., 2015).

4.4.3. Limitations of Wired Communication

Incorporating traditional wired communication systems into Smart Grids brings to the forefront some inherent limitations that present barriers to optimal efficiency and flexibility. One of the largest disadvantages is the rigidity of wired infrastructure, which does not allow scalability with the integration of new devices and technologies. Smart Grids' need for bi-directional energy flow and lots of IoT devices for control and monitoring means wired communication is lacking in accommodating the evolving demands of a dynamic energy sector. This rigidity causes delays in network upgrades and hinder the timely launch of new solutions, ultimately affecting the grid's reliability and efficiency, as indicated by the necessity for greater connectivity and automation in Smart Grid operations (Copeland et al., 2019). Additionally, the enormous physical constraints of wired systems limit their application in remote and rural areas where they can possibly not be economically implementable. These geographical challenges broaden the digital divide, and infrastructure development for efficient energy management is difficult to construct in sparse areas. The inadequacy of conventional wired networks is also apparent when consolidating decentralized energy sources, as one finds in new architectures being mooted for rural distribution networks. By being established more or less on wired solutions, such systems may neglect the utility of wireless technologies that are more adaptive and cheaper to install, stunting the grids ability to innovate and develop as a reaction to emerging needs in energy production and consumption (Díaz González et al., 2018). Also, operational and maintenance problems of wired communication systems add to the expense and complexity of operating a Smart Grid. The need for regular physical inspection and repair of massive wiring can strain resources and distract efforts away from handling more serious issues that affect grid stability and performance. In addition, wired infrastructure is also prone to physical interference by the environment, and this could lead to prolonged outages—a situation unacceptable in the context of current energy needs. In contrast, wireless communication technologies can provide an alternative less prone to these interference effects with reduced maintenance needs, hence a more responsive, dynamic Smart Grid capable of meeting the current energy needs (Copeland et al., 2019).

4.5. DATA MANAGEMENT AND ANALYTICS

Data analysis and integration into intelligent grid communications technology are at the heart of energy efficiency and performance enhancement. Smart grid advanced data analytics enable large volumes of data gathered from a range of sources, including smart meters, sensors, and other IoT systems, to be analyzed. Beyond facilitating real-time monitoring of energy consumption patterns, analytics enable utilities to predict demand and plan for resources accordingly. Therefore, improved decision-making processes are achieved, leading to an energy response grid that can dynamically adjust to changing energy demands. As stated, the effective use of smart technologies allows energy management to be automatized, which allows one to monitor and maintain energy performance at a high rate (Aliero et al., 2022). This scientific process not only assists in the encouragement of efficiency during operation but also enhances user engagement in energy management processes. Along with this, smart grid communications heavily utilize big data analytics to drive sustainability initiatives. The immense quantity of energy data accumulated demands robust mechanisms of data handling and monitoring to be able to generate their potential to the fullest extent. Without there being high levels of data handling facilities, the performance of smart grids can significantly be undermined. Big data analytics is a key solution in this area, and it enables better visualization of the grid as well as the optimization of energy consumption and distribution programs (Kumar et al., 2021). Analytics enhances one's ability for anomaly detection, such as energy theft, hence the integrity of the grids is maintained. Consequently, such data-driven strategy not only provides economic and technical advancements in energy management but also has more implications for achieving Sustainable Development Goals (SDGs) in the energy sector. Additionally, data handling in real-time plays an even greater role in the context of smart grids. As smart technologies improve with the potential to incorporate capabilities that allow users to track and optimize their energy consumption, the function of timely and precise data analytics remains squarely at the top of the list of priorities. The future of smart building energy management systems largely depends on enhanced privacy and security regulations, alongside cross-platform compatibility (Aliero et al., 2022). In this dynamic environment, machine learning and artificial intelligence solutions stand to transform energy management using effective data analysis. These solutions have the potential to predict patterns of demand for energy, paving the way towards a sustainable future and an energy-efficient world. The future will ultimately depend on being able to meet the challenges related to the integration of such sophisticated data management methods as smart grids move forward to dominate energy management strategies in the world (Figure 4.11).

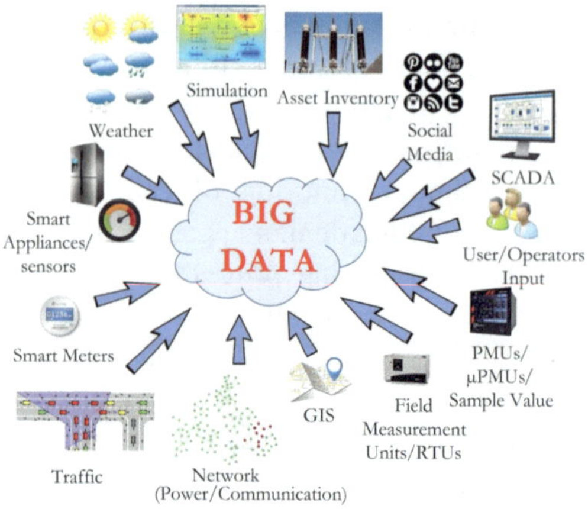

Figure 4.11. Data analytics in smart grids.

Source: https://miro.medium.com/v2/resize:fit:4800/format:webp/1*fEqcSobeBJhpVMifxIDlxQ.png.

4.5.1. Importance of Data Collection and Analysis

There are fundamental aspects of efficient data collection and analysis in the development and optimization of communication technologies in smart grids. The merging of traditional energy systems into advanced technologies such as IoT and AI in smart grids introduces a deluge of data, which is a double-edged sword. In-depth data analysis can enable utility providers to monitor energy consumption patterns in real time and optimize distribution, saving resources. This data improves resilience by providing a means to proactively manage possible interruptions and ensure a more stable power supply. Furthermore, by using gathered data, decision-making processes can proceed while being guided by the broader goals of efficiency and sustainability. The rapid evolution of communication technologies, and particularly in the case of smart grids, indicates that data is not ancillary but, on the contrary, it is central to effective operational planning and grid management, specifically with new technologies trying to balance energy needs and sustainability concerns (Jiří Jaromír Klemeš et al., 2020). In intelligent grid systems, the use of data collection and analysis is beyond critical applications such as predictive maintenance and energy consumption management. Advanced analytics of data has the ability to forecast failures in grid equipment, improving maintenance schedules and minimizing downtime. Maintenance is therefore proactive instead of reactive, thereby optimizing operating costs and extending the life of system elements.

In addition, through sensitive data analysis, power providers can tailor services to suit individual clients' needs, increasing user engagement and satisfaction. Pattern analysis aids demand response programs, which enable utilities to dynamically adjust prices according to consumption. Lastly, the impact of data utilization in smart grid communications technology is beyond mere day-to-day advancements; it directly initiates a transition towards a more efficient and more resilient energy tomorrow (Rasheed et al., 2020). In addition, data collection and analysis drive innovations in advanced services and regulatory needs, empowering smarter energy management solutions. The convergence of digital twins and advanced simulation tools is a paradigm shift in energy system design and operation. With virtual replicas of physical assets, stakeholders are now capable of monitoring performance, optimizing configuration, and foreseeing future operational problems. Data analytics provides effective resource allocation and allows for an integrated strategy across industries, promoting collaboration among utilities, consumers, and regulators. Through the provision of real-time observations for strategic energy projects, smart grid communication technologies can apply data-based methods for addressing system inefficiencies as well as environmental issues. Hence, the importance of quality data collection and analysis cannot be overstated, as not only does it raise the operational worthiness of smart grids, but also the sustainable development of modern energy systems (Jiří Jaromír Klemeš et al., 2020).

4.5.2. Tools and Technologies for Data Management

Effective data management is the cornerstone in the design of smart grids where the integration of power systems with information and communication technologies results in vast amounts of data. These include energy consumption patterns, availability of resources, and control signals in real-time from intelligent devices. Effective tools and technologies are thus required to process and manage incoming streams of data. Existing data management systems utilize optimized databases with specific functionalities to ensure easier data analysis and retrieval, which can lead to improved decision-making. Comprehensive visualization tools also enable stakeholders to efficiently analyze complex data sets. These technologies play a key role in transforming raw data into actionable intelligence, thereby improving operational efficiency in smart grids (Turck et al., 2013). Besides, pilot projects are also important proof grounds for new data management solutions. For instance, the Large Scale Smart Grid project called Linear deploys a robust infrastructure that can handle millions of data points coming from various sources, including smart meters and weather sensors. By decoupling data collection, processing, and display on specialized servers, the system maximizes performance and allows for real-time modifications to be made based on data received. The architecture not only enhances the accuracy of data but also distributed energy resource management, thus facilitating integration of renewable energy systems. The role of data management technologies in such efforts places them into perspective for their relevance in validating real-time experimental results and resolving operational problems appropriately (Turck et al., 2013). In addition to process optimization, innovative technologies also bring innovation in data management for smart grids. Advances in machine learning and artificial intelligence bring new perspectives to predictive analytics that can forecast consumption patterns and identify potential system disturbances in advance. With the ability to process historical data in conjunction with real-time inputs, such smart systems are able to provide actionable intelligence to utility suppliers and consumers alike, making the energy system more responsive. Additionally, with the integration of blockchain technology and smart grids, there is potential to offer better data security and create open and inviolable transaction records that will facilitate increased consumer participation and confidence. With newer technologies, their integration into management data systems is certain to play a central function towards addressing future energy requirements, including efficiency and sustainability.

4.5.3. Role of Big Data in Smart Grid Optimization

The integration of big data analytics within the optimization of smart grids is essential to improve energy efficiency and reliability. With rapid advancement in communication technologies, smart grids have evolved to leverage vast volumes of data generated from diverse sources like smart meters, IoT, and weather predictions. It makes utilities capable of learning about habits of consumption and health of the system, thereby making it possible for them to proactively manage the grid. For instance, real-time data analytics enable better predictability of demand, and as a result, energy firms pre-emptively make changes in their operations to prevent wastage and reduce expenses. Further, the integration of machine learning algorithms in big data platforms ensures predictive maintenance, which identifies potential failures ahead of time before they impact grid operations. All these optimizations not only enhance the reliability of services but also enhance the overall resilience of the grid, in accordance with the grand-scale visions of sustainability and efficiency embedded in modern energy systems (Meslouhi et al., 2025). Big data plays a significant role in making it possible to integrate renewable energy sources into smart grids. With the transition of the energy sector towards cleaner resources, variability and unpredictability in renewable resources call for advanced data management capabilities. Smart grids with big data analytics can handle the challenges posed by distributed energy resources cost-effectively by analyzing real-time consumption and generation information. This improves the flexibility and reliability of the grids, enabling operators to act quickly on supply and demand swings. With advanced data analytics tools, operators can maximize energy distribution so that excess energy from renewables is effectively harnessed and utilized. This Optimisation not only maximizes the value of renewable energy but also reduces the consumption of fossil fuels to bring about a transition towards a future of sustainable energy, an aspect that is emphasized in the narrative of Internet of Energy (IoE) applications (Zhou, 2025). Furthermore, big data maximizes the involvement of consumers within the realm of smart grid Optimisation. Through the application of advanced communication technologies, utilities can provide consumers with information regarding their usage patterns in detail, making them able to make quality decisions on what they consume. Technologies such as mobile applications allow consumers to track real-time energy data, adjust consumption habits, and participate in demand response initiatives. All this interaction translates to wiser energy consumption, which, in turn, leads to grid optimization. Also, big data enables utilities to deploy dynamic price models that are based on real-time demand, thereby encouraging consumers to shift their usage to off-peak periods, easing the grid. Therefore, the role of big data in encouraging consumer participation is central to achieving a more responsive and more efficient energy system, affirming the interdependence of technological innovation in smart grids (Meslouhi et al., 2025) (Figure 4.12).

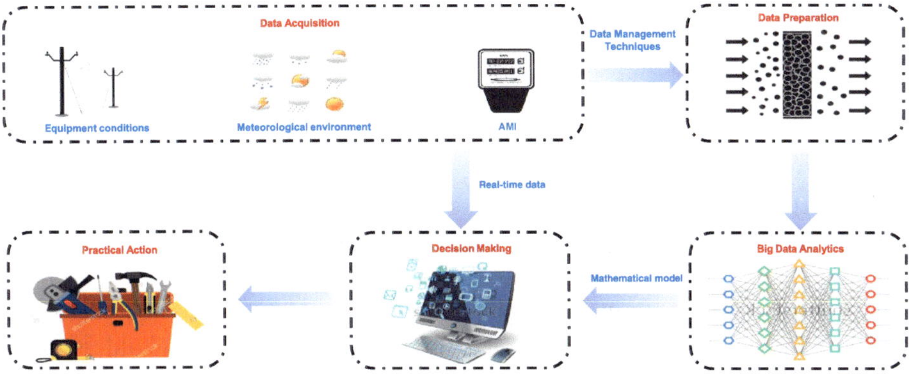

Figure 4.12. Example of big data analytic procedures in smart grid.

Source: https://media.springernature.com/full/springer-static/image/art%3A10.1186%2Fs42162-018-0007-5/MediaObjects/42162_2018_7_Fig4_HTML.png?as=webp.

4.6. REGULATORY AND POLICY CONSIDERATIONS

The transition to smart grids is not technological alone but also a regulatory problem of significant scale. Effective regulation must be established in order to handle a sophisticated interconnection of the actors involved in the production of energy, the consumption of energy, and the government. Because smart grids are composed of numerous assorted communication technologies, the regulatory environment must ensure interoperability among the systems along with security against cyber-attacks. This complexity is compounded by the need to maintain public trust in such systems, particularly for data handling procedures. Ensuring data security, privacy, and safety becomes a paramount concern as smart grids use internet-connected devices and two-way communication channels. Regulatory bodies must thus evolve existing frameworks to accommodate the new realities of smart grid technology while at the same time promoting openness and accountability among all stakeholders in this environment (Axon et al., 2014). In the European Union, regulatory compulsions over smart grid technology are pivotal towards the realization of a low-carbon energy future. Smart grids must be envisioned by the policymakers as integrated systems that facilitate renewable energy practices with clearly defined and transparent standards of operation for technology applications. Competition, pricing, and the integration of alternative sources are layers of complexity that are placed upon the realization of a unified regulatory approach. Notably, the deployment of smart grid technologies is highly reliant on public acceptance; regulators must hence be proactive in engaging stakeholders to reduce worries regarding technological threats and disturbances. It is important to develop regulatory policies that are accommodating for innovative development to ensure full utilization of smart grid technologies while being environmentally compliant (Borghetti et al., 2013). Furthermore, the convergence of policy questions and smart grid implementation means the significance of a multi-faceted approach to regulation. This includes not only defining responsibilities and regulations for energy providers but also establishing collaboration among sectors, such as telecommunications and data management. Having a platform for continuous communication among regulators, industry stakeholders, and customers can facilitate a more sustainable and dynamic regulatory system. In addition, regulatory innovation must be sensitive to the fast-changing nature of technology in the energy sector. By incorporating multiple perspectives and knowledge, regulatory systems can either be sustained or altered to deal with emerging challenges while facilitating the creation of smart grids, thereby making multiple environmental and economic benefits successful (Axon et al., 2014).

4.6.1. Government Regulations Affecting Smart Grids

Government policies play a central role in determining the development and deployment of smart grids, influencing everything from technology standards to consumer engagement. Governments have the responsibility of developing regulatory policies governing how smart grid technology is made available so that innovations align with overall energy policies of sustainability and efficiency. With such frameworks evolving, policymakers must consider an assortment of factors like public acceptance, data protection, and privacy considerations, all of which depend directly on consumer trust and participation in smart grid initiatives. Regulations thus don't merely stipulate requirements for compliance by energy firms but define the strategic environment where the technologies will function, establishing a platform that favors innovation while responding to resultant threats. The interplay between regulatory needs and the development of technology underscores the necessity of a tailored approach that solves both the complexity of energy systems as well as those of end-users. In addition, the impact of regulation on smart grid technologies is also underscored through its potential in establishing interconnectivity and an atmosphere of cooperation among various stakeholders. Effective policy frameworks can facilitate bringing together partnership possibilities between the energy sector, spanning utility companies to technology companies and regulatory bodies. This cooperation is essential to the operation of smart grids, which employ a mix of communication technologies to enhance two-way communication between suppliers and consumers of energy. The regulatory environment also serves to make investment in smart grid infrastructure appealing, encouraging the creation and integration of needed communication technologies to optimize operational efficiency and reliability. As smart grids can provide the significant carbon reduction objectives while ensuring the highest possible energy supply, the necessity to have encouraging and flexible regulations to unleash the whole range of benefits of smart grid systems is self-evident. Further, effective implementation of smart grids is also closely associated with continuous examination and adaptation of regulatory norms. As innovations in communication technologies advance, regulations are also anticipated to be dynamic in order to facilitate new practices and strategies in the evolving energy sector. The dynamic interaction that this requires implies that regulators ought to engage with industry players and stakeholders in order to develop regulations that both assure the public on issues like cybersecurity and privacy and encourage innovation in the energy sector. Indeed, such participation can lead to an accurate definition of what constitutes a smart grid, thereby regulatory efficacy and accuracy. As technical and functional issues are just some of the multiple facets present in smart grid deployment, regulatory tools must be not only receptive to these but also to socio-techno-economic drivers that shape consumer behaviors and uptake in this new energy paradigm (Axon et al., 2014; Debackere et al., 2025).

4.6.2. Standards for Communication Technologies

In the context of smart grids, progresses in communication technology have called for the establishment of strict standards to ensure effective and secure interfacing. Being a new technology in power supply, smart grids utilize various communication infrastructure to facilitate the two-way flow of electricity and information between the generation, transmission, and distribution sections. The application of information and communication technologies (ICT) in this setting is crucial to the optimization of efficiency, reliability, and security in energy management. With the complexity and multidimensionality of smart grids, it would be obvious that protocols need to be standardized so as to determine interoperability among heterogeneous systems and devices and therefore optimize the overall grid performance and integration with renewable energy systems. Adoption of such standards facilitates communication without any inconvenience, and this is the essence of real-time monitoring and control in very complex energy settings. Evolutions of such smart grid communication technologies also result in some serious security concerns that must be addressed through well-known standards. Smart grids are facing more cyber-attacks, highlighting the fundamental requirement of secure communication infrastructure that can resist potential attacks. The reliability of such systems hinges on robust standards that govern data encryption, access control, and key management procedures. Standards for communication technology must be cognizant of privacy, fault tolerance, and resilience of network infrastructure against cyber-attacks. This becomes even more important given the multi-plurality of endpoints and systems within the smart grid. As identified in prior work, i.e., Erol-Kantarci et al. (2015), the success of smart grid implementation rests heavily on the effectiveness of communication security measures, wherein robust standards become the norm to achieve the minimization of the possible vulnerabilities. While developing such communication technology standards, ongoing research is needed to keep pace with the evolving background of cybersecurity attacks against smart grids. Key management schemes, cryptography schemes, and secure communication protocols are the core parts of the framework to be used in securing such infrastructures. Such frameworks cannot be overstressed in that they not only provide fundamental security measures but also ensure the integrity and reliability of the communication within smart grid networks. As suggested by Alohali et al. (2025), there are many schemes that already exist but are still in need of continuous evaluation and adaptation for emergent threats. In the end, embracing far-reaching standards of communication technologies will be the solution to enabling the full potential of smart grids while providing protection against the inherent threats of increased digitalization of power systems (Figure 4.13).

Figure 4.13. Communication standard.

Source: https://cdn.coretigo.com/wp-content/uploads/communication-standard.jpg.

4.6.3. Future Policy Directions for Smart Grid Development

Smart grid development is a dramatic shift in the management of energy, that future directions in policy need to focus on highlighting advanced communication technology integration. While the conventional power systems are challenged to deal with increasing energy demands and diverse energy sources, policy initiatives must be targeted at enhancing two-way communication between the consumers and producers of energy. This two-way functionality is central to making possible the efficient flow of information and energy, which can further translate to minimized wastage of energy and enhanced reliability. Future policies should promote the installation of Internet of Things (IoT) devices, such as smart meters and sensors, which are instrumental in making energy distribution and consumption more streamlined. Through incentivizing their adoption, policymakers can enhance grid responsiveness and consumer engagement significantly, opening the doors to a more sustainable energy future (Copeland et al., 2019). Beyond communicating effectively between users and suppliers, future policies must respond to regulatory measures encouraging the development of smart grids. Current regulatory environments are not likely to keep up with the rapid technological advancements in Smart Grid solutions, and implementation is slowed down. Policy regulations must include clear guidance that supports innovation while establishing provisions for data security and consumer privacy. The introduction of Information and Communication Technologies (ICT) into smart grids provides a foundation that supports such innovation through enhanced management and optimization capabilities. Policymakers must collaborate with technology providers and industry stakeholders so as to identify the optimal practices that enable the unhampered integration of IoT technologies within the existing regulatory framework (Borghetti et al., 2013). Finally, future policy directions must also guarantee sustainability and resilience in smart grid development. As there is growing global interest in mitigating climate change and environmental health, there is a need for policies to ensure promotion of the development of renewable resources and integration into the smart grid infrastructure. This change, however, is reliant on effective communication and data sharing offered by future-generation ICT. Future frameworks need to facilitate research and deployment of emerging renewable technologies with guarantees that the smart grid is flexible enough to integrate emerging innovations in power generation. By giving priority to sustainability, resilience, and adaptive regulatory strategies, policymakers can facilitate the effective shift to smart grids, thus enhancing energy security and environmental sustainability for future generations (Borghetti et al., 2013; Copeland et al., 2019).

SUMMARY

In brief, the application of communication technologies in smart grids is a revolutionary energy management model, enhancing both efficiency and reliability. Deployment of advanced ICT enables real-time sharing of information and more sophisticated user-grid interaction. Such integration enables prosumers—residences that produce and consume energy—to actively engage in energy markets, assisting in load balancing and peak shaving. Smart grids not only maximize resource utilization but also facilitate users to contribute to energy conservation actively, leading to a greener industry landscape. As these technologies find wider use in urban settings, it is clear that smart grid innovations play a central role to counter the challenges of contemporary energy needs and environmental sustainability (Bonetto et al., 2017). Additionally, the dependency on advanced monitoring systems for improving the performance of the grid highlights the paramount need for cybersecurity. Although systems like wide-area monitoring can significantly improve operational effectiveness, they also pose dangers, most notably with regard to timing synchronization provided by global positioning systems. The potential for interference with these systems is in danger of undermining the safety and integrity of energy distribution grids. This means conducting an exhaustive analysis to identify and neutralize these threats is necessary to ensure the long-term viability of intelligent grid systems. Achieving a balance between technological advancement and security steps will characterize the smart grid's capability to resist outside interferences that would threaten their operations (Ahmed faheem Zobaa et al., 2014). Lastly, the future of smart grid communication technologies will be dependent on continuous innovation and stakeholder collaboration between government entities, private companies, and academic institutions. Standards and best practices will play the key role in creating interoperability between the technologies and devices. As the energy landscape shifts, even the installation of big data analysis and implementation of machine learning techniques can enhance the flexibility of smart grids by enabling real-time decision-making and predictive maintenance. Ultimately, a general philosophy encompassing technological advancements, robust security aspects, and stakeholder engagement will define the future of smart energy systems. By accentuating these aspects, societies can provide the foundation for a sustainable energy future that effectively responds to the issues of new power infrastructure (Ahmed faheem Zobaa et al., 2014; Bonetto et al., 2017).

In looking at communication technologies development in smart grids, the evolution from traditional electricity networks to improved smart grid systems is critical in addressing modern energy challenges. The legacy grid has been tested time and again to keep pace with the complexities ushered in by alternative sources of energy and fluctuating electricity demand. As explicated in Borghetti et al. (2013), the conventional systems prove inadequate for modern energy-saving policies, and hence there is a need for innovative solutions. ICT introduction offers a chance to enhance the reliability and flexibility of power networks. This transformation is not only technological; it is also about rethinking the management and optimization of electricity networks, necessitating a synergized understanding of technical and regulatory paradigms in designing smart grids. In addition, the role of ICT extends beyond connectivity; it is the foundation for dynamic energy management practices encompassing all the stakeholders, including prosumers. In modern smart grids, citizens are increasingly involved in energy consumption as well as generation, becoming active participants in markets rather than passive consumers. According to Bonetto et al. (2017), integrating signal processing and big data analytics into power grids' operations can significantly boost electrical stability and distribution. The advent of demand response paradigms exemplifies this transition since prosumers are now involved in load balancing and peak shaving, thus supporting the efficiency of the overall grid. Active involvement in this fashion not only helps to enable a more robust energy grid but also helps in empowering consumers to be in control of their energy consumption. In addition, the implications of smart grids with communication technologies have far-reaching ramifications even on environmental and social aspects. As urban centers leverage such technologies to optimize energy provisioning, they can actually reduce greenhouse gas emissions and enhance urban sustainability. Smart systems make it possible to evaluate massive data sets that are used to optimize better energy management and create energy-efficient behavior among end users. According to Bonetto et al. (2017), serious games are part of the new approaches that have been employed to encourage users toward energy-saving behavior. This multi-dimensional cooperation highlights the capability of ICT in not only improving grid performance but also encouraging a shared sense of responsibility toward sustainable energy use, bridging technology and community participation in the process.

REFERENCES

1. Aliero, M. S., Asif, M., Ghani, I., Pasha, M. F., & Jeong, S. R. (2022). Systematic review analysis on smart building: Challenges and opportunities. *Sustainability*, *14*, 3009–3009. https://doi.org/10.3390/su14053009 (accessed on 30th September 2025).
2. Alohali, B. A., Hurst, W., Kifayat, K., Shi, Q., et al. (2025). A survey on cryptography key management schemes for smart grid. https://core.ac.uk/download/42478732.pdf (accessed on 30th September 2025).
3. Al-Shetwi, A., Hannan, M., Al-Masri, H. M., & Sujod, M. Z. (2024). Latest advancements in smart grid technologies and their transformative role in shaping the power systems of tomorrow: An overview. *Progress in Energy, 7*. https://www.semanticscholar.org/paper/269840d777a75700903301263e2a92068549ec03 (accessed on 30th September 2025).
4. Alsuwian, T., Butt, A. S., & Amin, A. A. (2022). Smart grid cyber security enhancement: Challenges and solutions—A review. *Sustainability*, *14*, 14226–14226. https://doi.org/10.3390/su142114226 (accessed on 30th September 2025).
5. Ancillotti, E., Bruno, R., & Conti, M. (2013). The role of communication systems in smart grids: Architectures, technical solutions and research challenges. https://core.ac.uk/download/pdf/37831683.pdf (accessed on 30th September 2025).
6. Ancillotti, E., Bruno, R., & Conti, M. (2013). The role of the RPL routing protocol for smart grid communications. https://core.ac.uk/download/pdf/37831692.pdf (accessed on 30th September 2025).
7. Axon, C., Balta-Ozkan, N., Cipcigan, L., Connor, P., et al. (2014). Scenarios for the development of smart grids in the UK: Literature review. https://core.ac.uk/download/155776124.pdf (accessed on 30th September 2025).
8. Ben Rhouma, O., Rebai, C., Ben-Romdhane, M., Cara, D. D., Artale, G., & Panzavecchia, N. (2023). The environmental impacts of radio frequency and power line communication for advanced metering infrastructures in smart grids. *Sensors (Basel, Switzerland)*, *23*. https://www.semanticscholar.org/paper/1a923d3058c7708703e9f0b6fb0d467683d3add9 (accessed on 30th September 2025).
9. Biswas, P., Rashid, A., Al Masum, A., Al Nasim, M. A., Ferdous, A. S. M. A., Gupta, K., & Biswas, A. (2025). An extensive and methodical review of smart grids for sustainable energy management—Addressing challenges with AI, renewable energy integration and leading-edge technologies. *abs/2501.14143*. https://www.semanticscholar.org/paper/

adc4d876af05bcdaf9a2c3bdbdad30abc9accf82 (accessed on 30th September 2025).

10. Bonetto, R., & Rossi, M. (2017). Smart grid for the smart city. https://core.ac.uk/download/84922332.pdf (accessed on 30th September 2025).
11. Borghetti, G., Menga, G., Procaccianti, G., Kok, K., Ardito, L., Morisio, M., Mo, D., et al. (2013). Smart grid technologies in Europe: An overview. https://core.ac.uk/download/pdf/11431923.pdf (accessed on 30th September 2025).
12. Chin, W. H., Efthymiou, C., Fan, Z., Gormus, A., et al. (2011). Smart grid communications: Overview of research challenges, solutions, and standardization activities. http://arxiv.org/abs/1112.3516 (accessed on 30th September 2025).
13. Copeland, R., Crespi, N., Rehmani, M. H., et al. (2019). Internet of Things-aided smart grid: Technologies, architectures, applications, prototypes, and future research directions. http://arxiv.org/abs/1704.08977 (accessed on 30th September 2025).
14. Curea, O., Jiménez Verde, J., Llaria, A., et al. (2016). Application of wireless sensor and actuator networks to achieve intelligent microgrids: A promising approach towards a global smart grid deployment. https://core.ac.uk/download/157743960.pdf (accessed on 30th September 2025).
15. De Turck, F., Develder, C., Melis, S., et al. (2013). Design of a management infrastructure for smart grid pilot data processing and analysis. https://core.ac.uk/download/55867872.pdf (accessed on 30th September 2025).
16. Debackere, K., Monard, E., Raspoet, D., Ratinckx, et al. (2025). *Technologie En Innovatie In Vlaanderen: Prioriteiten.* https://core.ac.uk/download/pdf/6468460.pdf (accessed on 30th September 2025).
17. Develder, C., Haerick, W., & Mets, K. (2010). A simulator for the control network of smart grid architectures. https://core.ac.uk/download/55823207.pdf (accessed on 30th September 2025).
18. Díaz González, F., Gallart, R., Girbau Llistuella, F., et al. (2018). Smart grid architecture for rural distribution networks: Application to a Spanish pilot network. https://core.ac.uk/download/157811722.pdf (accessed on 30th September 2025).
19. Dong, L., Wu, J., Xu, X., Xiu, C., & Wang, Y. (2025). Enhancing grid reliability through Transformer-LSTM model in smart grids. *Proceedings of SPIE, 13486*, 134861K–134861K-7. https://www.semanticscholar.org/paper/110ed551855ce7d45287761b3af49ae8679f0dee (accessed on 30th September 2025).

20. El Meslouhi, S., El Fadil, H., Lassioui, A., Hasni, A., & Fhail, A. (2025). Internet of energy in microgrids and smart grids: State-of-the-art. *2025 5th International Conference on Innovative Research in Applied Science, Engineering and Technology (IRASET).* https://www.semanticscholar.org/paper/87f4b7cfa687494302bad50b103f8fdfbf38dbb5 (accessed on 30th September 2025).
21. Erol-Kantarci, M., Rehmani, M. H., Radenkovic, M., et al. (2015). Special section on smart grids: A hub of interdisciplinary research: *IEEE Access* special section editorial smart grids: A hub of interdisciplinary research. https://core.ac.uk/download/48318388.pdf (accessed on 30th September 2025).
22. Feng, C., Wang, Y., Chen, Q., Ding, Y., Štrbac, G., & Kang, C. (2020). Smart grid encounters edge computing: Opportunities and applications. *Advances in Applied Energy, 1*, 100006. https://doi.org/10.1016/j.adapen.2020.100006 (accessed on 30th September 2025).
23. Kaveh, M., Martín, D., & Mosavi, M. R. (2020). A lightweight authentication scheme for V2G communications: A PUF-based approach ensuring cyber/physical security and identity/location privacy. *Electronics, 9*, 1479–1479. https://doi.org/10.3390/electronics9091479 (accessed on 30th September 2025).
24. Klemeš, J. J., Fan, Y. V., & Jiang, P. (2020). COVID-19 pandemic facilitating energy transition opportunities. *International Journal of Energy Research, 45*, 3457–3463. https://doi.org/10.1002/er.6007 (accessed on 30th September 2025).
25. Kumar, P. V., Kasinathan, P., Elavarasan, R. M., Vinoth, R., Kumar, R. A., Subramaniam, U., Ghosh, A., et al. (2021). A comprehensive review on sustainable aspects of big data analytics for the smart grid. *Sustainability, 13*, 13322–13322. https://doi.org/10.3390/su132313322 (accessed on 30th September 2025).
26. Marwedel, P. (2021). *Embedded System Design.* https://doi.org/10.1007/978-3-030-60910-8 (accessed on 30th September 2025).
27. Mutlaq, K. A. A., Nyangaresi, V., Omar, M. A., Abduljabbar, Z., Ma, J., Al Sibahee, M. A., Aldarwish, A. J. Y., et al. (2025). Blockchain assisted signature and certificate-based protocol for efficient data protection and transaction management in smart grids. *PLOS ONE, 20.* https://www.semanticscholar.org/paper/ccd677e9cf9bc41bf05faff9febbba2a8ef6f50e (accessed on 30th September 2025).
28. Ogunsuji, Y. M., Amosun, O. R., Kumar, P., & Choubey, D. (2024). Integration of smart grid technologies: Implementing advanced

procurement strategies leveraging AI and machine learning. *World Journal of Advanced Research and Reviews.* https://www.semanticscholar.org/paper/1ac3fedfe1c91480533cc6f9e4347da3cea479cf (accessed on 30th September 2025).

29. Ojo, O. T., Habeeb, A. O., Adebiyi, O. E., Olanrewaju, S. A., Olabode, O. J., & Ezekiel, S. E. (2025). IoT-enabled energy storage systems for renewable energy grid integration. *Path of Science.* https://www.semanticscholar.org/paper/dd1e35630310eb32a22530a52c129e7f4084c77d (accessed on 30th September 2025).
30. Olatunde, T. M., Okwandu, A. C., Akande, D. O., & Sikhakhane, Z. Q. (2024). The impact of smart grids on energy efficiency: A comprehensive review. *Engineering Science & Technology Journal.* https://www.semanticscholar.org/paper/95786c54b5180741ea4a968c99d9f0101840b72f (accessed on 30th September 2025).
31. Peffer, T. (2010). UC Berkeley's Cory Hall: Evaluation of challenges and potential applications of building-to-grid implementation. https://core.ac.uk/download/288432727.pdf (accessed on 30th September 2025).
32. Rasheed, A., San, O., & Kvamsdal, T. (2020). Digital twin: Values, challenges and enablers from a modeling perspective. *IEEE Access, 8,* 21980–22012. https://doi.org/10.1109/access.2020.2970143 (accessed on 30th September 2025).
33. Tsitsimelis, A. (2020). Advanced signal processing techniques for robust state estimation applications in smart grids. https://www.semanticscholar.org/paper/66112bd7a093edb278406820b593d859626aca46 (accessed on 30th September 2025).
34. Zhou, X. (2025). Application analysis of wireless communication technology in China's smart grid. *Applied and Computational Engineering.* https://www.semanticscholar.org/paper/7e07ec99686b430a39c4062898091d07a85c8355 (accessed on 30th September 2025).
35. Zobaa, A. F., Vaccaro, A., Behrendt, K., Celano, T., Dickerson, B., Dion, A., Formato, G., et al. (2014). Vulnerability analysis of satellite-based synchronized smart grids monitoring systems. https://core.ac.uk/download/29204066.pdf (accessed on 30th September 2025).

CHAPTER 5

DEMAND SIDE MANAGEMENT (DSM)

Contents

INTRODUCTION

The last few years have seen the term Demand Side Management (DSM) being highly debated as energy systems are transforming towards greener and more efficient paradigms, particularly in the context of smart grids. DSM can be understood as technologies and strategies that are employed to optimize energy consumption patterns by influencing consumer habits and managing demand on an hourly basis. It not only assists utilities in balancing load but also allows customers to be in an active position to manage energy, leading to savings in energy and improved energy efficiency. As smart technology has evolved and renewable energy supplies have been added to the mix, DSM has become an integral part of energy policy strategies aimed at greenhouse gas emissions reduction and sustainability goals. Hence, understanding the fundamentals of DSM in smart grids is crucial for all the players involved in the energy sector. Smart grid revolution brings new opportunities and challenges to achieve DSM in a successful way. Smart grids make use of advanced communication technologies, analytics, and automation to create an interactive power system. This infrastructure facilitates real-time monitoring and control of energy usage, enabling consumers to modulate their use in response to price signals as well as to grid demands.

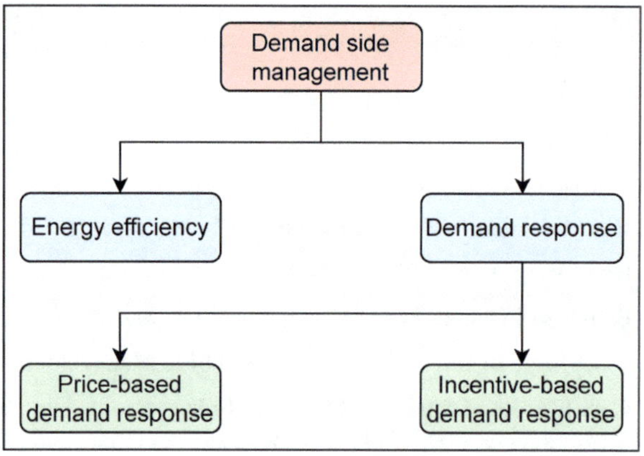

Figure 5.1. Demand-side management (DSM).

Source: https://www.mdpi.com/energies/energies-15-02863/article_deploy/html/images/energies-15-02863-g002.png.

The connectiveness of IoT devices and mobile edge computing also ramps up the responsiveness of DSM programs, providing the ability to integrate energy producers and consumers seamlessly. The promise of such innovations in technology makes DSM a key element in resolving the issues of modern energy systems. Hence, examining the interface between DSM and smart grid technology will highlight the directions for improving energy reliability and efficiency while encouraging consumer involvement. The incorporation of blockchain technology within DSM platforms also has the potential to revolutionize the energy sector by enhancing energy transaction transparency and security. As the energy sector is grappling with distributed energy resources and peer-to-peer energy trading, blockchain presents a probable answer that has the potential to fix some of the intrinsic challenges. By creating decentralized and tamper-resistant ledgers, blockchain facilitates new business models and secures trust for stakeholders. This is particularly applicable in DSM, where real-time monitoring and data exchange are important for effective demand response programs. With increasing interaction between DSM, smart grids, and blockchain, it will be essential to analyze how these technologies collectively enhance energy management policies. Overall, an interdisciplinary understanding of DSM in the context of smart grids and new technologies is essential in the development of sustainable energy systems that benefit consumers as well as the environment (Figure 5.1).

Definition of Demand Side Management (DSM)

DSM refers to the strategic redirection of domestic energy consumption patterns through a variety of signals, including price, incentives, and regulation. In a smart grid environment, DSM becomes a critical element to enhance energy efficiency and reliability by aligning supply and demand in real-time. This management approach allows consumers to become actively engaged in energy usage decisions, thereby achieving a more sustainable energy system. By using advanced technologies, such as smart meters and real-time data analytics, DSM facilitates a shift from passive consumption to dynamic engagement. This is imperative for grid operators because it assists in neutralizing peak load pressures and integrating renewable energy sources, hence enhancing the resilience of power systems while neutralizing the challenges posed by changing energy needs. DSM application is particularly valuable in its impact on peak reduction demand, which is of greatest concern for the majority of utilities. An entirely distributed system of DSM permits individual household consumers to manage their own energy consumption through dynamically changing pricing schemes. As illustrated in recent studies, this allows appliances to be timed in accordance with real-time levels of demand, which can result in significant peak energy absorption reductions. Research findings show that a completely distributed DSM architecture is capable of reducing peak demand by up to 55%, hence lowering the capital expenditure (CAPEX) required for meeting growing energy demands (Barbato et al., 2014). This capability is imperative to smart grids that aim to optimize operation while making sure energy resources are well distributed and utilized. Moreover, control measures that are the foundation of DSM systems are not confined to domestic consumers but are extended to other energy actors, i.e., electric vehicle charging stations. Coordination of distributed energy resources (DER) with load balancing capabilities is a demonstration of the multi-dimensional benefits of DSM. A high-end architecture allows for coordination of electric vehicle (EV) charging sessions with energy generation profiles in a manner that economic gains are achieved by all parties involved. With a two-level control system—made up of Load Area Controllers (LACs) at local levels and Macro Load Area Aggregators (MLAAs) at higher levels—DSM models can enhance operational efficiency and enable active load management through market equilibrium solutions. Such new methods emphasize the necessity of DSM to maximize energy usage and develop an integrated and efficient smart grid network (Giorgio et al., 2013) (Figure 5.2).

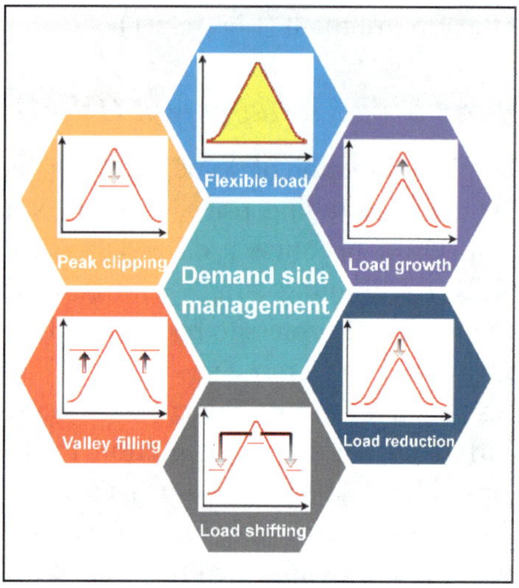

Figure 5.2. Demand-side management (DSM) techniques.

Source: https://www.mdpi.com/energies/energies-15-02863/article_deploy/html/images/energies-15-02863-g003.png.

Importance of Demand Side Management (DSM) in Modern Energy Systems

The integration of DSM in modern energy systems is important in terms of rendering grids more robust and sustainable. As more energy is demanded, there is a continually growing demand for an active interaction between consumers and utilities. DSM programs facilitate such an interaction by enabling users to change consumption patterns based on shifting supply conditions and price signals. Such a dynamic not only maximizes energy usage but also alleviates the pressure on power generation during peak times, ultimately contributing to a stronger grid. According to recent studies, incorporating DSM into the Smart Grids framework allows for the inclusion of smart technologies that enable real-time monitoring and management of energy usage, thereby improving system efficiency and reliability as a whole (Denysiuk et al., 2016). On top of that, DSM's role extends beyond direct load control to allow integration of renewable power into electricity markets. With the evolving energy system relying increasingly on intermittent solar and wind renewable resources, effective DSM methods are important to meet supply and demand volatility caused by the natural intermittency of solar and wind power generation. By engaging consumers to shift their consumption to conform with renewable generation patterns, DSM programs can efficiently maximize the utilization of the resources. Moreover, such coordination diminishes reliance on fossil fuel-based peaking facilities, enhancing a cleaner energy mix. Technological advances in Smart Grid enable such integration, with utilities capable of sharing real-time data and bringing consumers onboard for more dynamic patterns of energy consumption (He et al., 2015). Lastly, the relevance of DSM in today's energy systems is accentuated by its economic benefits, environmental sustainability, and enhanced grid reliability. Economically, DSM evaluations enable consumers to benefit from lower electricity costs through time-of-use prices, which equally fosters effective energy utilization. At a national scale, the implementation of DSM programs responds to national energy security by lowering peak load requirements and minimizing the risks arising from disruptions in energy supply. In addition, as the energy system continues to evolve, resilience from good DSM can significantly reduce the environmental footprint of energy production through the induction of energy savings and capping greenhouse gas emissions. Thus, incorporation of DSM into Smart Grids is not only beneficial but necessary to enable a future of sustainable energy (Denysiuk et al., 2016).

Overview of Smart Grids and Their Role in Demand Side Management (DSM)

The evolution of smart grids is a significant transformation in electricity distribution and management that directly affects DSM. Essentially, smart grids use advanced technologies, including smart meters and automatic networks, to present an interactive interface through which the energy providers and consumers are able to interact and share information in real-time. This interactivity makes the energy market more responsive, and customers are no longer passive recipients but participants in managing energy. By changing their consumption patterns in harmony with grid needs, consumers and businesses can each contribute their share towards end-to-end energy efficiency, underscoring the critical intersection of smart grid technology and DSM practices. With the stakeholders working in this new energy model, it is necessary that the user perspective be considered to utilize the best out of maximum capabilities of such technologies, as noted through the use of consumer attitudes and behavior in energy policy (Marez et al., 2010). Communication technology plays a crucial role in smart grids as it is the foundation for optimizing energy consumption and integrating demand-side resources. Real-time exchange of information between different segments of the grid—generation, transmission, and consumption—permits smarter control of energy and facilitation of coordinated reactions to changing conditions.

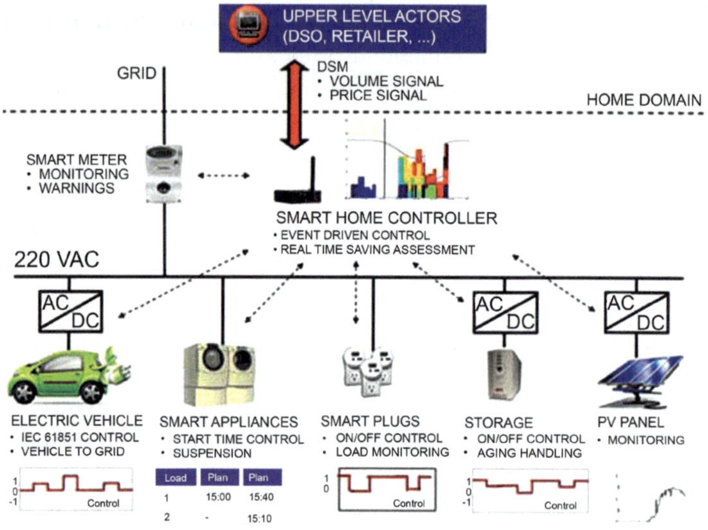

Figure 5.3. Smart grid and dynamic demand side management (DSM) environment.

*Source:*https://www.researchgate.net/profile/Zafar_Khan24/publication/277132053/figure/fig3/AS:667667597897732@1536195768336/Smart-Grid-and-Dynamic-DSM-Environmet.jpg.

Most critical, however, remains the development of interoperable and secure communications networks that can keep up with shifting energy requirements and minimize security concerns. Through investment in robust networking technologies and cross-industry standardization, electricity utilities are able to have smart metering infrastructure with coordinated activity across the various regions. These advancements in communications technology are the keys to future success with DSM, and it is these that are necessary to enable both utilities and consumers to gain the benefits of a smarter, more responsive grid system (Chin et al., 2011). Moreover, the consumer-oriented strategy of DSM and its coordination with smart grids highlights the significance of consumer participation in energy management. As individuals integrate smart appliances into their homes, consumer attitudes towards such devices are critical in shaping the efficacy of DSM programs. It is necessary to probe consumer attitudes, change willingness, and obstacles towards embracing smart technology to optimize energy consumption in home environments to the fullest. Studies and surveys, such as those conducted with mixed populations, can reveal how consumers understand and interact with smart devices. It is only through settling these perceptions that policymakers and energy suppliers can reasonably implement strategies promoting the adoption of sustainable energy practices, ultimately redefining the energy landscape with active participation in DSM (Marez et al., 2010) (Figure 5.3).

5.1. HISTORICAL CONTEXT OF DEMAND SIDE MANAGEMENT (DSM)

The history of DSM is traced to the revolution in the methods of load management from being simple and primitive to sophisticated, data-dependent approaches and has utilized the potential of smart grids. DSM initially gained interest and concentration in load shifting and mitigation of peak loads and was predominantly interested in mitigating stress on generation and transmission infrastructure. Early plans called for consumer incentive programs to reduce peak demand and typically were communicated directly by the utilities. The development of smart grid technologies shifted the DSM, with two-way communication between customers and utility companies now available. This shift is significant, as it emphasizes an ongoing effort to introduce renewable energy sources into action and increase system efficiency as a whole, propelling DSM into the limelight in energy management paradigms (Axon et al., 2014). Advances in information and communication technologies and smart metering have played key roles in driving the development of DSM in the past. With the rollout of sophisticated metering infrastructure, utilities have been able to tap into deep consumer insights and patterns of energy consumption. Such granularity enables the design of tailored demand-aware pricing plans, which encourage users to adjust their usage patterns in real time. The recent introduction of an "individualized" tariff policy represents a significant development in DSM, in which each customer is offered an individualized tariff regime that aims to discourage peak demand while ensuring reasonable access to energy. This new approach represents the shift toward a more customer-centric model that addresses not only energy efficiency but consumer fairness and involvement in demand management schemes (Hayes et al., 2016). Also, historical evolution of DSM has largely been influenced by policy and regulatory frameworks that aimed to enhance energy efficiency and reduce greenhouse gas emissions.

Governments and energy administrations across the world have increasingly understood the significant contribution that DSM can make towards attaining sustainable energy goals. Through the encouragement of smart grid installation and DSM programs, policymakers aim to enhance energy security, encourage economic development, and support the transition to a low-carbon economy. Increasing political and economic support is turning DSM strategies into primary drivers of achieving energy efficiency objectives (Axon et al., 2014). The constant interaction among stakeholders like consumers, regulators, and utilities in deciding these strategies reflects the dynamism of energy markets today, thereby building a more responsive and more resilient energy infrastructure.

5.1.1. Evolution of Energy Consumption Patterns

Shifts in energy consumption patterns have been achieved through phenomenal technological leaps accompanied by increasing awareness of sustainability. In the past, energy use was linear and fossil fuel-driven with minimal focus on efficiency or response to demand. However, with the introduction of smart grids and Demand Side Management (DSM) initiatives, consumer habits and energy consumption patterns have been revolutionized. This change is a move from passive consumption to active participation in energy management and the provision of users to modify their consumption according to real-time data and price signals. With more informed and engaged consumers, their choices not only define trends in demand but also drive technological innovation and infrastructure that aim to optimize energy transmission and reduce wastage (Axon et al., 2014). Through the actualization of smart technology in the energy sector, opportunities have been created towards a deeper understanding and analysis of consumers' energy consumption habits. The deployment of Internet of Things (IoT) technology and smart meters enables real-time monitoring and data collection and hence the creation of sophisticated analysis tools. The ability facilitates a level of responsiveness that was previously lacking in the energy market, since the utilities are able to match demand and supply effectively. For instance, big data analytics can unveil energy consumption patterns, the hours of peak consumption, and consequently offer tailored demand response solutions (Gharehpetian et al., 2019). Then, the technologies enhance the operational efficiency of smart grids while making consumers able to minimize their cost of energy consumption and environmental footprint. In addition, growth in energy consumption is closely intertwined with broader socio-economic conditions, including regulatory systems, economic advantages, and social opinions on renewable energy. Policymakers and business executives increasingly realize the importance of creating frameworks conducive to proper use of energy and environmental management. Political and financial backing for smart grid projects reflects a general understanding of their capability to make carbon reduction targets achievable and improve energy security (Axon et al., 2014). Furthermore, as societal attitudes change regarding the consumption of energy, so does the demand for transparent price structures and equitable access to the energy resources. Thus, the interaction of consumer trends, technological innovation, and regulatory structures will continue to drive the reformation of energy consumption patterns in the smart grids environment (Figure 5.4).

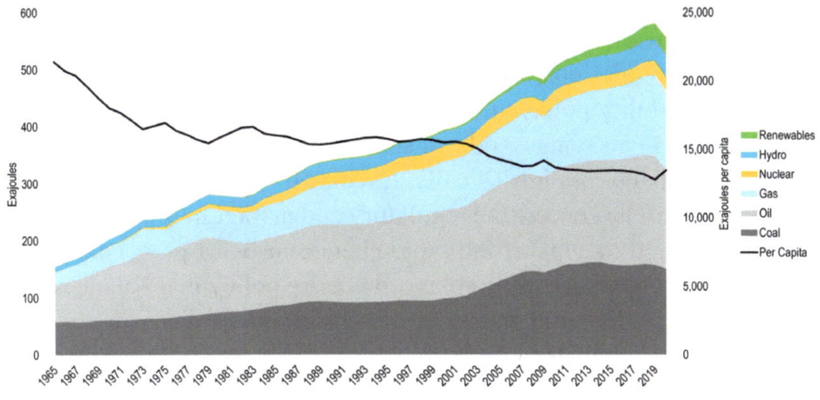

Figure 5.4. World energy consumption 1965–2018.

Source: https://i0.wp.com/transportgeography.org/wp-content/uploads/world_energy_consumption2.png?resize=1536%2C720&ssl=1.

5.1.2. Early Demand Side Management (DSM) Initiatives and Their Outcomes

Early DSM programs emerged in response to growing pressure for sustainable energy supplies under conditions of fluctuating energy demands. Early programs were focused on reducing peak load consumption, since it was understood by the utilities that effective demand management would relieve pressure on energy infrastructures. By means of time-of-use tariffs and load-shifting incentives, these programs sought to induce consumers to shift their electricity use to off-peak hours. Key findings were reduced cost of operations for the utilities and fewer outages, not only making the grid more reliable but also placing the energy in efficient use. This was also the beginning of a trend to adopt smart technologies capable of facilitating greater communication between the utilities and consumers, laying the foundation for the concept of smart grid that is core to current-day energy management principles. As the energy usage scenario evolved, DSM programs began to adopt newer technologies such as smart meters and real-time analytics. These facilitated deeper understanding and predictive analysis of usage patterns of energy, which helped in the effectiveness of demand-side interventions. Through the provision of real-time information regarding customer usage, the utilities managed to get the customers more involved and thus attain active participation in energy conservation activities. Also, the overlap of the technologies facilitated the development of segmented incentive schemes for different consumer groups, something that was helpful in managing demand well across different populations. The effects of the innovations created the realization of greater energy savings and reduced green gas emissions, which led to major climate goals that were related to energy sustainability policy and practice. While early DSM initiatives were successful in energy savings and consumer engagement, they also pointed towards certain challenges intrinsically associated with the transition towards smart grids. Most of these were the regulatory and market frameworks that were generally not geared to support the rapid change in DSM practice. As quoted in the literature, achieving a well-balanced energy policy that balances market efficiency, sustainability, and energy security is greatly reliant on cooperative action between stakeholders like regulators, producers, and consumers. The complexity of the situation dictates that an end-to-end approach must be adopted, one that draws learning from the policy as well as technical domains as we move along in the development of effective DSM schemes. The multi-dimensionality of this transition highlights the need for periodic review and revision of DSM frameworks to ensure conformity with wide energy goals and market forces (Axon et al., 2014; Jacques de Jong et al., 2025).

5.1.3. Technological Advancements Leading to Smart Grids

Smart grids represent a break-through in energy management caused primarily by technological advances in communication and automation. They provide an energy distribution framework that can react more dynamically, enabling real-time utility and consumer interaction. The milestone to all this is the advanced metering infrastructure (AMI), which enables precise observation of energy consumption patterns. The precision involved in data gathering ensures greater efficiency as it enables utilities to identify peak load times and prompt users to adjust usage patterns accordingly. Moreover, the implementation of such technologies makes Demand Side Management (DSM) initiatives easier, which support energy optimization and conservation. As society is moving towards sustainability, understanding how smart grids work not only improves the energy efficiency at large but also has a significant role to play in supply-demand equilibrium on the grid. Another important aspect of technological innovation in smart grids is the continuous optimization of algorithms that are intended to optimize energy consumption methods. The extant literature confirms that various approaches have been proposed for balancing consumer comfort and energy efficiency in domestic settings, pointing out the pressing engagement between such often conflicting demands. Implementing cutting-edge algorithms will lead to astounding reductions in energy loss without sacrificing user satisfaction.

Figure 5.5. Advancements in smart grid technology and its benefits.

Source: https://www.inventia.in/wp-content/uploads/2024/09/Smart-grid-blog-01.webp.

Despite this, there are challenges faced because most of the algorithms proposed recently, in promises of enhancing precision, have not been duly implemented in enhancing energy consumption in smart homes. By overcoming these challenges, researchers can also be able to make further contributions towards enhancing smart grid technology, which in turn will lead to implementations based on enhancing efficiency and comfort for users, as has been discussed in the existing debates in the field (Fayaz et al., 2019). The case of Smart Home Energy Management Systems (SHEMS) exemplifies the intersection of technological innovation and user-based design in the world of smart grids. While previous research focuses on technical developments, it is oblivious to the role consumer behavior can play in facilitating greater efficiency in energy consumption. Without it, solutions cannot actively engage users, hence limiting adoption and effectiveness of these systems. More integrated is an approach that combines technology innovations with insights into user interaction and incentives to take DSM efforts to the next level. By integrating technical and behavioral knowledge, SHEMS can significantly enhance household energy management, leading to major reductions in consumption and costs. The necessity of bridging this gap becomes apparent, as a balance strategy with simultaneous focus on technological advancement as well as user engagement will be instrumental in the effective change in demand-side management (McIlvennie et al., 2020) (Figure 5.5).

5.2. KEY COMPONENTS OF SMART GRIDS

The architecture of a smart grid entails several key components that form the foundation for optimal Demand Side Management (DSM). Centering this system is the advanced metering infrastructure (AMI) integration, where users are provided with smart meters to facilitate real-time data acquisition and two-way communication between customers and utilities. Two-way communication allows dynamic pricing mechanisms, where customers are incentivized to alter usage levels during peak times. Moreover, the built-in electronic control systems in smart grids enable the utilities to manage demand more effectively through the use of load balancing strategies and demand response programs. Not only do these minimize grid strain but also contribute towards a well-balanced operating space where supply and demand can be easily coordinated. Those kinds of offers are constituent elements that serve the overall themes of cost savings and sustainability, as emphasized in the smart grid innovation rhetoric (Axon et al., 2014). One of the most prominent aspects of smart grids is their adoption of Information and Communication Technologies (ICT), which significantly enhance operational capabilities. Utilities are able to make optimal energy supply and grid reliability a reality through the use of such technologies. The use of big data analytics allows for predictive modeling, through which utilities can forecast patterns of demand and respond to changes in energy consumption. Additionally, the development of prosumer networks—thereby consumers actively participate in energy generation—means a paradigm for the management of energy. In such instances, household consumers can participate in load balancing and peak shaving, which help in the overall efficiency of the grid. This marriage of ICT with energy production not only makes consumers more robust but also confronts energy management through collective approaches, leading to a more robust, flexible energy system (Bonetto et al., 2017). Regulatory frameworks also play a critical role in the smooth implementation of smart grids, considering their focus on demand-side engagement. Through setting policies that encourage the adoption of smart technologies and consumer engagement, governments can facilitate a more participative energy system. Such policies must address areas of data protection, cyber security, and price competition in order to secure people's trust and smart grid technology acceptance. As the level of interaction among various stakeholders increases, one must create comprehensive strategies that address all facets of smart grid operation. Lastly, innovation in technology, customer engagement, and regulation facilitation will converge to unleash the real potential of demand-side management in smart grids toward a sustainable future (Axon et al., 2014) (Figure 5.6).

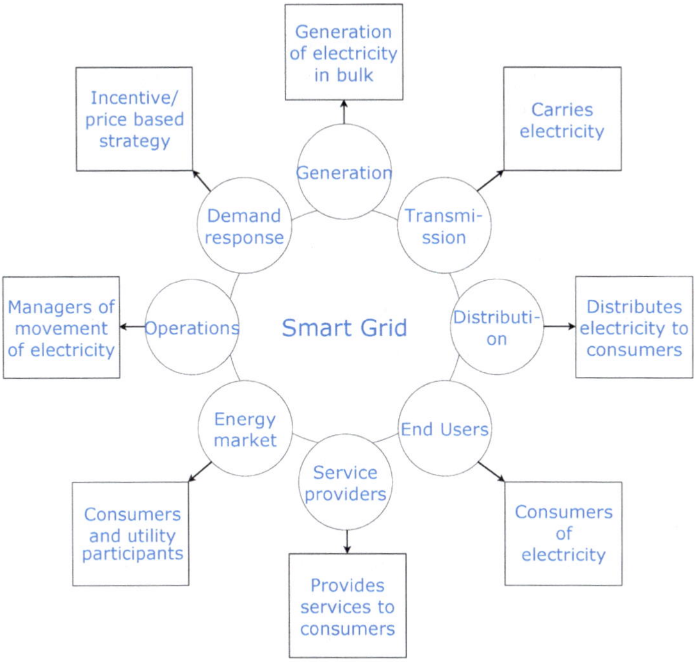

Figure 5.6. Architecture of smart grid.

Source: https://www.mdpi.com/energies/energies-13-04266/article_deploy/html/images/energies-13-04266-g001.png.

5.2.1. Smart Meters and Their Functionalities

The integration of smart meters into the modernized energy grid is an important advancement in Demand Side Management (DSM) in Smart Grids. Smart meters enable a two-way flow of information between consumers and utilities, with real-time monitoring and reporting of energy usage. Through this capability, consumers are in a position to know more about their patterns of energy usage, thus making smart choices that can lead to reduced energy bills and higher efficiency. By providing sophisticated usage data, smart meters shift the energy management paradigm from passive to active, wherein customers are more actively engaged with their own energy consumption, thereby maximizing consumption and encouraging more prudent behavior towards energy preservation. Furthermore, the functionality of smart meters transcends mere measurement; they play a significant role in the general communication protocols needed to optimize Smart Grid operation. These devices make use of advanced communication technologies to transmit information regarding the utilization of energy, thereby helping to balance supply and demand in the grid. As shown in descriptions of challenges and opportunities with smart grid communications, interoperability is key to attaining efficiency with the distribution and consumption of energy. Smart meters are at the forefront of such applications, as they must communicate coherently with other grid elements so that utilities and consumers are both offered information they need to make their operations more efficient and guarantee electrical grid reliability (Chin et al., 2011). Aside from their own direct purpose, smart meters play significant roles toward the higher goals of Smart Grids by enabling advanced levels of security and privacy. The shift from the traditional power grids to the Smart Grids necessitates robust data protection mechanisms and responses to privacy risks, with greater sensitive consumer information being shared more widely. The integration of the Internet of Things (IoT) technologies allows smart meters not only to report and log energy usage but also to provide crucial information to assess the security of the grid. As explained in IoT-enabled Smart Grid systems, smart meters' capability to leverage IoT enhances real-time monitoring and analytics functions, thus enhancing a more resilient and secure energy system (Copeland et al., 2019). Therefore, smart meters are not only demand management devices but critical components towards secure and efficient energy systems (Figure 5.7).

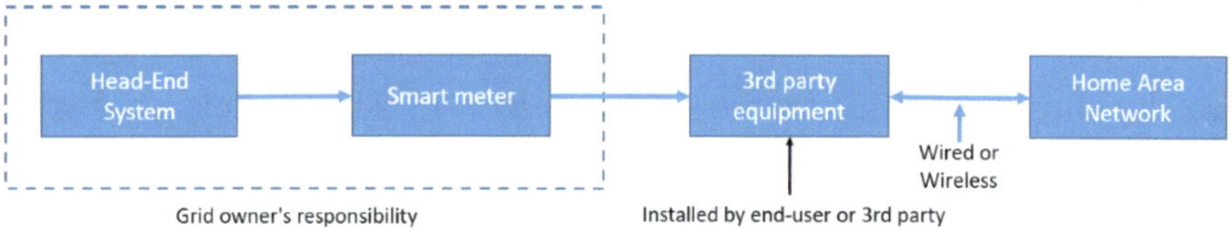

Figure 5.7. Smart metering architecture.

Source: https://www.mdpi.com/sensors/sensors-21-08087/article_deploy/html/images/sensors-21-08087-g002.png.

5.2.2. Communication Technologies in Smart Grids

Communication technology is the core of smart grids, providing real-time communication between producers and consumers of energy. Two-way communication is required in successful DSM as it helps consumers adjust their energy usage in accordance with fluctuations in supply and prices, thereby grid stability. Ideally, advanced communication protocols would use technologies such as smart meters and distributed energy resources to streamline energy flow and minimize wastage. Also, with the application of information and communication technology (ICT) in smart grids, it becomes easier to effectively manage renewable energy resources that are inherently intermittent. The efficient supply management, wherein these advanced communication technologies are used, is a cornerstone of what constitutes a smart grid, with significant consequences for energy efficiency as much as for sustainability (Axon et al., 2014). The alignment of these advanced communication technologies is needed in order to make smart grids robust DSM schemes possible.

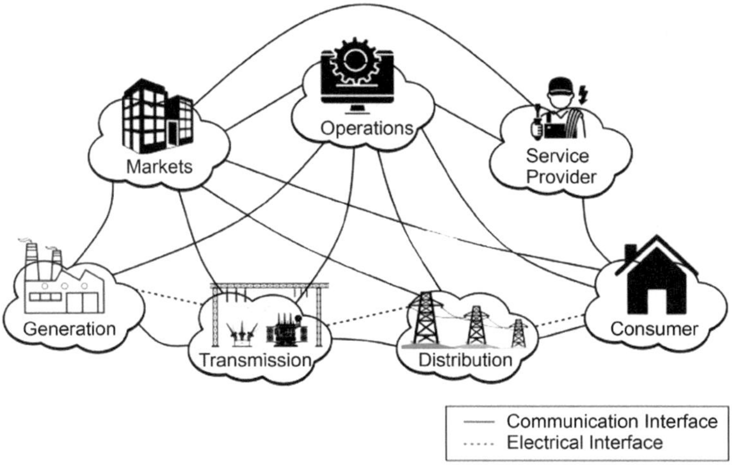

Figure 5.8. Smart grid communication.

Source: https://www.tutorialspoint.com/smart-grid-technology/images/smart-grid-communication.jpg.

By putting in smart meters and demand response solutions, the utilities can monitor patterns of consumption and notify the consumers to modify their behavior to support the balancing of peak load. This mutual dynamic not just makes saving for the consumer feasible but also helps make the grid efficient as a whole. For instance, communication technologies may facilitate the integration of distributed generation and storage resources that have the capability for decentralized management of energy. The approach can lead to optimized utilization of renewable energy resources with reduced reliance on additional infrastructure investments. The capacity to re-shape demand profiles, particularly by managing peak demand scenarios, can significantly enhance grid reliability and enable the achievement of environmental targets (Bakker et al., 2010). Furthermore, establishing trusted communication networks in smart grids brings significant challenges and opportunities. Along with greater dependence on data, so does worry about privacy, cybersecurity, and regimes of regulation for such technologies. Keeping consumer information secure and communication networks intact is critical to the proper deployment of smart grid technologies, since public trust is an important part of mass adoption. This multifaceted interface of technology, policy, and public acceptance demonstrates the complexity of selling smart grid solutions. Thus, while communication technologies hold huge promise to boost DSM potential, these must be synergistically integrated within a grand strategy that rationalizes not merely the technical but the regulatory and ethical dimensions of power management (Axon et al., 2014) (Figure 5.8).

5.2.3. Integration of Renewable Energy Sources (RES)

The integration of RES into existing power grids is fast becoming a focus area in the quest for sustainable power production. RES such as solar and wind are a cleaner alternative to fossil fuel but are marred by their intermittent nature, which causes significant problems to grid stability. To fight against such problems, Demand Side Management (DSM) policies are evolving to make energy consumption more responsive to supply fluctuations. Through the aid of advanced technologies and smart grid systems, potential exists to make the distribution of energy optimized in a way that renewable energy is used efficiently even though it is not predictable. Additionally, DSM allows consumers to be active in energy management and thereby make the grid stronger to provide for increased distributed levels of renewable generation without affecting stability or reliability (Barata et al., 2015). The role that smart grids play in this integration cannot be overstated. Smart grids look to incorporate a variety of distributed generation sources, like the energy of intermittent renewables, along with the necessary infrastructure to provide efficiently the necessary demand. A key component of this shift is implementing communications technologies that support real-time monitoring and control of energy consumption. Through enabling two-way communication between the consumers and producers of electricity, smart grids enable the capability to achieve dynamic real-time adjustments to unstable supply levels, aligning energy demand with supply. Greater interactions between consumers and energy infrastructure can lead to new behavioral patterns, leading to energy saving and load shifting, which are key to guaranteeing the optimal utilization of renewables integration (Axon et al., 2014). Furthermore, the effective integration of renewable energy into smart grids is largely determined by effective DSM measures with consumer participation. As emphasized by recent research, utilization of controllable appliances and energy storage facilities is vital for balancing supply and demand for energy. A DSM based on a multi-agent model can facilitate management of such distributed energy resources in a way that energy generation and consumption will be further aligned. For instance, customers can manage their energy usage in accordance with the forecast data generated by smart systems to make their operation optimal at peak renewable supply moments. This not only optimizes energy efficiency but also reduces total carbon emissions, thus enabling environmental sustainability. Hence, complementarity of DSM and integration of renewable energy is not only worthwhile but also important to achieve a low-carbon future (Barata et al., 2015) (Figure 5.9).

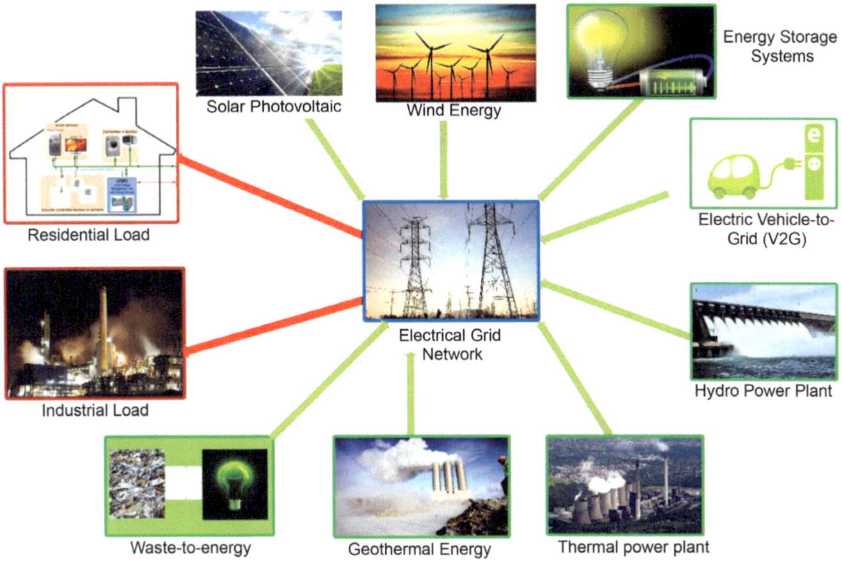

Figure 5.9. Various types of energy sources in the smart grid.

Source: https://www.mdpi.com/applsci/applsci-12-08914/article_deploy/html/images/applsci-12-08914-g004.png.

5.3. BENEFITS OF DEMAND SIDE MANAGEMENT (DSM)

DSM is responsible to maintain energy efficiency in the smart grid, and it has a wide range of benefits from consumers to utilities to the environment. Arguably, the most significant advantage of DSM is that it can relieve peak demand strains on the electric grid. Through the promotion of consumer behavior to modify their energy consumption patterns—e.g., operating appliances during non-peak periods—DSM load balancing is enhanced. This reduced peak demand not only postpones the need for upgrading costs but also lowers the risk of blackouts during peak hours of use. Besides, greater energy efficiency can offer consumers lower electricity bills, a scenario of mutual advantage that accommodates long-term energy management sustainability. In addition, DSM opens an arena for greater consumer education and awareness.

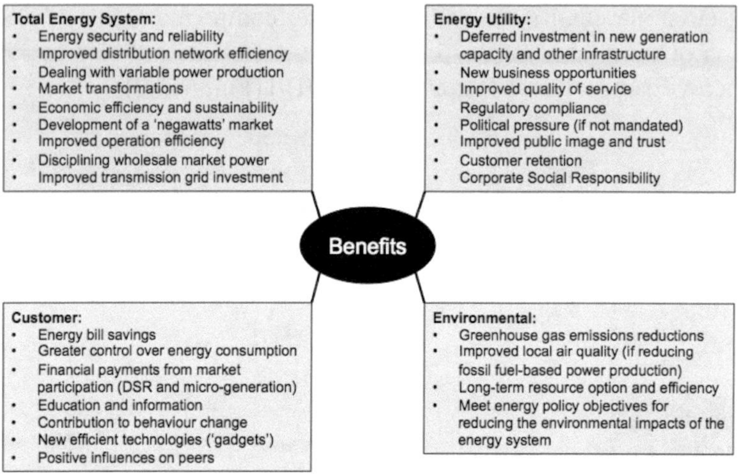

Figure 5.10. The benefits of demand-side management for utilities.

Source: https://www.researchgate.net/publication/332714483/figure/fig1/AS:1085864734138368@1635901735083/The-benefits-of-demand-side-management-for-utilities.ppm.

As the use of smart technology such as smart meters and real-time monitoring becomes part of everyday life, consumers are empowered with the control of their energy usage. Direct feedback from such advanced systems allows consumers to be aware of their usage patterns and adapt accordingly. With higher sensitivity, both household consumers and industries can make informed decisions leading to lower electricity bills and reduced environmental impact. The data made available by DSM programs can ultimately lead to a cultural shift in which conservation of energy becomes a part of consumer behavior, creating a stakeholder-focused process of energy efficiency that benefits everyone. In addition, the integration of big data analytics into DSM programs represents a paradigm shift in energy consumption management. By leveraging massive amounts of data gathered through smart grids, utilities can make their processes more efficient and design more effective DSM programs. Big data analytics brings out consumption trends, enabling utilities to make high-accuracy predictions on demand volatility and focus their programs to meet specific needs in a more efficient manner. This also makes the power supply more secure and reliable, which is illustrated by issues experienced with current research (Gharehpetian et al., 2019). Hence, the assimilation of DSM with big data solutions not only supports energy providers in resource management but also accelerates the use of sustainable techniques in smart grids to provide a robust energy system in the end (Figure 5.10).

5.3.1. Enhanced Energy Efficiency and Conservation

The urgency for increased energy efficiency and saving is increasingly important in the realm of Demand Side Management (DSM) of smart grids. With residential energy consumption on the rise, the traditional approaches to demand management fall short, and innovative approaches emphasizing the participation of consumers come into play. One such approach is the Automated Load Scheduling (ALS) algorithm, which facilitates dynamic appliance scheduling from real-time information and consumer preference. By lowering Peak-to-Average Ratios (PAR) and electricity costs, ALS not only aims to reduce usage, but rather maximizes the allocation of existing energy resources. Consequently, smart grids employing this technology can balance consumer needs and induce collective efficiency, offering a scalable approach to ensure maximum user satisfaction and grid stability. The implications of such innovation extend beyond dollar cost savings, encouraging cultural transformation towards household-level responsible use of energy (Ashraf et al., 2025). In exploring intensified conservation measures under DSM, it is crucial that one acknowledges the psychological dimensions of behavior regarding the use of energy. The implications of an empirical investigation into the licensing effect have been found to suggest that environmentally friendly individuals could paradoxically increase energy usage, especially in the presence of exceptional weather conditions. This behavioral impact is contrary to the traditional hypothesis of reduced consumption always being interpretable as sustainable self-identification. As a result, DSM initiatives which overlook these cognitive dimensions are less effective. In order to deploy energy efficiency and encourage conservation on a larger scale, utilities need to develop tailored interventions acknowledging the different drivers of consumer behavior. By addressing energy consumption complexities in today's homes, DSM programs are capable of creating more active and informed user populations, ultimately achieving substantial energy conservation outcomes in smart grid systems (Stenstadvolden et al., 2025). Technology and behavior meet to become a key force for promoting greater energy efficiency and conservation through

DSM. Interconnecting smart meters with houses not only provides consumers instant feedback regarding their energy consumption but also empowers them to take control of their electricity usage. Instant information allows responsive DSM strategy since it empowers consumers to synchronize behavior with consumption patterns and motivations. Besides, advances in technology for energy management systems enable home owners to monitor and control appliances, employing energy-saving options when grids become overloaded. The systems demonstrate how technology can enable an environment for conservation and efficiency. Accordingly, as smart grids evolve, there should be a call for adaptive and customer-centric approaches that ensure mutual responsibility in energy management, yielding a healthier sustainable energy environment (Ashraf et al., 2025; Stenstadvolden et al., 2025).

5.3.2. Reduction of Peak Demand and Grid Stress

The concept of Demand Side Management (DSM) lies at the heart of peak demand reduction as well as stress elimination on the grid, and particularly with respect to smart grids. Through the active participation of consumers in managing their energy consumption, DSM programs not only optimize the utilization of energy but reduce the load on the electric grid during peak times. With programs that provide monetary incentives for lowered energy consumption during peak periods, utilities are able to modify patterns of use. Not only does this stabilize the grid, but it also promotes general efficiency, allowing operators to get the maximum out of current infrastructure without calling for new construction. These strategies demonstrate the interconnectedness of consumer use and grid reliability, highlighting the central importance of DSM to sustainable energy management. Where renewable energy sources are increasingly integrated into grids, uncertainty from these resources can be compounded with peak period stress. The application of smart grid technologies facilitates improved monitoring and control of energy flows to make demand-side management more flexible. Active network management (ANM) is a newer strategy where flexibility in consuming energy and generating energy is offered. For instance, in times when maximum renewable energy generation is realized, demand can be redirected likewise. Such interactive equilibrium actually counters network congestion and promotes an amicable supply of energy consistent with prevailing demands imposed on infrastructure. With such a gap between transmission and generation plugged, therefore, DSM approaches directly contribute to lowering peak intensity of demand as well as bringing stability to the grid, as noted in recent studies (Gu et al., 2013). Furthermore, new technologies such as electric vehicles (EVs) have shown themselves to be good tools when dealing with stressed grid management. EVs can be harnessed as flexible demand resources, absorbing surplus renewable energy during low-demand periods and returning it to the grid when needed. This two-way function assists in enhancing the resilience of the grid through de-congestion and less use of fossil fuels, promoting ecological sustainability. Studies have suggested that the integration of such demand-responsive strategies can lead to a tremendous increase in renewable energy utilization, emphasizing the effectiveness of integrating demand management with generation strategies (Gu et al., 2013). As smart grids evolve, coordination of consumer behavior, integration of technology, and proactive energy management will be instrumental in addressing the challenges of peak demand and ultimately result in a more sustainable and reliable future for energy.

5.3.3. Economic Advantages for Consumers and Utilities

Implementation of Demand Side Management (DSM) in smart grids provides numerous economic advantages to consumers and utilities. Underpinning these advantages is the increased energy efficiency promoted by DSM. With customers being offered information in real time and flexible tariff packages, they can adjust their rate of consumption to respond to increasing and decreasing demand and supply prices. This active involvement not only equates to lower energy bills for consumers, but it also relieves peak-hour stress off the grid, allowing the utility to function more efficiently. Furthermore, as utilities optimize their energy delivery through demand response programs, they reap reduced costs of doing business, ultimately being transferred to economic savings for consumers as well. Besides, the smart grid migration enables a transformation of the roles of energy market participants in order to establish a less adversarial environment between consumers and utilities. With retail customers having the choice of engaging with markets in more interactive manners, there is room for them to buy demand response assets that will deliver substantial fiscal savings. This evolution of market interaction enables the formation of innovative pricing structures that align customers' electricity consumption with concurrent price signals. This not only encourages energy conservation but also maximizes the efficiency of the overall market, which is critical when it comes to ensuring the reduction of peak load and associated costs. With customers participating actively in managing energy, the utility companies have a more stable grid operation, which further results in cost efficiency and sustainability. Finally, strategic DSM initiative alignment aligns to regulation mechanisms that are designed to stimulate technology adoption by utilities. Governments will have better chances of developing positive regulations that will compel utilities to invest in smart grid infrastructure, knowing that such investments will yield substantial economic gains to consumers as well as the overall energy market.

For instance, utilities with robust demand response programs can minimize risks associated with peak pricing and infrastructure strain and, in the process, lower costs that are traditionally transferred to customers. Not only does this interdependence direct us to the economic rationale for DSM but also to the necessity of regulatory encouragement as one of the principal driving forces of this transformation. As discussed in the findings, the two objectives for enhancing market efficiency and system resilience through smart grid technologies end up creating a win-win situation for consumers and utilities as well, ensuring development in energy management. (Bergaentzlé et al., 2013)

5.4. CHALLENGES IN IMPLEMENTING DEMAND SIDE MANAGEMENT (DSM)

Implementation of DSM within smart grids presents intricate challenges that are both technology and regulation based. One challenge is the available electricity infrastructure, which is often found to be deficient in its capacity to accommodate the needs of DSM programs. The traditional power grid is a unidirectional flow of electricity and does not have the flexibility needed for modern energy management methods. The integration of new energy sources and energy policy with higher energy efficiency requires massive upgrading of such infrastructure. Furthermore, as mentioned, Information and Communication Technologies (ICT) only came to the level of reliability and flexibility that smart grids require in recent years. Thus, it is essential to overcome legacy systems' limitations while establishing conditions favorable for rapid technological progress to successfully implement DSM (Borghetti et al., 2013). Another significant hurdle is the need for effective communication and coordination among various stakeholders within the smart grid framework. Energy usage optimization in smart networks relies on near-real-time communication between generation, transmission, distribution, and load components. Interoperability between these disparate components is not an easy process. Data standardization, security, and privacy issues must be handled holistically to facilitate unproblematic information exchange. As defined, the use of common communication protocols is crucial not only for efficiency in operations but also for attaining consumer trust in smart grid technologies. This involves cooperative efforts on the part of regulators and industry stakeholders in developing interoperable systems that drive innovations while providing protection to users (Chin et al., 2011). Furthermore, consumer participation in DSM programs is often prevented by behavioral and perceptual issues. Customers are also not entirely informed of the potential cost savings and reduction of environmental impacts that DSM can provide. Additionally, the complexity of smart grid technologies might intimidate non-technical customers, discouraging them from engaging in energy management activities. Consumer education programs and outreach programs aimed at specific customers are urgently needed to de-mystify these technologies and educate customers so that they could actively participate. By building an educated consumer base that can engage with DSM initiatives, utilities can achieve the highest program effectiveness and benefit from increased energy savings. Hence, addressing these behavioral challenges is a part of the overall strategy in order to achieve the maximum potential of DSM in developing smart grids (Figure 5.11).

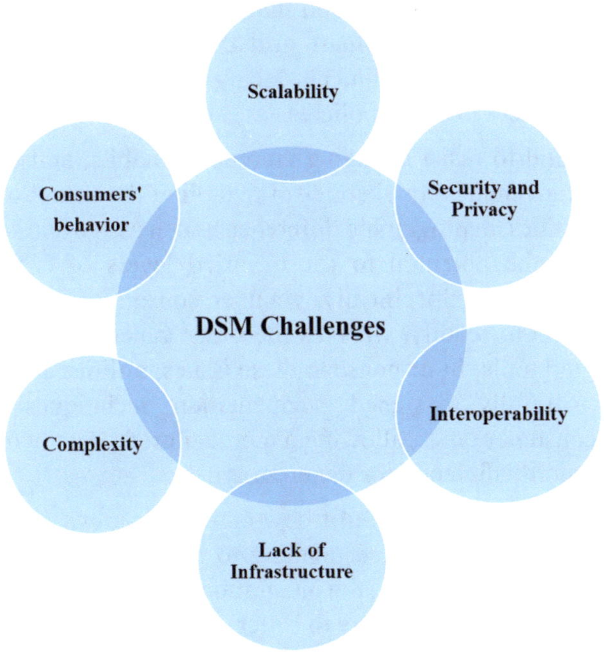

Figure 5.11. Demand side management (DSM) challenge.

Source: https://www.researchgate.net/profile/Ahsan-Raza-Khan/publication/283119722/figure/fig2/AS:667815275155461@1536230977225/LF-and-DSM-challenges-Integration-of-advanced-communication-infrastructure-is-considered.ppm.

5.4.1. Consumer Engagement and Behavioral Barriers

Consumer engagement is a critical element of Demand Side Management (DSM) in smart grids, as it contributes significantly to affecting the efficacy and effectiveness of energy consumption patterns. Engaged consumers are likely to alter their patterns of use in line with real-time data provided by smart meters and demand response efforts, which ultimately contributes to reducing energy outgoings and peak load management. However, behavioral barriers will discourage this participation, e.g., insufficient awareness of benefits from participation and insufficient knowledge about how to implement energy-saving technologies. Absence of knowledge may get in the way of consumers' inclination to play an active role in DSM programs and cause disappointing participation levels. Overcoming such impediments through targeted educational campaigns and intuitive technology interfaces is important to promoting a culture of participation that appreciates energy efficiency and embraces the overall goals of smart grid development. A familiarity with the distinctive behavioral impediments facing consumers is important to the successful enforcement of DSM policies.

Customers tend to resist changing entrenched habits, and this is because of perceived inconvenience or skepticism regarding the actual impact of their participation. Besides, demographic differences, such as age and socioeconomic status, also have the potential to cause varied levels of engagement with smart grid technologies. Specifically, younger consumers will tend to adopt new technology more readily, whereas the older generations will struggle to incorporate digital tools. In response to these issues, utilities and policymakers must utilize carefully designed segmentation techniques that engage heterogenous consumer types, allowing a more tailored strategy of engagement and a greater overall efficiency for the program.

This is supported by evidence of how communications specific in nature can lead to better results when responding to the specific needs of different consumer segments. Additionally, the integration of big data analytics in energy management presents an opportunity to better customize consumer engagement programs. By analyzing consumption habits and trends, utilities will learn about behavioral patterns that drive energy consumption.

For instance, combining water fixture stock data with smart meter readings has been shown to allow for personalized demand management policies that accurately segment consumers into groups based on their own behavior (Castelletti et al., 2019). Not only is a method like this able to segment groups of consumers with similar actions, but it provides valuable feedback that can be utilized to inform education campaigns. As the energy landscape shifts toward distributed energy systems with increasing reliance on distributed energy resources, constructing participation on the basis of strong data-informed insights will be the key to overcoming existing challenges and enabling the intended transformation of energy consumption patterns on smart grids (Hippolyte et al., 2017).

5.4.2. Technological Limitations and Infrastructure Costs

The integration of Demand Side Management (DSM) into smart grids is confronted with a range of technological limitation and infrastructure costs that must be addressed to realize their full potential. Though smart grids are designed to enable efficient energy management, the enabling technology still lags behind in terms of interoperability and scalability. For instance, the lack of standardized communication interfaces can cause real-time data exchange between producers and consumers to be complicated, leading to inefficiency in energy distribution. Additionally, the initial investment in advanced metering infrastructure, including smart meters and communications infrastructure, is a significant financial impediment for both utility companies and consumers, hence influencing the overall feasibility of DSM projects in the context of smart grid implementation (Axon et al., 2014). Infrastructure costs are among the most difficult challenges surrounding the installation of smart grids and the best DSM strategies. Replacing old energy distribution systems with new technologies can be extremely expensive. This does not just involve the replacement of physical plant but also the addition of new software and IT infrastructure required for two-way communications and managing data. These capital investments could strain utility coffers and end up increasing the cost of energy for consumers. Secondly, public acceptance of these costs can also be a limiting factor; people will resist efforts perceived to be costly, making the push for DSM in smart grids even harder (Axon et al., 2014). Therefore, being cautious in balancing investment vs. stakeholder requirements is key to success. Balancing technological limitations and cost of infrastructure in DSM is representative of a broader set of socio-technical challenges stakeholders will have to navigate. Regulators and policymakers play a central role in this setting, influencing investment alternatives and building systems that have the power to enhance or cripple the achievement of smart grid technology. Scenario-based methodologies offer a strategic method of exploring how various factors—regulatory decisions and market design—intersect and influence technological decisions available to energy producers and consumers (Axon et al., 2014). By cooperation between industry players and integration of public views, such scenarios can make possible the determination of potential directions for DSM strategies in which expenses and advantages can be balanced in the context of a constantly changing power environment.

5.4.3. Regulatory and Policy Hurdles

The integration of Demand Side Management (DSM) into smart grids entails numerous regulatory and policy issues that have a deep impact on its effectiveness. One of the main issues is the current frameworks that tend to be unable to offer full signals for long-term decarbonization. The EUs Climate and Energy Package is a perfect example of this incoherence, given the absence of coherent policies that complement energy efficiency and renewable energy sources (RES) with emissions trading schemes (ETS). The lack of complementarity makes low-carbon infrastructures' investments more difficult, extending the time frame of indispensable developments in smart grid technologies.

Thus, without robust DSM policy cues, the prospects for innovative demand response solutions could be jeopardized, leading to higher costs and a less dynamic move towards sustainable energy systems. Governments must remove these inconsistencies so that effective DSM strategies can be synchronized with broader decarbonization (Colombier et al., 2025).

Equally troubling are issues regarding the interoperability of those technologies that are essential to effective DSM implementation. The integration of information and communications technology (ICT) with current building systems is crucial to enabling the required responsiveness in electricity demand. Buildings, which use a considerable part of the world's electricity, have not been well incorporated into the smart grid system.

Legacy heating, ventilation, and air conditioning (HVAC) systems in older buildings tend to conflict with newer control technologies, suppressing possible contributions to grid stability. Successful DSM must be informed by a paradigm shift toward smart buildings that are capable of being active energy market participants. Yet, public policy has not been able to catch up with this abrupt change, and there are no incentives available for building retrofits capable of significantly increasing demand-side responsiveness (Boudreau et al., 2016). Finally, there is the absence of clearly stated regulatory structures that recognize the prosumption role of buildings as both energy consumers and producers that must be addressed.

Current policies have a tendency to overlook the potential of buildings to be "prosumers," generating energy as well as managing their demand according to grid needs. For effective DSM to be achieved, regulation bodies must provide standards that facilitate incorporation of renewables at the building level, thereby enhancing overall grid stability.

These regulatory enhancements, combined with retrofit incentives for existing infrastructures, will empower facility managers to optimize energy utilization without breaching occupant comfort or operating performance. Therefore, there is a need for an efficient regulatory regime in transforming the existing obstacles into opportunities and ultimately attaining the optimal potential of DSM in smart grids (Boudreau et al., 2016).

5.5. CASE STUDIES OF SUCCESSFUL DEMAND SIDE MANAGEMENT (DSM) PROGRAMS

Successful applications of DSM programs can be observed through various case studies analyzing their efficiencies in maximizing the consumption of energy and enhancing grid reliability. For instance, several municipalities have introduced smart home energy management technologies (SHEMS) to improve user engagement in energy use. These programs guarantee active participation by households, making it possible for users to manage their habits of using energy through real-time feedback as well as automatic control.

The active participant process has been determined to provide significant energy saving while making it possible for users to adopt greener habits. Therefore, SHEMS are a key solution for enabling energy efficiency and reducing overall demand, hence showcasing an ideal combination of technology and user-solutions in DSM programs (McIlvennie et al., 2020). Another effective case study showcases the implementation of DSM programs in smart grid systems, where utilities implement high-level communication technologies to dynamically manage energy loads. With two-way communication, utilities can learn better about consumer demand patterns and therefore respond accordingly. Demand response programs, for example, have helped utilities drop peak loads without additional generation capacity. This kind of strategic management stabilizes the grid but at the same time returns economic advantages to participating consumers with load-shifting strategies. The blend of DSM programs and advanced grid attributes highlights the significance of collaborative strategies among providers and consumers of power, thus determining the feasibility of successful DSM applications in smart grid systems (Axon et al., 2014). Finally, case studies have also determined the regulatory and policy frameworks as the determinants of successful DSM programs. For instance, some regions have established incentives to encourage energy efficiency activities at both the commercial and residential levels. The incentives have a high take-up because consumers seek to benefit from rebates and savings in cost. Encouragement can also help the integration of renewable energy sources into DSM programs to improve the sustainability of the power sector. The interaction of effective policies and best-designed DSM programs is capable of creating a conducive environment to achieve overall energy targets, which emphasizes the significance of coordinated efforts in the form of integrated strategies that balance technological innovations with policy interventions to exploit the potential of demand-side management initiatives fully (McIlvennie et al., 2020) (Figure 5.12).

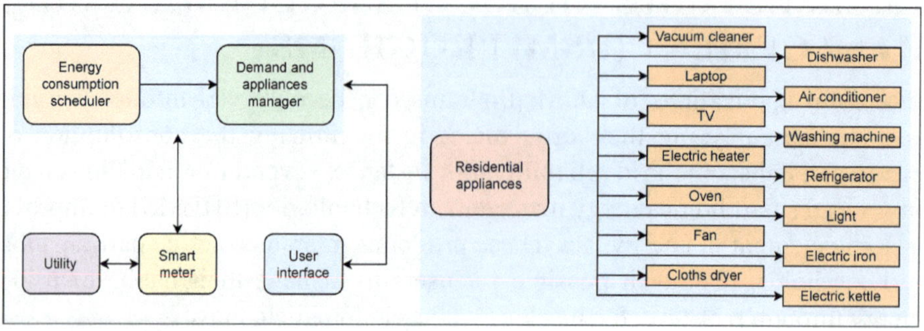

Figure 5.12. Block diagram for a home energy management system.

Source: https://www.mdpi.com/energies/energies-15-02863/article_deploy/html/images/energies-15-02863-g005.png.

5.5.1. Overview of Notable Demand Side Management (DSM) Initiatives Worldwide

Within the context of DSM, several exemplary initiatives across the globe testify to the revolutionary nature of such approaches in smart grid construction. For example, the United Kingdom has adopted DSM by establishing programs for increasing energy efficiency and enhancing consumers' ability to manage their energy use proactively. These efforts are especially important in view of the UK's mission to achieve ambitious targets for carbon reduction. Smart grids here are being constructed with smart metering and two-way communication between energy users and providers, which enhance the efficiency of operating energy systems. With financing and political support from the European Union as well as the UK government, the push for DSM is not only regarded as a priority for the environment but also as an economic stimulus to stimulate growth in terms of employment in the new clean technology sector (Axon et al., 2014). Similarly, cities all across the globe are taking major strides towards adopting DSM practice, particularly with increasing demand for electric vehicles (EVs). DSM-focused programs are particularly paramount in cities since infrastructure needs to be adjusted along with increasing energy consumption from EVs. Leadership in such cities is important because they implement systems that enable the installation of EV charging stations, and it is imperative in responding to current and future needs. Such efforts are typically based on a synthesis of city managers, developers, and utility entities to create robust responses that encourage sustainable energy-using habits. By utilizing urban innovation, such DSM initiatives promote energy efficiency and facilitate cultural shift towards embracing renewable and cleaner energy sources in cities (Fox-Penner et al., 2019). Another key characteristic of major DSM initiatives worldwide is congregating technical, policy, and regulatory systems and mechanisms to facilitate smart grid implementation. Policy makers and manufacturers recognize that the complexities of DSM necessitate a multi-faceted approach that combines supply-side variables with demand-side measures. This requires the institution of collaboration among disparate sectors, including public organizations, private companies, and consumer groups that represent consumers. By addressing dependent variables such as data protection, technological security, and market rivalry, stakeholders can develop a holistic system that facilitates consumers' engagement in energy management. Finally, successful DSM projects are those that integrate multiple stakeholders and forge articulate plans capable of navigating the intricate landscape of energy consumption and sustainability (Axon et al., 2014).

5.5.2. Analysis of Outcomes and Lessons Learned

The use of Demand Side Management (DSM) for Smart Grids has thrown into relief many fundamental outcomes contributing to the knowledge base on energy efficiency and consumer behavior. A review of these findings creates a mix of success and failure, and which are significant to future program planning. For instance, programs funded both by private and public means considerably demonstrate increased consumer involvement and peak-demand savings. However, variations in results from projects between the United States and Europe necessitate setting up a more comprehensive evaluation method to determine an extensive understanding of the implications and performance of these DSM methods. The findings highlight the importance of establishing comparable assessment structures for facilitation of cross-border coordination, as illustrated in discussions under the US-EU Energy Council (Steven et al., 2012). Among key lessons from DSM implementation in Smart Grids is the importance of data collection and exchange protocols. The heterogeneity of energy resources and consumer demands makes it extremely complex to manage, requiring robust data infrastructure for timely decision-making. In the absence of robust data analytics, utilities are less likely to implement DSM plans successfully, thus slowing down their momentum in keeping up with regulatory and consumer calls for change. EU and US expert sessions demonstrate compliance in strengthening methodological methodologies that not only develop the project evaluation but also uplift stakeholder transparency. Coordination in setting collective terminologies and methodologies will enrich critically the knowledge base and align appropriately DSM strategy (Steven et al., 2012). Moreover, creating consumer awareness and confidence is essential in order to make DSM programs viable in the long run for smart grid environments.

The findings of a number of DSM projects indicate that consumers are more likely to participate actively in energy conservation schemes when they are thoroughly informed and engaged. Conclusions emerging from such projects highlight the utility companies' need to develop appropriate communication strategies that will de-mystify the complexities of energy consumption and smart technologies. Creating open channels of information sharing can help bridge the gap between utilities and consumers, thus guaranteeing maximum levels of participation and effective management of energy. The conclusions reached from analyses of successful and less successful deployments highlight the necessity of continuous debate and strategy evolution in the changing Smart Grids context.

5.5.3. Comparison of Different Approaches to Demand Side Management (DSM)

In the context of comparison of multiple approaches to DSM in smart grids, it is essential to recognize the major role played by evolving technologies in optimizing grid efficiency and reliability. Multiple DSM methods not only aim for reducing energy usage but also focus on coordination between renewable sources of energy, storage facilities, and consumer behavior. For instance, the introduction of advanced Energy Management Systems (EMS) provides an efficient platform which optimizes grid operation and addresses varying patterns of demand and forecasted renewable energy output. The systems are able to dynamically update controllable resources based on real-time data as posited in Díaz González et al. (2017), consequently effectively lowering participants' energy costs while improving power quality in general. This multi-dimensional approach points to the potential for technology to make the energy system more sustainable and responsive in smart grids. In addition, a communication strategy analysis is vital in making out the effectiveness of DSM strategies in smart grids. Interconnectivity between different components of the grid relies on near-real-time communication protocols that can increase efficiency at each step from generation to consumption. Effective implementation of DSM necessitates overcoming these issues of communication, as noted by Chin et al. (2011). Via interoperability and security problems, the stakeholders are capable of delivering stability and effectiveness in the functioning of smart grids. Additionally, measures to harmonize communication standards across regions through regulation can promote an integrated system, thus promoting the efficiency of DSM programs. By enhancing communication dynamics in such systems, various DSM approaches can be more easily combined to result in smarter coordination and performance of smart grids. Finally, it is necessary to examine the economic benefits of various DSM approaches. As stated by

Díaz González et al. (2017), utilizing a robust EMS not only enhances technical performance but also enjoys significant economic advantages in terms of reducing energy imports along with associated costs. This distinction between cost reduction measures and traditional energy management practices is a quintessential example of how smart grid technologies can enable long-term sustainability and economic viability. The marriage of technical efficiency and economic sensibility is a paradigm shift in energy management planning. By offering an environment for DSOs that encourages the adoption of RE sources and DSM program execution, stakeholders are able to disentangle effectively the complexities of modern power networks, eventually leading to smarter and more resilient smart grids.

5.6. FUTURE TRENDS IN DEMAND SIDE MANAGEMENT (DSM)

With the changing energy world, DSM trends in smart grids in the future will radically revolutionize how energy is consumed and integrated into daily life. One of the top significant trends is increasing reliance on smart technologies, such as smart meters and home energy management systems, that allow two-way communications between consumers and utilities. This technology advancement not only allows for efficient utilization of energy but also enables real-time data-sharing to enable consumers to make logical decisions on energy utilization. As stated, the promise of smart grids is that they are able to manage supply intermittences while simultaneously making the entire system more stable and efficient (Axon et al., 2014). Its deployment is likely to democratize energy management to enable more homes and businesses to actively participate in DSM programs, thereby decongesting the traditional energy grid from pressure. Another trend installed is the use of demand response programs that compensate customers for lowering their peak load by making behavioral changes. These programs encourage customers to change their electricity consumption at times when demand is high to promote grid stability and cost savings. Involvement of the consumer through educative campaigns and incentive-based schemes is critical to the success of these DSM initiatives. The role of smart home energy management systems (SHEMS) has been highlighted, even though prospective research estimates that an integrated approach needs to be adopted. Current research reveals a general technical solution orientation without sufficient consideration of user-centric strategies that call for consumer involvement (McIlvennie et al., 2020). By embracing a balance of dialogue between technocentric innovations and consumer-driven initiatives, the effectiveness of demand management programs can be significantly enhanced. Finally, regulatory policies will continue to hold the key to DSM in the future as these adjust to the shifting needs of the energy sector under advancing smart grid technologies. Policy makers must be cognizant of the complex interplay among supply and demand, and the imperative of aligning interests among stakeholders like consumers, industry players, and policy makers. Balanced policies that encourage investment in smart grid infrastructure without sacrificing privacy, data security, and reliability will be key. It is essential that future policy allows innovation to occur while, simultaneously, it guarantees that the potential risks in switching to smart grids, e.g., the possibility of cybersecurity threats, are adequately addressed. As noted, successful progress in DSM relies on collective action by all the DSM stakeholders, thus stressing the importance of a holistic approach to technology and policy (Axon et al., 2014).

5.6.1. Role of Artificial Intelligence (AI) and Machine Learning (ML)

The integration of AI and ML into demand side management (DSM) for smart grids is a paradigm shift towards greater energy efficiency and sustainability. With the implementation of these advanced technologies, smart grids are able to process vast data in real time and facilitate intelligent decision-making in the distribution as well as consumption of energy. AI algorithms may forecast energy loads based on past trends and ongoing consumption patterns, managing the balance between supply and demand. This ensures efficient utilization of energy resources, reducing waste and improving the stability of the grid. Furthermore, with growing complexity as renewable sources are introduced, AI-driven analysis becomes imperative in managing variability and stability in energy supply, addressing one of the key challenges delineated in current literature in a direct manner (Mahmud et al., 2025). In particular, the role played by AI and ML towards DSM can be ascertained by observing their impact on consumer behavior and participation. Smart grids that have smart algorithms deployed can provide individualized energy consumption data and prompt consumers to behave differently and participate in the energy-saving cause willingly. For instance, with predictive analytics, AI systems can inform users of when they used the most energy and suggest changes that equal savings and reduced energy usage. This approach not only enables citizens to make educated decisions regarding energy consumption but also promotes sustainability culture at the grassroots level. Moreover, the elaborate modeling of household behavior, as recently emphasized in critical reviews, emphasizes the ability of machine learning in individualizing demand response schemes to be more appealing to user interests, further driving participation rates in DSM programs (Rehman et al., 2025). Despite the promising developments in utilizing AI and ML for DSM, several challenges, including data privacy concerns and the need for skilled personnel, must be addressed to maximize their potential. The reliance on vast datasets raises significant issues around the security and confidentiality of consumer information, necessitating robust frameworks to protect personal data while still enabling effective analysis. In addition, the high cost of deploying AI-based solutions creates obstacles, most significantly to consumers and small-scale utilities. There is a need for a collaborative approach among policymakers, technologists, and energy suppliers in order to develop a culture conducive to innovation and safe deployment of such technologies. By overcoming such challenges, stakeholders can effectively leverage AI and ML to enhance DSM strategies in smart grids, towards a more sustainable and more resilient energy future (Mahmud et al., 2025).

5.6.2. Potential for Increased Consumer Participation

Greater consumer involvement opportunities in Demand Side Management (DSM) in smart grids are a revolutionary revolution in the energy sector. Traditionally, consumers have been quite passive, reacting to energy prices instead of taking energy management decisions on a proactive level. With new technologies and smart grid infrastructures, it becomes feasible for consumers to be more proactive stakeholders. Through the delivery of near-real-time feedback on energy usage and prices to users, DSM programs are able to cause home households to modify consumption patterns in a manner that is beneficial to their economic interests as much as the stability of supply of the energy grid. Platforms such as mobile apps and smart thermostats facilitate this interaction, allowing customers to customize their energy consumption and, in so doing, become actively involved in grid operation, thus achieving energy efficiency and cost savings (Ashraf et al., 2025). The central point towards improving the participation of consumers is incorporating automated systems that honor user preference while promoting energy saving. Advanced Demand-Side Management platforms, particularly with Automated Load Scheduling (ALS), are equally potent tools in this connection. Besides optimizing energy consumption allocation, these platforms also facilitate individualized schedules according to individual home needs. Through dynamically lowering peak loads through equipment scheduling, customers can maintain their energy expenditure low while enhancing their comfort (Ashraf et al., 2025). The dynamic classification of devices and scheduling their use on the basis of feedback from actual consumption results in higher chances of consumer engagement, facilitating the balancing of demand and supply of energy, particularly in peak periods. Second, the integration of renewable energy sources into smart grids has an important impact on the involvement of consumers in DSM activities. As homes become increasingly involved in managing energy, they can respond in real-time to price uncertainty caused by the intermittency of sources like solar and wind. The employment of energy management systems which aggregate distributed energy resources allows the consumer to navigate the market signals aptly, thus becoming a steward of the environment and reduced energy costs. Through these mechanisms, consumers are able to align their energy usage with cleaner production, ultimately building a more robust grid. So, the prospect of increased consumer involvement is not so much about individual profit but creating a collective feeling of responsibility towards energy efficiency and conservation in the larger scheme of smart grid operation.

5.6.3. Predictions for Demand Side Management (DSM) Evolution in Smart Grids

As smart grids evolve, incorporation of sophisticated technology ensures significant DSM enhancement. Future of DSM is that it increasingly will depend on real-time analysis of data to facilitate improved energy usage. Advanced algorithms and machine learning are employed by utilities to facilitate improved consumer-supplier interaction. This shift towards a more responsive power grid is not simply a technological upgrade; it is a paradigm shift in the habits of energy consumption. The stakeholders, whether consumers or otherwise, will be able to witness more of their habits and thus be motivated towards being more proactive in energy management. This evolution is in line with the worldwide aims for a low-carbon economy as outlined within the various policy proposals and business strategies, with smart grids being identified as central to this transition (Axon et al., 2014). Among the attributes of the evolving DSM landscape is the inclusion of demand response programs that use behavioral incentives together with new price structures. As more sophisticated DSM measures are adopted by utilities, customers will be provided with options that reward them for reducing or shifting their usage during peak hours. This not only reduces pressure on the grid but also raises overall efficiency and dependability. Newer technologies such as smart devices and energy management solutions will allow consumers to have greater control over their energy decisions through automated energy choices. By integrating systems like model predictive control (MPC), as depicted in recent work, appliance management, including that of refrigerators, can be optimized to steer consumption based on real-time pricing and forecasted loads. Such innovations mark a significant step towards decentralized energy control and highlight the potential for new solutions in the replacement of traditional consumption patterns (Bindner et al., 2015). The future way for DSM in smart grids will also involve increased collaboration between various stakeholders like utilities, technology vendors, regulators, and consumers. For successful implementation of such advanced DSM platforms, a combined strategy with respect to regulatory, technological, and social aspects is necessary. In this collaborative model, policies are formulated for ensuring privacy and security of the consumer data but for actively promoting user involvement in DR programs. In addition, public acceptance of smart technologies and perception will decide the level and speed of DSM implementation. A combined analysis of the dynamic interaction between supply and demand patterns supported by robust scenario analyses can allow for effective use of smart grids. These anticipatory measures will be crucial to achieving the overall goals of energy efficiency and sustainability that form the basis of the ongoing development of DSM as part of the smart grid framework (Axon et al., 2014).

SUMMARY

In general, the implementation of DSM in Smart Grids is a critical innovation in the energy sector that can enhance energy efficiency, sustainability, and overall grid reliability. Whereas utility firms grapple with fluctuating energy usage and increasing demands for renewable sources of energy, DSM offers strategic paradigms that make consumers participants in energy management. The implementation of technology such as SHEMS is imperative because not only does it facilitate the saving of energy, but it also invites active use by consumers. However, research indicates a disparity in existing research where technical parameters receive greater emphasis while user-based paradigms are taken for granted. This points to a gap that needs to be addressed in order to be in a position to effectively utilize the potential of DSM programs to promote consumer engagement and achieve substantial energy gains (McIlvennie et al., 2020). Moreover, leveraging the potentials of Smart Grids facilitates two-way communication between energy consumers and producers, fundamentally transforming traditional power delivery frameworks. With the deployment of advanced technologies such as the IoT, Smart Grids establish a dynamic environment under which real-time monitoring, data analytics, and advanced control schemes can thrive. Not only does this assist in the optimization of energy utilization, but the predictability of fluctuating demands is also enhanced, ultimately translating to reduced operating costs as well as minimizing wastage of energy. The seamless integration of DSM measures in Smart Grids becomes a call for resolving the technical intricacies and operation openness necessary to facilitate this transition. Upcoming research must strive to develop holistic solutions that combine technological and behavioral elements to create effective demand-side interventions (Copeland et al., 2019). The effective implementation of DSM in Smart Grids largely relies on effective coordination between technology developers, utility providers, and consumers. With a collaborative setting, stakeholders can share information and resources in a bid to enhance energy efficiency practices and innovation. Additionally, if fully embracing the potential of DSM, an interdisciplinary approach that values the inter-linkage between technology, user behavior, and policy measures needs to be developed. As the energy environment continues to evolve, its focus on integrated DSM strategies will be critical to addressing climate change challenges and coping with an ever-increasing global energy demand. Emphasis on consumer engagement and awareness of the diverse motivations underlying energy consumption will be pivotal in remaking patterns of energy usage and ushering in a more sustainable future (Copeland et al., 2019; McIlvennie et al., 2020).

Successful DSM in smart grids depends on enhanced communication systems that enable the real-time communication of all grid elements. Most debates reveal that optimizing energy consumption depends on embedding advanced communication technology for synchronizing generation, transmission, and distribution. This correlative effect provides a platform where energy loads can be easily controlled, contributing significantly to energy production modes as well as consumer habits. Literature indicates that resolution of the communication problems that are caused by interoperability and security problems is important to future-proof smart grids. The necessity for a concerted effort towards standardization in communications helps to highlight the industry's interest in the development of a robust DSM framework with the ability to accommodate shifting energy needs and customer interests. These technological developments will not only improve operational efficiency but also be part of the overall resilience of smart grid infrastructure (Chin et al., 2011). Furthermore, scholarly debate is centered on the idea of smart grids as agents for low-carbon energy transformation, with DSM processes central to energy supply and demand planning. Evidence indicates that while basic definitions and operation models of smart grids are still being developed, the integration of various components of technology—involving IT solutions and smart meters—is promising. These components enable two-way communication between producers and consumers of energy and more dynamic energy management practices. Further, DSM discussions enhance visibility of smart grid contribution towards carbon reduction objectives, employment generation, and minimizing the cost of energy generation and distribution. This indicates the expansiveness of DSM practice impact and stakeholders, from policymakers to consumers, on shaping the evolution of smart grid technology (Axon et al., 2014). Finally, the sophisticated nature embracing DSM in smart grids calls for consideration of socio-technical elements impacting their deployment. The coupling of supply and demand sides, regulatory demands, and new emerging technology dynamics is a holistic context that must be examined in its entirety. Scenarios have been proposed as viable tools for incorporating variables into harmonious stories, improving our understanding of potential smart grid deployments. Literature reports that economic factors have been the conventional topic of historical examinations, and the integration of other socio-technical factors is essential to enable holistic scenario building. With the integrated approach, not only will the effectiveness of DSM strategies be enhanced but also responsiveness of such strategies to evolving consumer, regulator, and overall energy market needs. The intersection of these disputes is the sophisticated challenges and opportunities involved in the establishment of smart grids and their capacity to offer sustainable energy management practices (Axon et al., 2014).

REFERENCES

1. Andoni, M., Robu, V., Flynn, D., Abram, S., Geach, D., Jenkins, D., McCallum, P., et al. (2018). Blockchain technology in the energy sector: A systematic review of challenges and opportunities. *Renewable and Sustainable Energy Reviews, 100*, 143–174. https://doi.org/10.1016/j.rser.2018.10.014 (accessed on 30th September 2025).
2. Andreadou, N., Covrig, C. F., Fulli, G., Olariaga-Guardiola, M., Poncela Blanco, M., & Prettico, G. (2015). *Smart Grids Laboratories Inventory 2015.* https://core.ac.uk/download/38629517.pdf (accessed on 30th September 2025).
3. Ashraf, M., Arshad, M., Zaidi, S. H., & Shaukat, K. (2025). Balancing comfort and conservation: Dynamic programming algorithm for appliance scheduling in residential demand-side management. *Mehran University Research Journal of Engineering and Technology.* https://www.semanticscholar.org/paper/bec10dc920667d39da6996800fb50d2b0da4f621 (accessed on 30th September 2025).
4. Axon, C., Axon, C., Baker, P., Baker, et al. (2014). *Scenarios for the Development of Smart Grids in the UK: Synthesis Report.* https://core.ac.uk/download/161104070.pdf (accessed on 30th September 2025).
5. Axon, C., Balta-Ozkan, N., Cipcigan, L., et al. (2014). *Scenarios for the Development of Smart Grids in the UK: Literature Review.* https://core.ac.uk/download/155776124.pdf (accessed on 30th September 2025).
6. Bakker, V., Bosman, M. G. C., Hurink, J. L., et al. (2010). Demand side load management using a three-step optimization methodology. https://core.ac.uk/download/pdf/11478024.pdf (accessed on 30th September 2025).
7. Barbato, A., Capone, A., Chen, L., et al. (2014). A distributed demand-side management framework for the smart grid. https://core.ac.uk/download/55248842.pdf (accessed on 30th September 2025).
8. Bergaentzlé, C. (2013). *From Smart Technology to Smart Consumers: For Better System Reliability and Improved Market Efficiency.*
9. Bindner, H., Costanzo, G. T., Douglass, P. J., Lakshmanan, et al. (2015). Grey-box modelling of a household refrigeration unit using time series data in application to demand side management. https://core.ac.uk/download/43252872.pdf (accessed on 30th September 2025).
10. Bonetto, R., & Rossi, M. (2017). *Smart Grid for the Smart City.* https://core.ac.uk/download/84922332.pdf (accessed on 30th September 2025).
11. Borghetti, G., Menga, G., Procaccianti, G., Kok, L., Ardito, M., Morisio,

M., et al. (2013). Smart grid technologies in Europe: An overview. https://core.ac.uk/download/pdf/11431923.pdf (accessed on 30th September 2025).

12. Bossart, S., & Giordano, V. (2012). *Assessing Smart Grid Benefits and Impacts: EU and U.S. Initiatives.* https://core.ac.uk/download/38627650.pdf (accessed on 30th September 2025).
13. Boudreau, M. C., Helsen, L., Henze, G. P., et al. (2016). Ten questions concerning integrating smart buildings into the smart grid. https://core.ac.uk/download/129148665.pdf (accessed on 30th September 2025).
14. Castelletti, A., Castelletti, A., Cominola, A., et al. (2019). Data mining to uncover heterogeneous water use behaviors from smart meter data. https://core.ac.uk/download/287885431.pdf (accessed on 30th September 2025).
15. Chin, W. H., Efthymiou, C., Fan, Z., et al. (2011). Smart grid communications: Overview of research challenges, solutions, and standardization activities. http://arxiv.org/abs/1112.3516 (accessed on 30th September 2025).
16. Colombier, M., Guerin, E., Marcey, C., et al. (2025). Decarbonizing the EU power sector: Policy approaches in the light of current trends and long-term trajectories. https://core.ac.uk/download/pdf/6633745.pdf (accessed on 30th September 2025).
17. Copeland, R., Crespi, N., Rehmani, M. H., et al. (2019). Internet of Things-aided smart grid: Technologies, architectures, applications, prototypes, and future research directions. http://arxiv.org/abs/1704.08977 (accessed on 30th September 2025).
18. de Barata, F. A. S. F. (2015). *Control in Distribution Networks with Demand Side Management.* https://core.ac.uk/download/157632111.pdf (accessed on 30th September 2025).
19. De Jong, J., Glachant, J. M., Hafner, M., & Grant, R. (2025). *Toward a Smart EU Energy Policy: Rationale and 22 Recommendations.* https://core.ac.uk/download/pdf/6380049.pdf (accessed on 30th September 2025).
20. De Marez, L., Hauttekeete, L., & Stragier, J. (2010). Introducing smart grids in residential contexts: Consumers' perception of smart household appliances. https://core.ac.uk/download/55823187.pdf (accessed on 30th September 2025).
21. Denysiuk, S., Opryshko, V., Strzelecki, R., et al. (2016). Implementation of the Smart Grid concept through the use of demand response programs and modern power electronics systems. https://core.ac.uk/download/81630897.pdf (accessed on 30th September 2025).

22. Di Giorgio, A., Mercurio, A., & Purificato, F. (2013). Optimal fully electric vehicle load balancing with an ADMM algorithm in smart grids. http://arxiv.org/abs/1305.1044 (accessed on 30th September 2025).
23. Díaz González, F., Girbau Llistuella, F., & Sumper, A. (2017). Optimization of the operation of smart rural grids through a novel energy management system. https://core.ac.uk/download/157811591.pdf (accessed on 30th September 2025).
24. Ekins, P., & Watson, J. (2014). UK energy in a global context: Synthesis report. https://core.ac.uk/download/20391336.pdf (accessed on 30th September 2025).
25. Fayaz, M., Lajis, A., Nasir, H., et al. (2019). A review on energy consumption optimization techniques in IoT-based smart building environments. https://core.ac.uk/download/300473894.pdf (accessed on 30th September 2025).
26. Fox-Penner, P., Hao, J. (R.), Hatch, J., et al. (2019). The critical role of public charging infrastructure. https://open.bu.edu/bitstream/2144/39112/1/MeltingIceBook_Web_FINAL.pdf (accessed on 30th September 2025).
27. Gharehpetian, G. B., Marzband, M., Moradi, J., et al. (2019). Attributes of big data analytics for data-driven decision making in cyber-physical power systems. https://core.ac.uk/download/305121394.pdf (accessed on 30th September 2025).
28. Gu, C., Hu, Z., Le Blond, S., et al. (2013). Cost/benefit assessment of a smart distribution system with intelligent electric vehicle charging. https://core.ac.uk/download/610712792.pdf (accessed on 30th September 2025).
29. Hayes, B., Mancini, T., Melatti, I., et al. (2016). Residential demand management using individualized demand aware price policies. https://core.ac.uk/download/80309937.pdf (accessed on 30th September 2025).
30. He, X., Kuo, C. C. J., Leung, K. C., et al. (2015). A survey on smart grid communication system. https://core.ac.uk/download/80964229.pdf (accessed on 30th September 2025).
31. Hippolyte, J. L., Howell, S., Jayan, B., et al. (2017). Towards the next generation of smart grids: Semantic and holonic multi-agent management of distributed energy resources. https://core.ac.uk/download/82895612.pdf (accessed on 30th September 2025).
32. Kumar, B. S., Devi, A., Sharma, S., Bhavana, M., Babu, B. S., & Singh, N. (2024). Integrating renewable energy sources into smart grids with an aggregator-based energy management system for efficiency and resilience. *2024 7th International Conference on Contemporary Computing and Informatics (IC3I)*, *7*, 621–626. https://www.semanticscholar.org/pap

er/119fad928689021cb7e571d8bbb2e96ac134c6b5 (accessed on 30th September 2025).

33. Mahmud, R., & Waheduzzaman. (2025). The role of artificial intelligence in advancing smart grid technologies. *EJSMT.* https://www.semanticscholar.org/paper/69d586a71183473fae7a5455c61634b988465f7f (accessed on 30th September 2025).
34. Mao, Y., You, C., Zhang, J., Huang, K., & Letaief, K. B. (2017). A survey on mobile edge computing: The communication perspective. *IEEE Communications Surveys & Tutorials, 19*, 2322–2358. https://doi.org/10.1109/comst.2017.2745201 (accessed on 30th September 2025).
35. McIlvennie, C., Pritoni, M., & Sanguinetti, A. (2020). Of impacts, agents, and functions: An interdisciplinary meta-review of smart home energy management systems research. https://core.ac.uk/download/323062198.pdf (accessed on 30th September 2025).
36. Rehman, U., Faria, P., Gomes, L., & Vale, Z. A. (2025). Future of energy management models in smart homes: A systematic literature review of research trends, gaps, and future directions. *Process Integration and Optimization for Sustainability*. https://www.semanticscholar.org/paper/af6bb2ad7b9bb33b1696ee09ba8f00e18a022106 (accessed on 30th September 2025).
37. Stenstadvolden, O., Yang, Z., Zhao, L., Bestgen, S., & Lee, W. J. (2025). Study of residential power demand behavior from a consumer psychological perspective. *2025 IEEE Industry Applications Society Annual Meeting (IAS).* https://www.semanticscholar.org/paper/2f42f7415ef3fff38b42b9dc9399c35a581764cc (accessed on 30th September 2025).

CHAPTER 6

RENEWABLE ENERGY INTEGRATION

Contents

INTRODUCTION

With the backdrop of climate change and its associated challenges, the transition away from traditional sources of energy to renewable energy systems has never been more imperative than before. Integration of renewable energy is a matter of technologies and policies that make it possible to incorporate renewable resources, such as solar, wind, and hydroelectric power, into existing energy systems. This overlap isn't merely a technologic shift; it is a significant paradigm shift when it comes to energy management, requiring innovative solutions for grid stability and reliability. This shift towards renewables necessitates solid understanding of the complexities involved, from variability of resources to grid planning, underlining the crucial significance of policies and cooperative frameworks between countries to make serious progress here. Regionally, for Southeast Asia, regional efforts demonstrate a resolve to increase the share of renewable energy in the energy mix. ASEAN member states have outlined aggressive intentions to raise their share of renewable energy, with the target of bringing the share to 13%–23% by 2025.

Figure 6.1. Renewable energy integration.

Source: https://media.licdn.com/dms/image/v2/D5612AQGNWDBgA8pxcw/article-cover_image-shrink_600_2000/article-cover_image-shrink_600_2000/0/1707575803254?e=2147483647&v=beta&t=Jyn7AS8NJzUqNdHmvCmXvQu7HsODYMnah-uOIMvlFd4.

That pledge is an indication of the necessity of regional coordination to overcome challenges in the use of renewable energy. Grid flexibility is thus the first consideration to enabling such a transition; without appropriate adaptation measures and infrastructure, efforts to integrate renewable resources may collapse. Moreover, the application of tools like a grid flexibility evaluation can establish strengths and limitations of grid systems across various ASEAN nations, and hence tailored strategies for enhancing the overall performance are crucial (Huang et al., 2019). In their accomplishment of these challenges, smart grids have a vital part to play in transforming the move towards a low-carbon energy future. Smart grids integrate advanced technologies to attain efficient management of energy, demand response, and bidirectional communication between the consumer and producer. This promotes the effective integration of intermittent renewable sources while ensuring grid stability and security. Critical to smart grid development are public acceptability considerations and regulatory structures able to adapt to the changing energy environment. The socio-technical dynamics of economic expansion, policy development, and consumer activism influence the deployment of smart grids and must be taken into account through strategic planning. Successfully harnessing the potential of smart grids would eventually usher in the successful integration of renewable energy systems, guiding towards a sustainable energy future (Axon et al., 2014) (Figure 6.1).

Definition of Renewable Energy

The definition of renewable energy captures various sources of energy that readily renew themselves, thereby offering sustainable substitutes for fossil fuels. As world attention turns to sustainable utilization of the environment, renewable energy has expanded beyond privileged forms like solar, wind, hydro, and geothermal to fresh forms that are highly scalable and affordable. Renewable energy derives its strength from its ability to generate power without depleting non-renewable resources or emitting greenhouse gases. This automatically makes renewables significant in reducing climate change and facilitating energy security. Besides, renewables are characterized by their non-dispatchability, with the generation depending on the weather, and this presents challenges to integration into existing supply systems, necessitating the innovation of energy storage technologies as part of a coordinated approach to deploy renewable energy (Villanueva et al., 2019). Another defining characteristic of renewable energy is its role in the development of a low-carbon future, particularly through tools like net zero energy buildings (NZEBs). These structures are models of the application of energy efficiency and use of renewable energy, supporting a building design transition towards minimizing energy consumption and maximizing local energy generation. To become NZEB entails striving to make systems optimal based on local climatic conditions, affirming the definition of renewable energy as not only the source but also the effectiveness of its application. Case studies show that well-designed NZEBs are likely to utilize renewable energy technologies in combination with passive building principles such as natural ventilation and daylighting to optimize overall energy performance. Surprisingly, some NZEBs even turn out to be net-positive energy houses, generating more energy than they consume, thereby redefining traditional measures of energy efficiency and injecting renewable energy into mainstream building practice (Feng et al., 2019). The integration of renewable energy into overall energy systems is both an opportunity and a threat that must be dealt with cautiously. Intermittence of renewable resources demands robust energy storage means to harness supply-demand mismatches. Transcritical carbon dioxide cycles combined with thermal energy storage are some of the emerging technologies poised to contribute to the reliability as well as the efficiency of renewable energy systems. Through the development of alternatives to conventional storage options like pumped hydro, these technologies represent the evolving picture for renewable energy storage. Moreover, advancements toward economic feasibility indicate a significant potential for integration across power levels, as illustrated by estimates for a similar levelized cost of electricity (Villanueva et al., 2019). The seamless integration of renewable energy not only contributes towards the achievement of sustainability targets but also represents an essential transformation into novel energy solutions critical to future energy systems (Figure 6.2).

Figure 6.2. Definition of renewable energy.

Source: https://images.takeshape.io/86ce9525-f5f2-4e97-81ba-54e8ce933da7/dev/9d498117-3179-4718-81f4-f1c7fe8e892b/Pillar%201%20Energy%20Transition%20wind%20energy%20solar%20panels%20dreamstime.png?auto=compress%2Cformat&w=1440.

Importance of Renewable Energy Integration

Integration of renewable energy sources into the existing energy system is significant for mitigating climate change and sustainable development. With conventional sources such as coal and natural gas increasingly criticized for their carbon content and environmental impact, the alternatives in the form of solar, wind, and hydroelectric power emerge as a green option. Transitioning towards renewables not only reduces greenhouse gases but also renders energy more secure through energy diversity in the mix. Moreover, the use of renewables spurs economic growth as the sector creates jobs in manufacturing, installation, and maintenance. The application of cutting-edge sustainable technologies (IST) is at the core of this transition, offering efficient solutions that can potentially optimize energy use and lower the operational cost. According to the literature, IST selection is accompanied by several challenges, including threats posed by rapid technological advancement, which necessitates long-term strategic planning in the energy sector (Ankrah et al., 2014). Promoting the use of renewable energy sources also promotes the building of strong infrastructures that will be able to cater to future energy needs. The process of integration requires upgradation in grid systems to handle the intermittent characteristics of renewable energy, and this will need massive investment in smart grid systems.

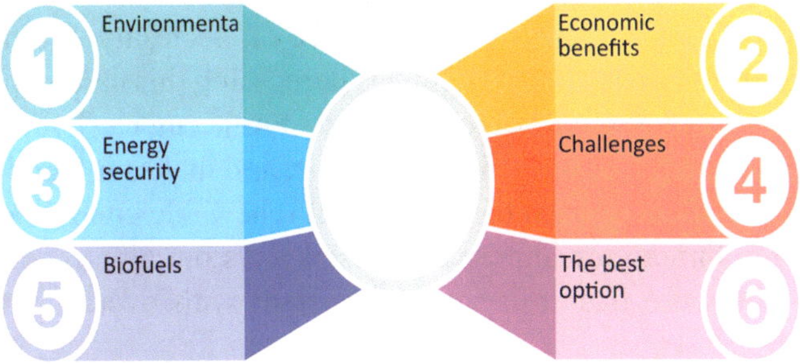

Figure 6.3. Importance of renewable energy.

Source: https://fastercapital.com/i/Renewable-Energy--Revolutionizing-Renewable-Energy--The-Role-of-Biofuels--The-Importance-of-Renewable-Energy.webp.

As demonstrated models such as the C2C BIZZ project show, effective planning and management of energy systems can enable the uptake of sustainable processes in business activities. Combining renewable energy and existing infrastructures can be seen as a central effort towards achieving energy resilience. By making use of sophisticated tools in measuring financial performance and potential investment returns, firms have the potential to investigate developments that implement sustainable energy concepts, ultimately benefiting the overall economy and environment (Ankrah et al., 2014). Second, the use of renewable energy is at the heart of innovation in the energy sector. While firms focus on IST development and application, they are faced with opportunities for improvement that can lead to increased efficiency and performance. The issue is to ensure stakeholders possess the knowledge and skills necessary to make sound decisions regarding such technologies. Although uncertainties and risks may face contractors, incorporation of systemized decision-making practices can enable appropriate IST selection for low-carbon transitions. This emphasis on new solutions puts the imperative relationship between economic payback and environmental stewardship into sharp relief, as firms seek to balance profitability and sustainability goals. Successful integration of renewables and supporting technologies will therefore not merely alleviate current energy problems but provide a direction for a more sustainable and low-carbon future (Fuertes et al., 2014) (Figure 6.3).

6.1. TYPES OF RENEWABLE ENERGY SOURCES

The landscape of renewable energy is rich and varied, with a variety of main sources each serving a unique purpose in the provision of sustainable energy. Solar power, for instance, uses sunlight in photovoltaic cells or solar hot water systems to provide a clean and limitless source of electricity. Wind power gets kinetic energy from wind flows as turbines, harnessing this natural phenomenon into mechanical and then electrical energy. Furthermore, hydropower also utilizes flowing water, like from rivers or dams, and makes it feed into turbines in the generation of energy. All these forms of renewable energy not only offer a solution to the limitations imposed by fossil fuels but also encourage efforts at curbing greenhouse gas emissions, consistent with overall environmental objectives.

Biomass energy is also a significant form of renewable energy, harvesting energy from organic material, like plant and animal waste. This process can effectively convert putrid organic refuse into renewable energy in the forms of burning or biological breakdown. Not only can this reduce waste, but it also facilitates local economies in terms of waste management and agricultural initiatives. Geothermal energy, which is derived from the internal heat of the Earth, is another critical source, particularly in regions with high tectonic activities. This energy source provides a stable and consistent energy supply that is less susceptible to fluctuation than solar and wind energy. This diversification of renewable energy sources is crucial towards providing an efficient and sustainable energy mix. The incorporation of these varied renewable energy sources into the mainstream energy systems is an opportunity as well as a challenge. As shown in Californias vision of achieving a 33% penetration by 2020, there are substantial constraints surrounding how these energy sources interact within distribution systems. Integrating technologies like photovoltaic systems and electric vehicles requires shifting strategies in voltage control and grid stability. Studies have highlighted gaps in current approaches, particularly in Volt/VAR Optimization and Control (VVOC) technologies, and these need innovations to facilitate greater integration of renewables (Arghandeh et al., 2013). Therefore, full awareness of renewable sources of energy not only enhances energy independence but also echoes the importance of modifying infrastructure to tap their full power, ultimately benefitting sustainable energy integration (Deckmyn et al., 2013) (Figure 6.4).

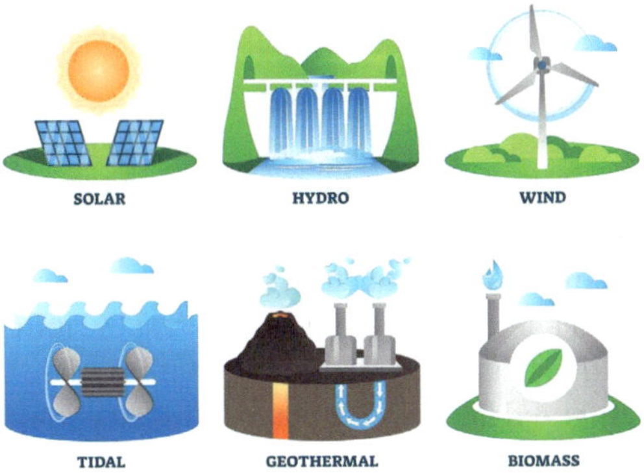

Figure 6.4. Types of renewable energy sources.

Source: https://www.inspirecleanenergy.com/_next/image?url=https%3A%2F%2Fimages.prismic.io%2Finspirecleanenergy%2Ftypes-of-renewable-energy-sources.jpeg%3Fauto%3Dcompress%252Cformat%26rect%3D0%252C0%252C1840%252C1036%26w%3D1840%26h%3D1036&w=1920&q=50.

6.1.1. Solar Energy

The assimilation of solar power within the realm of renewable energy systems is central to overcoming the challenges of climate change and the depletion of fossil fuels. Solar power, which is harnessed by the use of photovoltaic cells or solar thermal systems, is an enduring source of energy that is productive as well as recyclable. As urban communities grow, so does the importance of energy-efficient technology. By incorporating solar technology into skyscrapers and urban infrastructure, cities can reduce their carbon footprint without sacrificing energy demand. The application of smart energy management systems plays a crucial role in optimizing solar energy efficiency, with the capabilities of real-time monitoring and predictive analytics which modulate energy consumption to corresponding conditions, thereby realizing the optimal potential of this renewable energy source (Yan, 2025). Furthermore, the volatile nature of solar energy necessitates the integration of advanced integration techniques to ensure a stable supply of energy. Unlike conventional energy resources, the generation of solar energy is intermittently variable in nature and thus introduces variability in energy supply. For the proper handling of such variations, hybrid energy systems combining solar with other renewable energy sources are essential. These systems provide stabilized as well as predictable output by utilizing complementary energy sources and thus enable a more balanced supply of energy. Specifically, the incorporation of energy storage systems, such as solid-state batteries, allows for the storage of excess solar energy to draw upon in fluctuations, thus ensuring energy reliability and security (Kumar et al., 2025). Not only does this address the volatility of solar generation but also enhances resilience in the overall energy infrastructure. Aside from technological progress, policy frameworks and incentives contribute significantly to effective integration of solar energy into broader renewable energy policies. Governments and local authorities bear a crucial responsibility for creating pro-environmental conditions supportive of the adoption of solar technologies. It could range from grants and tax credits to less stringent permitting processes that stimulate domestic and commercial installations. Secondly, public education and information regarding the benefits of solar energy contribute to a cultural shift towards the embracement of renewable energy. The better-informed communities are regarding the economic and environmental advantages of solar integration, the greater the support for policies encouraging the shift towards sustainable energy. Therefore, the encouragement of collaboration between the government, industry, and public is crucial in unleashing the full potential of solar energy in ensuring renewable energy integration (Yan, 2025) (Figure 6.5).

Figure 6.5. Solar energy.

Source: https://greenhomesystems.com/wp-content/uploads/2023/09/blog-cover-photo-98.jpg.

6.1.2. Wind Energy

Wind energy importance as being one of the cornerstones of renewable energy integration cannot be overemphasized. Wind energy is presently one of the fastest growing sources of green energy, driven by technology and environmental awareness. That it can produce electricity without adverse emissions is a source of optimism to supplant fossil fuels. Also, its modularity allows for application in many contexts, from megascale wind farms to microscale community projects. Since it is difficult for renewable sources of energy to be intermittent and variable, integrating wind power in the energy matrix is essential. Besides ensuring stability of energy supply, it minimizes the risk of excessive reliance on conventional energy sources, thus opening the doorway to an energy future that is sustainable. Challenges in integrating wind energy with conventional systems are varied. These are due to the unstable nature of wind, which can lead to fluctuations in power generation. To mitigate these problems, novel techniques, such as hydrogen energy storage (HES) systems, can be employed. HES technology turns excess wind energy into hydrogen, stored, and then turned back into electricity when there is maximum demand. This approach enhances the efficiency of the overall power system, particularly in combined heat and power (CHP) applications.

Figure 6.6. Wind energy.

Source: https://renown-electric.com/wp-content/uploads/wind-turbine-500x699.jpg.

A day-ahead scheduling approach with CHP-HES systems has been shown to manage wind energy integration efficiently. With cost savings and reduced wind power curtailment, the technology demonstrates the potential for maximizing renewable energy penetration in electric networks (Marzband et al., 2019). Deployment of wind-diesel hybrid systems in actual applications has accelerated, particularly in rural areas where conventional sources of energy are limited. Increased application of wind-diesel systems in rural Alaska, for instance, reflects both the need for a stable supply of energy as well as availability of financing through the Renewable Energy Fund. The projects are significant because of their two-way contribution to energy security and local capacity-building. It becomes imperative to learn about the determinants of success for these installations with increasing communities adopting the utilization of wind energy technologies. Research aiming to support such projects highlights the need for sustainable management and long-term care strategies. The investment of wind projects in 27 communities at more than $107 million confirms the sustained growth of the venture to integrate renewable sources of energy (Fay et al., 2011) (Figure 6.6).

6.1.3. Hydropower

As the world is getting more concerned about the usage of renewable energy, hydropower is a vital component because it can provide a consistent and reliable supply of electricity. Holding the kinetic energy of flowing or falling water, hydropower power plants convert the potential into electrical energy at a high efficiency rate. This technology, apart from saving the environment in that it does not emit anything when utilized, also offers grid stability through provision of energy storage facilities. In nations like Brazil, where massive investment into energy infrastructure is projected, the deployment of hydropower is at the core of a sustainable future. Decisions today about energy development can either place countries into carbon-intensive tracks or propel them toward a low-carbon future, and therefore hydropower is a critical part in strategic planning (Lucon et al., 2015). Incorporation of hydropower into more advanced energy systems also reflects the need for robust policy frameworks and institutions. As the nation's work towards meeting their renewable energy targets, the role played by multilateral development banks (MDBs) assumes growing importance. The MDBs can help facilitate the reforms by providing the essential funds and establishing the right regulation that encourages public and private sector investments in renewable energy. By breaking down the specifics of the requirements for sustainable development of hydropower, the MDBs can create a more conducive environment for innovation and development in renewable energy. This comprehensive approach is required to gain maximum benefit from hydropower while addressing various challenges associated with energy transition and climate change (Ballesteros et al., 2010). In addition, incorporating hydropower into the renewable energy palette requires weighing its environmental and social impacts. Although hydropower is extensively promoted as being environmentally friendly, the impact on local society and environment must be weighed. Large dams can disrupt natural environments and displace people, sparking social justice and environmental sustainability concerns. In addressing such challenges, stakeholders must engage in participatory planning processes that involve local community decision-making. By creation of a discourse that addresses the varied implications of hydropower projects, societies can aim at seeking solutions that reconcile both energy needs and environmental integrity towards the advancement of renewable energy integration and sustainable development initiatives (Figure 6.7).

Figure 6.7. Hydropower.

Source: https://www.innovationnewsnetwork.com/wp-content/uploads/2020/08/Hydropower.jpg.

6.2. BENEFITS OF RENEWABLE ENERGY INTEGRATION

Integrating renewable energy into the existing energy system has several environmental benefits with main contributions being green gas emission reduction. With countries fighting against the pervasive and increasing challenge of global warming, the transition from fossil fuels to renewable energy is imperative. Wind, solar, and biomass energy reduce carbon emissions to a great extent, thereby easing the negative impacts associated with climate change. Secondly, utilization of innovative sustainable technologies (IST) enhances efficiency in energy utilization in various sectors, particularly buildings. In accordance with Fuertes et al. (2014), choosing the right IST makes it possible to attain low carbon emissions as well as maximum utilization of energy. The transition to renewables not only helps meet worldwide climate targets but also triggers sustainable urban development, making it vital to incorporate overall strategies that increase the integration of renewable energy. Apart from environmental considerations, the economic benefit of renewable energy integration cannot be ignored. Renewable energy offers enormous employment opportunities, starting from manufacturing to installation and maintenance. By investing in clean technology, governments are able to generate jobs within their local economy, developing and maintaining their economy. Additionally, lowered costs of using clean energy technology are making it more affordable and competitive within the market compared to traditional energy. By embracing these technologies by governments and companies, they enhance innovation and lower the cost of energy for consumers. This change not only resuscitates ailing job economies but also ensures energy security via fuel supply diversification. Economic returns for the adoption of renewable energy measures are often able to reap a positive return on investment, offering a long-term sustainable path to economic stability and sustainability. Social justice and development are also fields where the integration of renewable energy is advantageous, particularly for developing countries. As climate change effects hit worst among vulnerable communities, mainstreaming renewable energy could become a pathway towards climate-compatible development. Policies and programs that deliver "triple wins" of adaptation, mitigation, and development advantages must be forged to increase resilience in such communities, as exemplified in Bood et al. (2013). Mainstreaming renewable energy can increase access to energy, fostering local development through the empowerment of local communities and relief from energy poverty. Greater access to renewable energy can lead to enhanced health outcomes, learning, and economic growth. While nations strive for equitable solutions, integrating renewable energy is an important aspect of policies to tackle climate challenges as well as inclusive growth and development (Figure 6.8).

Figure 6.8. The benefits of renewable energy integration in smart home systems.

Source: https://www.knx.org/wAssets/img/news/homeowners/The-benefits-of-renewable-energy-integration-in-smart-home-systems.png.

6.2.1. Environmental Impact Reduction

There has never been a greater need to minimize impacts on the environment as nations implement initiatives to transition to sustainable energy systems. At the heart of it all is the fact that GHG emissions in the energy sector need to be addressed to achieve global climate targets. As highlighted in recent literature, countries such as Brazil are at a crossroads, and they need to make decisions that will either fix energy carbon habits or go towards sustainable architecture that increases economic competitiveness and addresses climate change. The possible investment of around 0.5 trillion U.S. dollars in energy investment in the next decade can basically define Brazil's path to a low-carbon economy, highlighting the paramount importance of policies that advance energy goals while reducing emissions at the same time (Lucon et al., 2015). These combined strategies will play an important role in guaranteeing a sustainable future for energy that serves current as well as future needs. Moreover, integrating renewable energy sources into existing systems offers distinctive advantages for environmental impact reduction. By utilizing innovative technologies, such as solar microgeneration, organizations can enhance energy efficiency significantly. For example, employing standards like Power over Ethernet helps in minimizing transformer losses during energy distribution, thereby maximizing efficiency from generation to consumption. This strategy not only achieves optimal power management of IT devices but also diminishes reliance on traditional grid power, which contributes more to the environment (Counsell et al., 2009). Implementation of such renewable technologies can also result in lower carbon emissions, promoting further the intended ends of incorporating sustainable energy. Thus, the focus on technological innovation and efficiency improvement assumes a pivotal position in mitigating the challenges caused by climate change. Lastly, the engagement of various industries, particularly high-impact industries such as transport and power generation, is essential to reduce environmental imprints. These subsectors can be focused upon to find major abatement possibilities that align with overall energy and climate objectives. For instance, the proposals outlined for Brazil emphasize the need for a combined approach addressing the transport sector emissions and industrial process emissions. These pro-inclusive policies can range from investing in cleaner technologies and subsidization to sustainable practices that not only fulfill the international climate commitments but also contribute to economic development (Lucon et al., 2015). By enforcing the position of stakeholders in renewable energy, there is a big chance to apply systemic reforms that minimize environmental effects while facilitating a stronger and more sustainable future.

6.2.2. Economic Advantages

The economic advantages of renewable energy integration are not only short-term cost savings but also long-term sustainability and resilience of the energy system. As increasingly popular technologies like electric vehicles (EVs) and vehicle-to-grid (V2G) systems gain ground, they have various advantages over traditional metrics of economic efficiency. In a large-scale experiment investigating the interfacing of EVs and V2G technologies, experts discovered a number of advantages, including reduced operational costs and improved integration of renewable energy into the grid. These figures emphasize the economic return opportunities beyond mere emission and fuel savings, illustrating the pay-offs of enhanced performance and noise reduction as integral parts of enhanced urban lifestyles and living standards (Kester et al., 2018). In addition, renewable power schemes catalyze massive job generation within the local economies due to new employment opportunities related to green energy transitions.

Figure 6.9. Economic benefits of renewable energy.

Source: https://www.certrec.com/wp-content/uploads/2024/09/Exploring-the-Economic-Benefits-of-Renewable-Energy-and-Its-Impact-on-Global-Markets-Certrec-1024x636.jpg.

For instance, the implementation of bioenergy production from landfill gas has been found to have extensive economic benefits as well as waste management issues. In Crete, the development of a power plant using landfill gas can produce a capital cost between 2.4 and 3.2 million euros with payback periods estimated to be 3.5–6 years. Their profitability level is indicated by the net present value of such projects, proving that local communities have much to benefit from waste-to-energy projects (Skoulou et al., 2016). The combined impact on the economic and environmental dimensions adds weight to the role played by integrating renewable energy in triggering local economies while, at the same time, lowering climate change. Finally, the beneficial impacts of integrating renewable energy extend beyond economic indicators to encompass total socio-economic benefits, enhancing community health and environmental health. By investing in renewable technology, communities acquire waste reduction and greater energy resilience, leading to more sustainable urban environments. Moreover, the shift towards renewable energy sources increases the share of clean energy in the overall energy production basket, a change that is crucial in achieving national and global climate goals. The interplay between economic growth and sustainable development is uncovered whereby investments in renewable energy contribute to job opportunities and stimulating local economies but also contribute to a cleaner environment resulting in enhanced quality of life (Skoulou et al., 2016). These advances raise the argument that the economic gains of embracing renewable energy are much broader than financial, with growing concern being a holistic view of community resilience and sustainability (Figure 6.9).

6.2.3. Energy Security and Independence

In contemporary debate concerning energy security and independence, the transition towards renewable sources of power emerges as an essential policy towards minimizing reliance on fossil fuels. Diversification of energy systems through the integration of renewable resources enhances resilience against geopolitical tensions and supply interruptions historically associated with traditional energy markets. For instance, while developing economies and developed nations alike bristle with the threat of climate change, a robust shift to renewable energy sources such as solar, wind, and biofuels can provide long-term energy security. Through strategically diversifying energy portfolios, nations can better deal with the limitations of energy supply and demand, ultimately to realize greater insulation from volatile fossil fuel markets, commonly beset by price volatility and geopolitical tensions, a concern underlined by recent global events (Harvey et al., 2010). A comparison of policies for biofuels across different countries illustrates the varied directions that define energy security. Brazil's early transition to sugarcane-based ethanol represents a model in which renewable energy not only assists energy independence but also connects with agricultural economic policy. By contrast, the U.S. emphasis on corn-to-ethanol production represents another dynamic set that influences energy policies as well as market structures. The same initial conditions of each nation, including resources and political institutions, define their trajectory to energy independence. Therefore, the cross-case experience within these countries presents an intriguing case study in how the incorporation of renewable energy varies and is driven by overarching drivers such as energy security, rural development, and economic market potential (Harvey et al., 2010). In regards to the dependability of renewable power systems, and more notably, the electricity supply, it is crucial to consider implications for energy security. The United Kingdom's electricity sector demonstrates, for instance, how the inclusion of low carbon objectives results in an increased diversity of fuel mix, providing protection against disruption. An examination of wind generation reveals that while intermittency cannot be avoided, deliberate diversification of source can offset reliability issues. This adaptability of the electricity supply infrastructure underscores the importance of renewable energy toward securing long-term security under unfavorable conditions generated by climate change and resource scarcity. The achievement of different electricity generation capacities supports not only strategic security goals but also sustains environmental stewardship, thus reaffirming the central role of incorporating renewable energy toward securing energy independence (Figure 6.10).

Figure 6.10. Energy independence.

Source: https://static.scientificamerican.com/dam/m/4fc157e97635e01d/original/GettyImages-2064884603.jpg?m=1714411438.234&crop=16%3A9%2Csmart&w=1920.

6.3. CHALLENGES IN RENEWABLE ENERGY INTEGRATION

Integration of renewable energy sources into the current energy infrastructure poses several significant issues that must be given serious thought and innovative solution. The initial significant issue is the intermittence of renewable energy resources such as solar and wind power, whose power supply will affect the stability of the supply of electricity. This periodicity in power generation makes it difficult to coordinate energy production with varying consumption patterns, ultimately resulting in inefficiencies and potential grid instability. Cellular networks, for instance, increasingly rely on stable provision of energy in order to cope with the soaring wireless data traffic; thus, matching the supply and demand is crucial in maintaining operating efficiency as well as reducing costs. The incorporation of advanced technologies in smart grids, like energy collaboration among base stations, is a potential path of recourse to mitigate such challenges, although the implementation of such cooperative mechanisms is an aggregated endeavor (Duan et al., 2014). An issue of integrating renewable energies is the need for robust energy storage mechanisms with the potential to accommodate variabilities inherent in green power generation. Energy storage plays a vital role in bridging the gap between supply and demand by storing overproduction energy during times of excess generation for use in times of sluggish production. Traditional storage methods are prone to capacity and efficiency problems, thereby compromising the efficiency of renewable energy systems. Recent projects like the INGRID European project demonstrate the potential of high-capacity solid-state hydrogen storage systems to overcome such limitations. With the deployment of advanced-level information and communication technologies (ICT), these systems would be able to upgrade the intelligent integration of different energy networks, ultimately making renewable energy not only effectively generated but also efficiently consumed (Zaluska et al., 2015). Lastly, the integration of renewable energy, particularly in the smart grid, is very significant regulatory and logistical obstacles. Policy-makers are faced with the daunting challenge of developing policy that will encourage investment in renewable technologies without ensuring existing installed energy infrastructure goes to waste. Grid stability and energy distribution demand the entire gamut of regulatory solutions encompassing health, safety, and environmental considerations. As stakeholders work towards developing a harmonious strategy for integrating renewables, they are faced with the consequences of market forces and supply chain constraints. The path to integrating renewables seamlessly calls for technological advancements but also a responsive regulatory framework that keeps pace with innovative energy solutions in order to provide an efficient and sustainable energy future.

6.3.1. Intermittency and Reliability Issues

The addition of renewable energy sources into the grid raises enormous challenges in terms of intermittency and reliability. Intermittency is largely because of the uncertain nature of sources such as solar and wind, which are extremely sensitive to weather patterns and time of day. This volatility leads to enormous variations in electricity generation, and grid operators find it difficult to maintain real-time supply of electricity on a consistent basis. For instance, research has shown that unforecastable of intermittency can cause welfare losses, making renewable energy generation less economically desirable in some situations, particularly when compared with more predictable conventional energy sources like natural gas. Such unreliability necessitates the establishment of advanced forecasting methodologies and grid management strategies to offset the adverse effects of intermittent power supplies (Gowrisankaran et al., 2025). Besides that, the utilization of a diversified mix of renewable resources is considered important to enhance grid reliability. The fact that different generations are added offers balancing of the imbalances generated by the randomness of individual resources. For example, whereas solar energy production will be most during daytime, wind power will supply the gaps during late night or low-solar hours. The strategic mix of various renewables not only supports achieving low carbon emissions targets but also enhances the robustness of the system. A study in the UK shows that a balanced fuel mix is crucial to achieving long-term electricity supply security while, simultaneously, achieving carbon reduction objectives. This confirms the assumption that a diversified strategy for renewable energy can alleviate most of the reliability problems with sole source reliance (Butler et al., 2025). Mitigating the intermittency of renewable energy entails comprehensive policy regimes and technological advancements aimed at ensuring grid operations are reliable. Enhanced investments in energy storage technologies, such as batteries, have a significant role to play in supply stabilization by storing surplus energy produced at high-generation hours for usage at low-generation hours. Additionally, demand management measures can aid in coordinating patterns of energy consumption with accessible supply, hence minimizing possible reliability issues. The policymakers will have to weigh the economic expenses of such measures while ensuring that inflexible renewable mandates that have socially costly welfare-boosting social costs do not generate additional welfare losses. Ultimately, a diversified approach relying on both efficient energy production technologies and innovative management techniques will be important to overcoming the intermittency challenges on the path to a future powered by alternative energy (Gowrisankaran et al., 2025).

6.3.2. Infrastructure Limitations

The integration of renewable energy systems in established infrastructure has severe limitations that hinder their worldwide use. The traditional power grids were designed primarily for the centralized production of electricity and thus give rise to uni-directional energy transfer from central power plants to end users. The paradigm not only allows challenges in adapting to the vagaries of renewable resources such as solar and wind but also exacerbates issues on energy wastage and grid stability. There is an inherent necessity of modernization, with the aging infrastructure not being able to sustain the bi-directional flow required for effective integration of renewable energy. An approach in this direction is the shift to Smart Grids (SGs), which can manage energy more effectively through more responsive and adaptive systems that are able to cope with the demands of a heterogeneous energy mix including renewable energy-driven sources. The systems need to develop in response to these infrastructure constraints for a sustainable and resilient energy future. In addition, the challenges of infrastructure constraints are reflected in the geographical conditions that define the deployment of renewable energy. For example, though rooftop solar sites offer a great hope for clean energy production, they also pose complexity with respect to grid operations. It is highlighted by a recent study that, even though there is a latent potential for large-scale on-site electricity generation in cities like Los Angeles, the majority of the buildings consume more energy than what they produce. The infrastructure is generally not sufficient to facilitate the integration of the excess energy onto the grid and limit the net positive impact of decentralized renewable sources. Cumulative potential sales to the grid indicate a considerable unused energy output, which indicates the imbalance between renewable energy output and grid capacity. This is an example illustrating that upgrading of infrastructure is required to maximize the benefits of renewable energy and to facilitate a more sustainable urban electric environment. Upgrading of infrastructure bottlenecks involves strategic investment and technology, as well as policy reform. Smart Grids technology, as mentioned earlier, needs financial and technical resources to implement the necessary upgrades capable of handling new technologies such as IoT devices for real-time monitoring and control. Furthermore, identification of potential from distributed generation systems such as rooftop solar is necessary for effective energy planning. Rules for interconnection of distributed systems need re-evaluation so as not to miss available renewable resources. An energetic approach, one that targets the identification of net electricity generation opportunity and supports policies for poor people, can facilitate moving towards a more integrated system of renewable energy. Overall, overcoming these infrastructure limitations can create the potential for a more sustainable, resilient, and equitable energy future.

6.3.3. Policy and Regulatory Barriers

The integration of renewable energy into existing energy systems is often thwarted by a complex policy and regulatory framework. The governments play a pivotal role in establishing systems that facilitate or impede the use of renewable energy, which has a direct influence on the industry's potential to achieve its maximum capacity. A review of ASEAN's effort to boost its proportion of renewable energy finds the member nations to be lacking the essential components such as proper forecasting facilities and open electricity market access needed to have such resources being integrated in an optimum way. The outcomes indicate that weaknesses in grid flexibility emerge as a critical concern, particularly for nations with excessive variability in the supply of renewable energy. As ASEAN transitions to meet its renewable energy goals by 2025, bridging such policy gaps matters since it shows policy settings can be employed to spur or stifle the path to a clean energy future (Huang et al., 2019). At the same time, transitioning to smart grids has emerged as a key factor in bridging policy obstacles associated with incorporating renewable power. While smart grids have the promise to optimize power distribution through newer, improved technology, their adoption is stunted by legislative uncertainties and regulative systems that lag behind their requirements. Implementing smart grids is not only a matter of technology but also of an enabling policy environment. As was underscored, public acceptability and the dynamics between various stakeholders—namely, regulators, industry stakeholders, and consumers—are pivotal in providing an environment under which smart grid technology can thrive. Improved regulatory clarity is required to guarantee that issues regarding data security, confidentiality, and vulnerability to cyber-attacks are handled, with all relevant stakeholders in agreement in order to facilitate a transition to a low-carbon energy system (Axon et al., 2014). Further, the intricacies of regulatory systems engender unintended impacts that also increase obstacles to merging renewable energy. For instance, energy regulations that promote traditional sources of energy can unintentionally discourage investment in renewable energy, creating a competitive imbalance that decays market fairness. Rigidity in policies generally prevents adaptive measures that would promote greater flexibility in energy systems from being put into place, particularly as demand for renewable sources rises. Regulatory bodies need to take an active step towards harmonizing policies across jurisdictions so that the transition to a greener energy paradigm is a problem-free one. Through collective action at a regional level with the sole emphasis placed on innovative policy solutions, nations can work together to overcome such obstacles and create consistent energy markets that are holistically inclusive of renewable energy potential, paving the way for cost-effective solutions for the grid as a whole (Huang et al., 2019).

6.4. TECHNOLOGICAL INNOVATIONS

Technological innovations are imperative in the adoption of renewable energy to enable the shift of traditional energy systems to more sustainable alternatives. Technological advancement in alternative energy technologies has been at the forefront of recent decades in response to outside shocks such as oil price volatility and shifting environmental regulations. As explained in the literature, they not only become more energy efficient but also make the green technology more competitive by providing a platform where businesses can experiment, innovate, and innovate (Bhuyan et al., 2025). Supporting the shift from fossil resources to renewable resources, innovation in energy storage technology, management of the grid, and efficiency is most essential. Collectively, these innovations constitute the basis of a modern energy system that is based on resilience and sustainability, effectively reducing greenhouse gas emissions and enabling large-scale adoption of renewable resources into the energy base. Renewable energy integration is supported by resource efficiency and new material use strategies. The cascading utilization of renewable resources, keeping in mind the multi-utilization before conversion to energy, realizes the maximum value derived from raw materials with the minimum waste. It prioritizes synergistic strategies that combine production, resource management, and technology (Geldermann et al., 2015). With the promotion of interdisciplinary studies in maximizing production and distribution networks for renewable resources, we see a direct impact on sustainability and the efficiency of resources. These technology-driven solutions not only support cost-effective production of energy but also favor guidelines of circular economy, which emphasize sustainability through regeneration of resources. Because of this reason, developments in material science and production methodology are important in driving the renewable energy sector towards resource-saving practices that are synchronized with both economic efficiency and environmental sustainability. Finally, consumer behavior influences the adoption and acceptance of renewable energy technologies to a large extent, emphasizing the need for innovation that is tailored to meet market needs. The role of information technology in enhancing consumer awareness and involvement cannot be overstated, as it equips people and communities to make informed decisions regarding their energy consumption. Technological innovations must then be focused on producing easy-to-use systems that embody low levels of energy consumption while promoting ecologically friendly behavior. This intersection of new technologies and public support shapes the future direction of renewable energy integration, towards a general movement towards cleaner energy products. As researchers and practitioners move through the interfaces of technology, policy, and market forces, the future growth of renewable energy systems will undoubtedly rely on new approaches that reconcile both environmental and economic factors.

6.4.1. Energy Storage Solutions

The effective functioning of renewable energy systems is greatly dependent upon effective energy storage solutions. Due to their natural variability in nature, as of renewable energy sources such as solar and wind, energy storage technologies must be robust enough to counter variability in generation. This variability renders it tough to maintain a steady energy supply, and therefore storage solution developments are imperative to make renewable energy integration a reliable option. Technologies such as lithium-ion batteries, pumped hydro, and emerging options such as flow batteries seek to offer the storage capacity to store such surplus energy produced during peak hours to be used during low times of generation. The use of diversified energy storage devices not only stabilizes the grid but also optimizes the overall efficiency of energy consumption, which is a priority area for sustainable development work. The incorporation of energy storage technologies in the energy system requires deep innovation and flexibility, particularly with the emergence of smart grid technologies.

Figure 6.11. Energy storage trends.

Source: https://knowhow.distrelec.com/wp-content/uploads/2024/11/GettyImages-2155976732-1024x652.jpg.

The integration of Information and Communication Technologies (ICT) has become feasible to a larger extent to deal with the intricacies related to renewable energy systems. Smart grids allow for efficient transmission of energy with better distribution along with enhancing storage and retrieval capacity with greater proportions as per real-time demand and supply situations. In addition, the crossroads of energy storage technologies and smart grid infrastructures provide a seamless transition to a flexible network that can accommodate various sources of power. Through the integration of storage solutions on these advanced systems, energy firms can better manage the utilization of resources as well as improve response time to changes, hence improving grid resilience. This convergence is leading to an energy ecosystem that can be attuned to both consumer needs and sustainability issues, thereby becoming sustainable. Also, the convergence of energy storage technologies with technology innovations like Generative Artificial Intelligence (GenAI) holds favorable prospects of streamlining renewable energy operations. With the use of MLOps frameworks, energy storage systems can continuously enhance their efficiency and effectiveness, immensely supporting applications such as predictive maintenance and resource optimization. GenAI potential to expand to automating the assessment of energy production scenarios can lead to improved decision-making processes for energy storage deployment and utilization. This new methodology not only enables operational excellence but also sustainability goals by reducing energy reliance on traditional sources. As policymakers advocate for standardized regulations and investments in infrastructure, the application of GenAI in energy storage practice will undoubtedly be an engine for the transformation of our energy systems, which will result in a greener future for all concerned stakeholders ((Borghetti et al., 2013; Gade et al., 2019) (Figure 6.11).

6.4.2. Smart Grid Technology

The deployment of smart grid technology is pivotal in the enablement of higher rates of renewable energy integration into electrical grids. With the shift towards more distributed and intermittent power sources, like wind and sun, traditional grid infrastructures are gravely strained to provide reliability and efficiency. Smart grids deploy advanced digital communications technology to enable real-time monitoring and control of the energy flow, essential for supply-demand balancing under fluctuating output of renewables. This dynamic approach not only enhances the grids' robustness but also makes room for embracing new-generation energy solutions such as energy storage and demand response, paving the way for a greener energy landscape. Emerging technologies in Volt/VAR Optimization and Control (VVOC) specifically hold the potential for filling prevailing lacunas in prevailing methodologies to increase voltage profile stability in distribution systems with high renewable energy penetration (Arghandeh et al., 2013). In addition, electric vehicles (EVs) are on the verge of becoming both a threat and opportunity in the context of smart grid technology. With the mass introduction of EVs also another pressure on low voltage power distribution networks, energy usage can surge erratically and perhaps overload the current infrastructure. But through smart grid technology, coordinated charging schemes can also be utilized to mitigate issues. With real-time data analysis and advanced grid management systems, DNO can normalize charging schedules to reduce peak demand and enhance load management. Such an approach not only averts the risk of transformer overload but also the whole efficiency of renewable energy utilization. Smart charging of EVs is, therefore, a quintessential example of the synergistic cooperation between smart grid systems and integration of renewable energy, demonstrating an example of sustainable energy infrastructure harmonizable with contemporary environmental and economic aspirations (Butans et al., 2016). Furthermore, evolution of smart grid technology is crucial to enhance consumer participation in energy management. With the advent of smart meters and networked home appliances, customers can monitor and control their energy usage on a continuous basis. Not only does this increased participation result in more efficient usage of energy, but it also allows users to take advantage of time-of-use pricing and reward programs, particularly when there is excess renewable energy. This consumer-centric energy management framework leverages the natural fluctuation of renewable resources in a manner that permits efficient distribution and consumption patterns along the grid. By providing active residential and commercial participation in energy markets, smart grid technology makes the integration of renewable energy more viable, thus enabling a transition to a cleaner, more secure energy future (Figure 6.12).

Figure 6.12. Smart grid technology.

Source: https://yourelectricianwinnipeg.ca/wp-content/uploads/2023/12/view-bio-engineering-advance-with-human-hand.jpg.

6.4.3. Advanced Forecasting Methods

The integration of advanced forecasting methods is crucial in the optimization of renewable energy systems to offer optimized avenues for enhancing the reliability of the power supply in response to varying generation trends. Traditional forecasting methods are unsuitable when applied to the complexity of renewable energy sources since they are ineffective in handling big data. On the contrary, advanced techniques, such as big data analytics, utilize artificial intelligence and machine learning concepts to analyze large data sets and yield valuable insights. These methods help combat the limitations of conventional forecasting, such as divergence in results and mistakes, by detecting trends that would otherwise be obscured within typical analyses. Using sophisticated forecasting, energy producers can make more accurate predictions of energy consumption and generation, resulting in enhanced grid stability and the effective integration of renewable sources into energy systems (Gharehpetian et al., 2019). Additionally, sophisticated forecasting methods can greatly boost forecasting energy production, particularly in solar photovoltaic (PV) systems where data accumulation over time is vast. Smart optimization modeling techniques leverage control system information, operating information, and maintenance event information to enable predictive analytics. These techniques are more advanced than mere forecasts and instead offer intelligent information that facilitates condition-based maintenance (CBM) strategies and operating efficiency optimizations. With these sophisticated techniques, power system operators can take raw data and convert it into actionable intelligence that translates directly into return on investment (ROI) through reduced production losses. As renewable energy systems become more advanced, the demand for inclusion of such sophisticated forecasting techniques grows even further in designing efficient management strategies (Crespo Márquez et al., 2019). Moreover, the applicability of sophisticated forecasting techniques is measurable in terms of enhancing microgrid and smart grid implementation performance. With the application of big data analytics along with predictive analytics, grid operators can not only manage the integration of intermittent sources of renewable energy but also the increasing complexity of power distribution. Such processes facilitate the generation of active distribution networks that can respond in real-time to variations in energy supply and demand, making them resilient as well as sustainable. The advancements in forecasting also further impel innovation in energy market strategies, making the economic viability of renewable energy projects further enhanced. Ultimately, the merging of improved forecasting methods builds a solid foundation for cultivating active management of renewable energy systems and creating avenues towards sustainable energy transitions (Crespo Márquez et al., 2019; Gharehpetian et al., 2019).

6.5. CASE STUDIES OF SUCCESSFUL INTEGRATION

A detailed analysis of Indias renewable energy integration plan shows significant advancement coupled with equally significant issues, particularly against the backdrop of its ambitious target renewable capacities. The country has a goal of achieving 10% of its electricity from renewable sources by 2020, increasing to 165 GW installed capacity by 2022 with significant contributions from solar and wind energy. Nevertheless, these goals require a robust policy environment that can address techno-economic issues as well as socio-political ones. Currently in place, such as Renewable Purchase Obligations (RPO) and Feed-in Tariffs (FiT), are meant to encourage economic incentives but are insufficient to create the regulatory environment for long-term infrastructure expansion. Current estimates point to the fact that India will miss its ambitious targets if more all-encompassing efforts that consider many political, social, and economic factors are not taken (Ramamurthi, 2016). An examination of the evolution of smart grids as a prime facilitator for the transition towards renewable energy systems points to the fact that their development is multi-dimensional. Smart grids enhance efficient energy management by allowing room for intermittent sources of power using advanced technological designs, including two-way communications and data analysis. Investment in smart grid technology receives widespread political and financial backing as a path towards the achievement of carbon reduction goals because it enhances energy stability and reduces costs. The addition of smart grids, though, is accompanied by the mitigation of cybersecurity, data privacy, and regulatory compliance concerns of the public. Therefore, the successful integration of smart grids not only requires the improvement of technology but also robust consumer confidence and participation because these play an important role in the social acceptability and the resultant installation of such grids (Axon et al., 2014). Success stories regarding new energy integration across the globe underscore the need for inclusive approaches involving various stakeholders. For instance, regions that have been able to successfully integrate renewable technologies will often exhibit cooperative frameworks for the integration of government policy, private investment, and community engagement. Such models of success demonstrate not only the utility of technological innovation but also social acceptance in pursuing renewable energy initiatives. Furthermore, they suggest that leadership and coordination across various sectors are necessary to overcome challenges like infrastructure build-out and regulation issues. The energy scenario of today demands a fusion of economic, technological, and social factors in order to achieve an integrated, sustainable energy system. On the basis of these case studies, policymakers in India and other places can more easily understand the challenges of renewable energy integration, which will result in more efficient and sustainable energy solutions (Axon et al., 2014; Ramamurthi, 2016).

6.5.1. Germany's Energiewende

Germanys Energiewende, or energy transition, has become a leading template for the integration of renewable energy into national-level energy systems. In its pursuit of an unprecedented increase in the share of renewables in its energy system, Germany has become an internationally known leader for its ambitious green vision. However, the execution of this ambitious project has revealed both successes and setbacks. Observing the complexities that are involved, the German government has emphasized a number of challenges, which touch on the volatility of renewable energy and the need for additional energy storage and grid capacities.

Figure 6.13. Germany's energiewende.

Source: https://www.agora-energiewende.org/fileadmin/Sections/Home/AEW_home_header.jpg.

As much as observed by energy specialists, the challenges point towards the need for an integrative vision that is aligned with the broader priorities of sustainability and security of energy, not just for Germany, but the entire European Union (Hedberg et al., 2017). The implications of Germany's Energiewende are universal, to the extent that they have lessons for other member countries in the European Union. According to the thesis on energy security in Europe, the policy-making framework within Germany can serve as a yardstick for nations that are confronted with energy dependency and the risks that come with it. Other nations can minimize their carbon footprint while simultaneously improving energy self-reliance by emulating similar measures. A replication of the steps that reflect the failures and achievements of Germany's model might offer useful lessons regarding the prospects and challenges of embedding renewable energy. Since the environment is under severe pressure in Europe, lessons from these dynamics might inform policies to create more sustainable energy systems in Europe (Millsap et al., 2018). Germanys Energiewende also points to the significance of coordinated policy steps that amalgamate economic and environmental considerations. A comprehensive strategy with stakeholder engagement, technological innovation, and facilitation of rules is required for effective transformation to renewable energy. The findings from Germanys experience bring to the limelight that ambitious targets rely on collective effort among different tiers of government and sectors of society. In the absence of adequate planning and coordination, the way to a low-carbon economy can be contradictory and inefficient. Therefore, Germany's experience is a crucial reminder of the importance of not just visionary goals but pragmatic approaches towards setting up robust renewable energy integration (Hedberg et al., 2017) (Figure 6.13).

6.5.2. California's Renewable Energy Initiatives

California, leading the way in renewable energy projects, has initiated an ambitious move from fossil fuels to cleaner sources of energy. This is supported by the state's broad policies aimed at curbing the emission of greenhouse gases and boosting the proportion of renewables in the energy mix. The enforcement of such legislations as the Renewable Portfolio Standard (RPS) mandates that utilities obtain 60% of their electricity from renewable resources by 2030 and, in the end, 100% by 2045. These all mirror Californias strong commitment to not only rendering its energy more secure but also to avert climate change. Moreover, the state has also seen a tremendous influx of investment in wind, solar, and other green industries, which not only fulfill environmental objectives but also drive economic progress through the creation of green jobs. The introduction of renewable energy into Californias power grid both has potential and difficulties. Complex smart grid technologies are a key component to managing the variability of renewable sources such as solar and wind and ensuring a stable supply of energy. The state has adopted various technologies to achieve maximum consumption and delivery of energy and improve the stability of the grid. With the utilization of smart meters and advanced analysis, California can enhance energy efficiency and overall decrease of consumption during peak periods. These innovations reflect a visionary approach to meeting consumer needs with renewable energy production. The challenge remains, though, with how to enable existing infrastructure to accommodate the growing transmission of renewable energy resources without compromising reliability and affordability to consumers (Alava et al., 2017). Public engagement and policy activism have been key to mobilizing public support for California's renewable energy policies. Grassroots campaigning and cross-sector collaboration between the government, private sectors, and community organizations have been responsible for driving awareness of climate issues and advancing legislative changes. Public support is evident in high participation levels of clean energy projects and community renewables schemes, showing a collective push by the population towards sustainable energy futures. Additionally, involving the stakeholders in planning helps in the development of an integrative model of energy policymaking, in terms of addressing economic disparities, and ensuring equal access to renewable substitutes for various populations. Through such complementary momentum, California continues to be at the edge of renewable energy initiatives statewide and globally in establishing benchmarks, inspiring other places to revisit their energy policy (Kada et al., 2004) (Figure 6.14).

Figure 6.14. Solar and wind power projects have been booming in California, like the Pine Tree Wind Farm and Solar Power Plant in the Tehachapi Mountains, but that doesn't mean fossil fuels are fading away quickly.

Source: https://opb-opb-prod.cdn.arcpublishing.com/resizer/v2/RMEY5CDQJI-44JOJJRVJGXTTMKY.jpg?auth=e23591bc4bc2872217185ed4ef138af2ea776d3fb92d16178ae2233a41138459&width=767.

6.5.3. Denmark's Wind Energy Strategy

Denmark's aggressive wind energy plan is a mainstay of its overall renewable energy incorporation strategies. As a global leader in wind power, the country reaped generously from its advantageous geographical location, with robust and steady winds. The government of Denmark initiated its journey with wind power in the 1970s, as a result of the oil crisis and further encouraged by subsequent policies to reduce the consumption of fossil fuels. A technological innovation promise in wind turbines has provided for the deployment of onshore and offshore wind farms, enabling Denmark to make more than 47% of its electricity usage accessible through wind power by 2019. This is a reflection of the success of policy-driven innovation since Denmark has continued investing in research and green technology development, setting a precedent for other nations that are undertaking the same objectives for sustainability and energy independence. At the core of Denmark's wind power policy is a robust regulatory framework that encourages competition and innovation in the energy sector. The Danish government established market structures promoting investment in renewable energy technologies while simultaneously minimizing entry barriers for new firms. This has encouraged a good climate for the development of innovative solutions, particularly wind energy, as indicated by the increase in the number of patents generated in the industry. Evidence has indicated that the relationship between policy stimulus and innovation is strongly positive for low-barrier deregulated energy markets, and it is to this extent that Denmark's regulation has driven the growth of renewable technology (Nicolli et al., 2025).

Figure 6.15. Offshore wind energy tender to power Denmark's entire electricity demand.

Source: https://www.innovationnewsnetwork.com/wp-content/uploads/2024/04/shutterstock_Tony-Moran_109075526-1024x576.jpg.

Secondly, Denmark's wind power integration is supplemented by its focus on decentralized production, complementing a local empowerment and energy security culture. The use of waste heat recovery in combination with wind power is another innovative aspect of Denmark's energy policy. Using waste heat in conjunction with district heating systems, Denmark is in a position to mitigate the inherent fluctuating nature of wind power generation and still be capable of producing sufficient cooling capacities to support industry processes. This unification creates a dynamic energy production model, where the excess power generated through renewable sources can be effectively utilized to meet demand and stabilize the grid. Moreover, a Virtual Power Plant (VPP) concept facilitates effective management of renewable sources by utilizing local capacities to match consumption and production (Eetvelde et al., 2013). Taking such steps not only optimizes the use of the resources at hand but also demonstrates Denmark's passionate dedication towards energy efficiency and sustainability, eventually translating into its strategic initiative towards a renewable energy future (Figure 6.15).

6.6. FUTURE TRENDS IN RENEWABLE ENERGY INTEGRATION

As the world shifts towards a sustainable energy-driven future, the integration of renewable energy into existing grids will be the driver behind achieving climate goals. New trends suggest that there will be increased reliance on grid flexibility because of the dominance of intermittent renewable resources like solar and wind in harvesting new generation capacities. Grid managers must deal with problems associated with the reliability of such sources; therefore, grid flexibility must be enhanced to accommodate variable output. Grid flexibility assessment indicates that nations, such as ASEAN, may experience radical variation in readiness. There will be a need for concerted effort among nations to create a stronger, more functional system capable of being serviced by higher percentages of renewables, fostering cost savings and environmental sustainability in energy markets, as highlighted by evidence that points towards the necessity of improved forecasting tools and access to ASEAN electricity markets (Huang et al., 2019). The economic impact of greater variable renewable energy (VRE) penetration is already seen within the traditional power markets. There are indications that the larger VRE rise, the more that it draws wholesale energy prices and alters operation conditions for established thermal generators. Importantly, already the trend has shown to decrease prices, eventually benefiting the consumer, but not without some detriment to traditional baseload generators. Areas with high levels of VRE generally face sophisticated market conditions, including instances of adverse pricing in low-demand scenarios. Yet, even though the effect of VRE has been obvious in some areas, its overall influence on historical average wholesale price levels in wider markets is restricted when set alongside other influences like varying natural gas prices (Botterud et al., 2020). This complexity will need continued analysis and rebalancing of the regulatory frameworks in existence to guarantee electricity markets are robust. In the future, the relationship between emerging technologies and integration with renewable energy will continue to shape the profile of energy. Technologies like energy storage devices, smart demand response, and decentralized models of generation are expected to enhance the resilience and efficiency of the grid. Energy storage technologies will be especially useful by lowering renewables' supply volatility, thus making it more reliable when demand peaks. Decentralized models will also be empowering consumers since they will have an active role to play in energy generation and control. As utility firms adapt with these technologies and structures in the markets change, the energy sector will likely transition towards higher interconnection and coordination between participants. Achieving synergy between technological innovation and renewables integration will not only maximize resource utilization but also reconfirm the ultimate vision of a sustainable energy future where integration of renewables is seamless and effective (Figure 6.16).

Figure 6.16. Future trends in renewable energy integration.

Source: https://entrapeer.com/wp-content/uploads/2024/08/united-state-independence-day-background-scaled.jpg.

6.6.1. Increasing Investment in Renewable Technologies

Transition to renewable technologies demands massive investment to fund their integration into the existing energy infrastructure. History in the past illustrates that government expenditure has a vital role to play to facilitate development in renewable energy sectors. For instance, President Obama's fiscal year 2015 budget proposed enormous funding for the Department of Energy (DOE) to aid clean energy programs to address climate change. A nice touch, the budget put aside $1 billion in a Climate Fund that works to help communities cope with the increasingly intense frequency of climate events, ultimately showing that financial investment not only spurs technology development but also raises infrastructure resilience to climate issues (N/A, 2014). All these investments are key to creating an atmosphere in which renewable technology can thrive, thereby driving overall energy transition activity. In addition to direct government investments, ensuring a stable market environment through the provision of differentiated subsidies also contributes significantly towards enhancing the potential of renewable energy sources. Research indicates that a well-designed regulatory environment can reduce the volatility of facing when incorporating intermittent renewable energies into power markets. With the implementation of technology-specific subsidies, policymakers can effectively minimize the level of excess production curtailment that typically accompanies additional renewable generation capacity. As seen, such tailored approaches can significantly lower the required investments in energy storage, addressing the issues of supply-demand imbalances haunting renewable energy efficiency (Abrell et al., 2019). In this manner, a strategic approach that aligns investment with policy incentives is able to enhance the viability and integration of renewable technologies within the energy system. In addition, increasing investments in renewable technologies can also enhance economic growth by creating employment opportunities and stimulating innovation in the targeted industries. As governments and private investors spur funding for renewables, the resulting transformation can stimulate employment opportunities in manufacturing, installation, and upkeep of renewable systems. Also, such investments stimulate innovation in allied technologies, like smart grids and energy storage, that are required for efficient energy management. This collaboration of factors results not just in a cleaner energy infrastructure but also in a more robust and resilient economy. Such economic multipliers strengthen the argument for increased investment in renewable technologies as a key component of environmental sustainability and economic development within the broader context of renewable energy integration.

6.6.2. Role of Artificial Intelligence (AI) and Machine Learning (ML)

In the pursuit of achieving maximum renewable energy integration, utilization of AI and ML has been a game-changer. These technologies have the ability to process huge datasets created by energy systems, facilitating more effective decision-making. Conventional approaches have frequently been challenged by the intricacies involved in big sets of data, causing inaccuracies and inefficiencies in managing energy. AI and ML methods can effectively process multiple variables that determine energy production and consumption, thereby providing a multifaceted insight into energy trends. With the use of such advanced analysis tools, energy decision-makers can improve forecasting accuracy, optimize resource usage, and maximize demand response actions, thus achieving a more stable and sustainable energy system in integration programs (Gharehpetian et al., 2019). Additionally, AI and ML use in renewable energy systems significantly enhances operational efficiency with maximum energy storage and load balancing. As renewable sources such as wind and solar power tend to be intermittent, the provision for sustaining grid balance becomes a serious issue. Energy output from these sources can be forecasted using AI programs, which allow grid managers to predict supply fluctuations and be prepared for them. Predictive analysis plays an important role in incorporating renewables into the existing grid network, reducing fossil fuel reliance, and making sustainability more overall. Furthermore, machine learning models can facilitate distributed energy resources and microgrid functionality to make supply and real-time demand precisely aligned. Such developments highlight the significance of AI-based solutions in the construction of smart grids and the energy sector as a whole (Ahmed et al., 2025). The use of AI and machine learning does not stop at optimization but also involves the development of new paradigms in energy production and consumption. The use of big data analytics within power systems promotes the emergence of collective operation trends and elevated microgrid autonomy. Based on IoT device behavior data and other data sources, AI can guide demand-side management strategies that boost energy conservation and peak load mitigation. Furthermore, machine learning applications have the ability to detect abnormalities, i.e., electricity theft and equipment malfunction, to enhance grid security and integrity. Overall, the integration of AI and ML in renewable energy not only addresses the challenges of variable source energy, but also constructs a more solid and adaptive energy infrastructure, at last fostering the establishment of a sustainable energy future (Gharehpetian et al., 2019) (Figure 6.17).

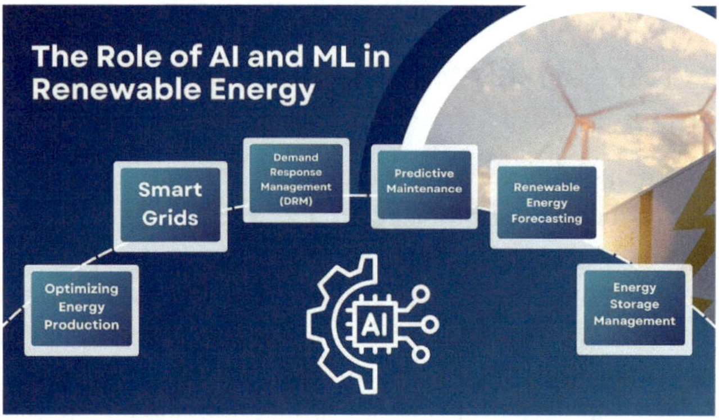

Figure 6.17. The role of artificial intelligence (AI) and machine learning (ML) in renewable energy.

Source: https://miro.medium.com/v2/resize:fit:4800/format:webp/1*XVimbZZMa3y13ttL3LcG7g.jpeg.

6.6.3. Global Collaboration and Agreements

In the context of renewable energy integration, global cooperation and agreements become major forces driving the transition towards sustainable energy systems. Energy landscapes' increasing complexity, characterized by interplays of various technologies and interests of various players, necessitates countries' cooperation together. Global norms, such as the Paris Agreement, help encourage cooperation through establishing joint goals to reduce greenhouse gas emissions and build climate resilience. Such agreements do not only encourage national policies but also aid in sharing best practices and technology, hence accelerating the spread of renewable sources of energy worldwide. With participation in these global initiatives, countries can align their plans for energy so that they advance mutual environmental objectives while addressing their unique economic and social conditions. The requirement for this coordination cannot be overstressed since it lays the foundation for a unified attempt at combating climate change.

The integration of renewable energy into existing infrastructures involves major challenges that can be adequately addressed only through global cooperation. All three sectors, governments, research institutions, and the private sector, must work together to devise new solutions and share information to overcome systemic limits. For instance, attempts to pair water supply systems with renewable energy have shown regional cooperation to be capable of great efficiency and sustainability gains. The cross-regional comparative analysis of initiatives across different regions, as illustrated in Ani et al. (2024), offers key lessons in addressing technological, financial, and infrastructural challenges. Further, such case studies highlight the necessity of sharing resources and knowledge to ensure scalable and resilient solutions that travel across contexts. All this collaboration goes beyond technicalities to foster partnerships that further foster shared interests and advance the global clean energy agenda. Cybersecurity elements of the renewable energy infrastructure also highlight global cooperation and agreements.

As highlighted by Ogunmoye et al. (2025), the increasing reliance on distributed energy resources (DERs) further increases the landscape's vulnerability, with cybersecurity being a pressing concern. Nations are recognizing that cross-border cyber threats should be tackled through multilateral efforts to create efficient security mechanisms. Building cooperative agreements to develop uniform cybersecurity processes can enhance the overall security of energy infrastructures such that utilization of renewable resources will not come at the cost of security. This requires cooperative efforts that focus on sharing threat intelligence, best practices, and regulatory policy harmonization. Finally, the creation of cooperative alliances that transcend borders is key because it not only promotes energy security but also produces mutual confidence between nations towards a sustainable energy future.

SUMMARY

In general, the incorporation of renewable energy sources into present power networks is a crucial step towards achieving a sustainable energy future. As we shall see in this chapter, shifting towards renewals is not a blue-sky idea but rather something that has to be implemented practically using processes involving a number of parties and integrating the latest technologies. In particular, solar microgeneration solutions are excellent examples of best ways of transmitting power efficiently while significantly reducing the use of conventional electricity infrastructures. This model promotes the importance of energy maximization through minimizing losses by employing systems, thereby playing its part in overall carbon emissions and energy use reduction in technology-oriented industries, namely media technology use. The expediency of green programs emphasizes the importance of such projects in planning future practices of managing energy, which ultimately benefits more significant goals of environmental conservation and energy security. Moreover, the role of stakeholder cooperation in renewable integration cannot be overstated. Involving multiple organizations—ranging from government agencies and business enterprises to communities—is vital in developing comprehensive strategies that include mutual goals and address local energy needs. For instance, paying attention to backcasting within stakeholder meetings demonstrates the utility of process-oriented cooperation in planning for the long run while being linked to short-term policy initiatives (Bärlund et al., 2010). Through harmonization of short-term goals with long-term visions for sustainable development, stakeholders can create implementable strategies that result in successful energy conversions. Long-range planning enables a proper response to new challenges through ensuring that renewable energy integration results in technological success as well as social and economic success. It is definitely a complex journey towards sustainable energy systems, but through technological, regulatory, and social challenges. Nevertheless, ever-evolving technological innovations such as those that integrate solar power into low-power IT devices show the potential of new solutions to overcome such obstacles (Counsell et al., 2009). As the clean energy landscape keeps evolving, policymakers, businesses, and communities must remain steadfast in integrating such innovations. By establishing a culture of ongoing innovation and implementing policies favorable to the use of renewable energy, we can ensure our future is built on sustainable grounds. In summary, the use of renewable energy presents a revolutionary opportunity to re-conceptualize our energy systems towards a more resilient, efficient, and greener future.

Integration of renewable energy sources into current power grids poses both tremendous opportunities and severe challenges. Of particular significance is the comparison of several Principles of Access (POA) implemented in other regions of the globe, e.g., the United States, Great Britain, and Ireland. The referred principles, i.e., Last-in First-out (LIFO), Pro Rata, and Market-based approaches, specify how new power producers can enter and treat the electricity grid. Each region's experience bears the unique strengths and weaknesses that are characteristic of their preferred approaches, underlining the need for a tailored strategy to encourage integration and alleviate the hazards of curtailment and investment in generation plants. The examination of these experiences led to critical conclusions that could potentially help distribution network operators in the UK, among other locations, integrate distributed generation more efficiently and economically (N/A, 2013). Further, the achievement of the ambitious renewable energy target set by institutions such as ASEAN relies on the ability of the regions to ensure greater grid flexibility. The challenge lies in accommodating the inherent volatility of renewable energy sources that could lead to supply uncertainty to make grid management increasingly challenging. Notably, a grid flexibility assessment tool has been developed to measure the current level of grid reliability and other determinants of importance among ASEAN nations. The tool applies quantitative indicators to classify countries in three categories in terms of their preparation for renewable energy integration: better, moderately, and least prepared. The findings reveal a disconcerting lack of forecasting infrastructure and limited market access, both of which are essential in enhancing the grids response to shifts in supply and demand in the energy as penetration of renewables increases (Huang et al., 2019). Finally, encouraging improved regional coordination is an urgent way to bridge the challenges related to renewable integration. Grid flexibility analysis should not just provide current shortcomings but also inspire collective action among ASEAN member nations to combine collective capabilities and resources. Such collective effort is necessary in addressing the common issues of integrating renewables and improving overall system performance. By prioritizing investment in grid infrastructure, forecasting, and access to electricity markets, member states can enhance their readiness for higher penetration of renewable energy. The open approach guarantees cooperation at the regional level is harmonized with high-level environmental goals while at the same time maximizing the economic benefit derived from a transition towards a cleaner energy mix (Huang et al., 2019).

REFERENCES

1. Abrell, J., Rausch, S., & Streitberger, C. (2019). Buffering volatility: Storage investments and technology-specific renewable energy support. https://core.ac.uk/download/286703813.pdf (accessed on 30th September 2025).
2. Ahmed, A. A., Alarga, A. S. D., & Belrzaeg, M. (2025). AI-driven optimization of renewable energy grids enhancing efficiency and sustainability. https://core.ac.uk/download/663684328.pdf (accessed on 30th September 2025).
3. Alava, L. A., Arauz, W. M., Castillo, G. A., et al. (2017). The future of micro-grids in Ecuador. https://media.neliti.com/media/publications/268613-the-future-of-micro-grids-in-ecuador-466df48b.pdf (accessed on 30th September 2025).
4. Ani, E. C., Olajiga, O. K., Sikhakane, Z. Q., & Olatunde, T. M. (2024). Renewable energy integration for water supply: A comparative review of African and U.S. initiatives. *Engineering Science & Technology Journal*. https://www.semanticscholar.org/paper/4d8a0ac15dd1ea9d956ce88661200b89202b2f36 (accessed on 30th September 2025).
5. Ankrah, N. A., Baffour-Awuah, K., Hammond, F., et al. (2014). The valuation tool user guide: Monetizing Cradle to Cradle®. https://core.ac.uk/download/30650393.pdf (accessed on 30th September 2025).
6. Arghandeh, R., Brown, M., Cibulka, L., et al. (2013). Distribution system voltage management and optimization for integration of renewables and electric vehicles: Research gap analysis. https://core.ac.uk/download/288432751.pdf (accessed on 30th September 2025).
7. Axon, C., Balta-Ozkan, N., Cipcigan, L., et al. (2014). Scenarios for the development of smart grids in the UK: Literature review. https://core.ac.uk/download/155776124.pdf (accessed on 30th September 2025).
8. Ballesteros, A. R., & Nakhooda, S. (2010). Investing in sustainable energy futures: Multilateral development banks' investments in energy policy. https://core.ac.uk/download/71352408.pdf (accessed on 30th September 2025).
9. Bärlund, I., Dubel, A., Kok, K., et al. (2010). Pan-European back casting exercise, enriched with regional perspective, and including a list of short-term policy options. https://core.ac.uk/download/pdf/29235489.pdf (accessed on 30th September 2025).
10. Becerra Villanueva, J. A., Calderón, B., Carvalho, M., et al. (2019). Trans critical carbon dioxide charge-discharge energy storage with integration

of solar energy. https://core.ac.uk/download/322845447.pdf (accessed on 30th September 2025).

11. Bhuyan, R., & Mytelka, L. K. (2025). Changing configuration of alternative energy systems. https://core.ac.uk/download/pdf/6941347.pdf (accessed on 30th September 2025).
12. Bood, N., Dyer, J., Fish, M., et al. (2013). An investigation of the evidence of benefits from climate compatible development.
13. Borghetti, G., Menga, G., Procaccianti, G., Kok, L., Ardito, M., Morisio, M., et al. (2013). Smart grid technologies in Europe: An overview. https://core.ac.uk/download/pdf/11431923.pdf (accessed on 30th September 2025).
14. Botterud, A., Levin, T., Mills, A. D., et al. (2020). Impacts of variable renewable energy on wholesale markets and generating assets in the United States: A review of expectations and evidence. https://core.ac.uk/download/305124749.pdf (accessed on 30th September 2025).
15. Butans, E., Nieto-Martin, J., & Zafred, K. (2016). Electric vehicles–Effects on domestic low voltage networks. https://core.ac.uk/download/74409962.pdf (accessed on 30th September 2025).
16. Cheng, D., Federico, F., Fournier, E., et al. (2020). Net solar generation potential from urban rooftops in Los Angeles. https://core.ac.uk/download/322493841.pdf (accessed on 30th September 2025).
17. Copeland, R., Crespi, N., Rehmani, M. H., et al. (2019). Internet of Things-aided smart grid: Technologies, architectures, applications, prototypes, and future research directions. http://arxiv.org/abs/1704.08977 (accessed on 30th September 2025).
18. Counsell, J. M., Haughton-Brown, C., & Tuff, S. (2009). Renewable power for lean desktops in media applications. https://core.ac.uk/download/9022822.pdf (accessed on 30th September 2025).
19. Crespo Márquez, A., Ferrero Bermejo, J., Guillén López, A. J., et al. (2019). Review and comparison of intelligent optimization modelling techniques for energy forecasting and condition-based maintenance in PV plants. https://core.ac.uk/download/288003394.pdf (accessed on 30th September 2025).
20. Deckmyn, C., Timmerman, J., Van Eetvelde, G., Vandevelde, et al. (2013). Techno-economic energy models for low carbon business parks. https://core.ac.uk/download/55683188.pdf (accessed on 30th September 2025).
21. Duan, L., Xu, J., & Zhang, R. (2014). Cost-aware green cellular networks with energy and communication cooperation. http://arxiv.org/abs/1411.4139 (accessed on 30th September 2025).

22. Fay, G., & Udovyk, N. (2011). Factors influencing success of wind-diesel hybrid systems in remote Alaska communities: Results of an informal survey. https://core.ac.uk/download/162576116.pdf (accessed on 30th September 2025).
23. Feng, W., Hao, B., Ji, H., et al. (2019). A review of net zero energy buildings in hot and humid climates: Experience learned from 34 case study buildings. https://core.ac.uk/download/323310631.pdf (accessed on 30th September 2025).
24. Fuertes, A., Goodhew, S., Wei, P., et al. (2014). Contractors' perspective on the selection of innovative sustainable technologies for achieving zero carbon retail buildings. https://core.ac.uk/download/29819173.pdf (accessed on 30th September 2025).
25. Gade, P. K. (2019). MLOps pipelines for GenAI in renewable energy: Enhancing environmental efficiency and innovation. https://core.ac.uk/download/636358611.pdf (accessed on 30th September 2025).
26. Geldermann, J., Kolbe, L. M., Krause, A., et al. (2015). Improved resource efficiency and cascading utilisation of renewable materials. https://core.ac.uk/download/151157330.pdf (accessed on 30th September 2025).
27. Gharehpetian, G. B., Marzband, M., Moradi, J., et al. (2019). Attributes of big data analytics for data-driven decision making in cyber-physical power systems. https://core.ac.uk/download/305121394.pdf (accessed on 30th September 2025).
28. Gowrisankaran, G., Samano, M., & Reynolds, S. S. (2025). Intermittency and the value of renewable energy. https://core.ac.uk/download/pdf/6496966.pdf (accessed on 30th September 2025).
29. Harvey, M., & McMeekin, A. (2010). Political shaping of transitions to biofuels in Europe, Brazil and the USA. https://core.ac.uk/download/9065438.pdf (accessed on 30th September 2025).
30. Hedberg, A. (2017). Germany's energy transition: Making it deliver. *EPC Discussion Paper, 9 October 2017.* https://core.ac.uk/download/132578636.pdf (accessed on 30th September 2025).
31. Huang, Y. W., Kammen, D. M., & Kittner, N. (2019). ASEAN grid flexibility: Preparedness for grid integration of renewable energy. https://core.ac.uk/download/323310932.pdf (accessed on 30th September 2025).
32. Kada, N., & Kiy, R. (2004). *Blurred Borders: Trans-Boundary Impacts & Solutions in the San Diego-Tijuana Border Region.* https://core.ac.uk/download/71345897.pdf (accessed on 30th September 2025).

33. Kester, J., Noel, L., Sovacool, B. K., et al. (2018). Beyond emissions and economics: Rethinking the co-benefits of Nordic electric vehicles (EVs) and vehicle-to-grid (V2G). https://core.ac.uk/download/161102773.pdf (accessed on 30th September 2025).

34. Kumar, M. S., Sen, S., & Shiva, C. (2025). Recent advances in microgrid energy management using renewables and storage devices. *2025 IEEE 1st International Conference on Smart and Sustainable Developments in Electrical Engineering (SSDEE).* https://www.semanticscholar.org/paper/e1738630a4294e85c59d3d0333b7df4099fc1f1a (accessed on 30th September 2025).

35. Lucon, O., Fransen, T., & Romeiro, V. (2015). Bridging the gap between energy and climate policies in Brazil: Policy options to reduce energy-related GHG emissions. https://core.ac.uk/download/75782972.pdf (accessed on 30th September 2025).

36. Marzband, M., Mirzaei, M. A., Mohammadi-Ivatloo, B., et al. (2019). Day-ahead network-constrained scheduling of CHP and wind-based energy systems integrated with hydrogen storage technology. https://core.ac.uk/download/199234955.pdf (accessed on 30th September 2025).

37. Millsap, R. E. (2018). The German gamble: An analysis of the Energiewende policy and its implications for energy security in Europe. https://core.ac.uk/download/233951628.pdf (accessed on 30th September 2025).

38. N/A. (2013). Understanding best practice regarding interruptible connections for wind generation: Lessons from national and international experience. https://core.ac.uk/download/42337156.pdf (accessed on 30th September 2025).

39. N/A. (2014). Obama administration FY 2015 budget proposal: Sustainable energy, buildings, transportation and climate. https://core.ac.uk/download/71365211.pdf (accessed on 30th September 2025).

40. Nicolli, F., Vona, F., & Nesta, L. (2025). Determinants of renewable energy innovation: Environmental policies vs. market regulation. https://core.ac.uk/download/pdf/6441011.pdf (accessed on 30th September 2025).

41. Ogunmoye, K. A., Agupugo, C. P., Ejichukwu, E., Barros, P., & Hayden, M. D. (2025). Protecting the grid: Cybersecurity strategies for renewable energy integration. *World Journal of Advanced Research and Reviews.* https://www.semanticscholar.org/paper/753f3ac06cf8556be51950e04367fd0c9c60d4ff (accessed on 30th September 2025).

42. Ramamurthi, P. V. (2016). Political economy of renewable energy deployment in India: Case study of Karnataka. https://core.ac.uk/download/71338519.pdf (accessed on 30th September 2025).
43. Skoulou, V. K., & Vourdoubas, J. (2016). Case study of viability of bioenergy production from landfill gas (LFG). https://core.ac.uk/download/151162915.pdf (accessed on 30th September 2025).
44. utler, L., Grubb, M., & Sinden, G. (2025). Diversity and security in UK electricity generation: The influence of low carbon objectives. https://core.ac.uk/download/pdf/7093211.pdf (accessed on 30th September 2025).
45. Van Eetvelde, G., Vandevelde, L., Vandoorn, T., et al. (2013). Renewable energy balancing with thermal grid support. https://core.ac.uk/download/55732711.pdf (accessed on 30th September 2025).
46. Yan, B. (2025). Innovative integration and optimization of RE and smart energy management systems in skyscrapers: Enhancing sustainability and energy self-sufficiency in vertical cities. *E3S Web of Conferences*. https://www.semanticscholar.org/paper/ec6edb1822ebbe698a138bc41068a62d4e421718 (accessed on 30th September 2025).
47. Zaluska, A., Liang, G., Huot, J., Shulz, R., & Bowman, R. C. (2015). The INGRID project: Development of solutions for sustainable and highly interconnected grids. https://core.ac.uk/download/55252621.pdf (accessed on 30th September 2025).

CHAPTER 7

MICROGRIDS AND DISTRIBUTED ENERGY RESOURCES (DERS)

Contents

INTRODUCTION

The transition towards a decentralized energy infrastructure has spawned immense interest in microgrids and distributed energy resources (DERs). These emerging technologies portend a paradigm shift towards the production and consumption of electricity with the aim of enhancing resilience, efficiency, and sustainability in power systems. Microgrids facilitate local energy generation and control, reducing reliance on centralized grid systems often prone to outages and volatility. This regional approach enables communities to tailor energy solutions according to their own needs, leveraging the resources to the fullest. With more and more cities getting populous and the pressure building up on climate change, embracing microgrids and DERs can prove to be crucial in upgrading the energy infrastructure and embracing green applications. The use of advanced technologies, e.g., Information and Communication Technologies (ICT), facilitates this transition by improving energy distribution and communication security in such networks (Bonetto et al., 2017). Microgrids are further differentiated by their ability to operate independently and alongside main grid systems.

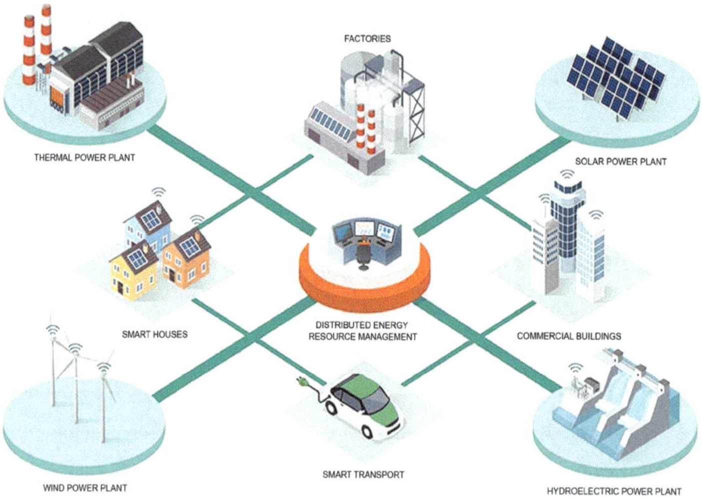

Figure 7.1. Distributed energy resource (DERs) management systems.

Source: https://www.nrel.gov/images/libraries/grid-images/distributed-energy-resource-management-systemsb4a96ab1-2f94-4167-a1a6-02321aecefa1.jpg?sfvrsn=c7f8bd6a_2.

The deployment of direct current (DC) microgrids, for instance, illustrates the increased application of power electronic converters to make processes more efficient. These systems can self-regulate and adapt to varying working conditions without the need for external communication networks. Based on local bus voltage information, decentralized methods make it possible for microgrids to balance generation capacities and load requirements with high efficiency. The new framework is particularly useful in addressing the Optimal Economic Dispatch problem, which is paramount to balancing supply and demand at reduced costs (Angjelichinoski et al., 2018). As microgrids gain popularity, it becomes essential to examine their function mechanisms and how they contribute to forging a resilient energy future. The role of the consumer cannot be overemphasized as the energy industry continues to develop. Prosumer communities, who both produce and consume energy at the same time, have a key role in maximizing energy utilization and encouraging demand response programs. By fostering participation and behavioral changes towards energy consumption, these communities are able to actually complement the performance of microgrids. The use of big data analytics for demand response planning also complemented this trend so that solutions can be tailored to synchronize the supply of energy with consumer behavior. Education and awareness programs can guarantee energy conservation measures, actively involving passive consumers in the energy economy. Lastly, the growth of microgrids and DERs also underscores the significance of decentralized systems but further emphasizes the collective responsibility of all stakeholders to play an active role in this energy transformation (Figure 7.1).

Definition of Microgrids

Microgrids are a significant innovation in energy distribution systems defined by their capability to operate independently or in the main grid. Generally described as localized DER networks, they consist of various power generation units such as solar panels, wind turbines, and batteries. Microgrids enhance energy resilience and reliability, particularly in power outage-susceptible areas or areas with high energy security requirements. By including several sources of power within a smaller spatial area, microgrids have the ability to manage energy loads in an efficient manner, optimize sources of generation, and improve sustainability through minimizing the use of fossil fuels, thus being of vital importance to the future energy infrastructure. The operation dynamics in microgrids are highly dependent on advanced control mechanisms, particularly the droop control technique. This system permits several energy sources to share a load proportionally along with controlling the voltage of the distribution bus. As discussed in subsequent studies, systems based on droop control can seamlessly switch between a mode of operation and another, being connected to a larger grid or being isolated in islanded mode. This flexibility not only enhances the resilience of microgrids but also aids in the integration of various energy sources, adapting to altering patterns of generation and load demand. Such types of capabilities become even more crucial with growing demands for sustainable energy solutions, rendering microgrids a critical piece of the contemporary energy systems landscape (Caldognetto et al., 2018). Additional advancements in control technologies, especially in decentralized systems, are crucial to the optimal operation of microgrids. Independent controllers utilizing local bus voltage measurements can make optimal decisions autonomously without relying on external communication networks. Decentralization can allow microgrids to estimate generation capacities, load demands, and distribution line conductance's more accurately and with faster responsiveness. Research shows that these innovations enhance microgrids' operating characteristics significantly, leading to ideal economic dispatch solutions with improved supply and demand matching (Angjelichinoski et al., 2018). The contribution of microgrids to enabling more sustainable, resilient, and decentralized energy systems becomes more important as development in the field of microgrids continues to progress, thus highlighting their role within the broader context of energy production and use (Figure 7.2).

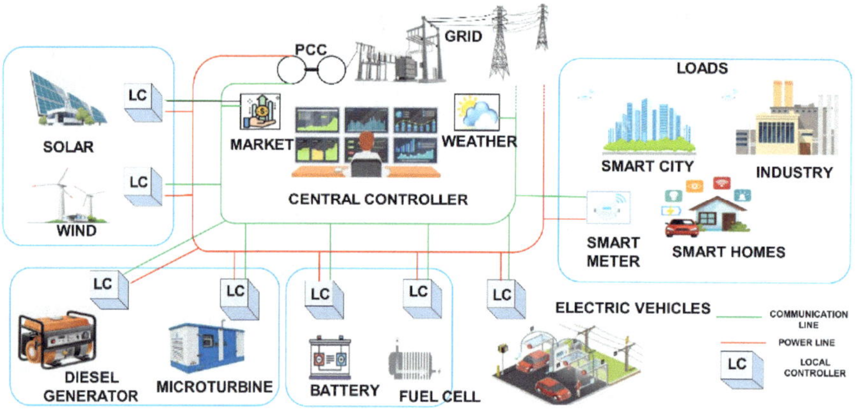

Figure 7.2. Structure of a microgrid.

Source: https://www.mdpi.com/sustainability/sustainability-15-06366/article_deploy/html/images/sustainability-15-06366-g001.png.

Overview of Distributed Energy Resources (DERs)

DERs entail a wide array of technologies that generate or store electricity at the site where it is consumed, representing a significant divergence from the traditional centralized power generation system. DERs entail solar panels, wind turbines, energy storage batteries, and combined heat and power systems among others, whose total advantage is improved system efficiency and environmental benefit. The incorporation of DERs into the power grid can facilitate not only increased resilience and reliability of energy supply but also facilitates the transition towards renewable energy resources. For instance, where there are recurring power outages, DERs can be installed to facilitate local power generation and thereby increase energy security and reduce loss due to long-distance transmission. Therefore, the integration of DERs is now seen as critical to preserving energy resiliency and independence in rural and urban regions. Sizing and optimization of DERs in microgrids have become central concerns in enhancing their effectiveness and reliability. Various sizing techniques, such as the load-profile based sizing method (LPBSM), have been developed to calculate and optimize DER capacities. This innovative approach utilizes real load data in conjunction with relevant electricity tariffs to maximize the overall performance of microgrids in a manner that makes the system closely align with the local energy requirement. LPBSM not only resolves operational issues but also minimizes consumers' costs by integrating an optimal mix of renewable energy sources.

Figure 7.3. Distributed energy resources (DERs).

Source: https://www.cummins.com/sites/default/files/styles/newsroom_hero_image/public/images/newsroom_article/windfarm.jpg?itok=0C9pCGSr.

Current studies emphasize the significance of such approaches, as they make it easier to deploy renewable resources in microgrids to tackle energy demands holistically while optimizing economic and environmental benefits (Jansen et al., 2025). Control and management of DERs in microgrids are complex interactions and advanced technologies for guaranteeing power quality and reliability. One of the key problems faced by microgrids working in islanded modes is generator output control and reactive power balance, which result in voltage oscillations and power quality issues if not managed. To combat such problems, technologies like adaptive virtual impedance droop control (AVIDC) have been researched that enable effective power sharing and voltage regulation among dispersed resources. This decentralized approach allows self-sustaining operation with assured strong coordination between DERs. Moreover, utilization of multiagent systems (MAS) enhances the capability of microgrids through facilitating individual devices to communicate and respond to real-time situations, optimize their operational status, and enhance overall grid stability (Qashqaie et al., 2025). Thus, effective control strategies play an indispensable role in unleashing the full potential of DERs in current energy networks (Figure 7.3).

Importance of Microgrids in Modern Energy Systems

Integration of microgrids in modern energy systems is a revolution towards more efficient and robust control of energy. Microgrids, founded on localized sources of energy and advanced information technologies, facilitate a decentralized approach to electricity supply. This system not only makes energy efficiency higher but also allows greater flexibility to counter swings in energy supply and demand. As cities grow and grapple with increasing energy needs, adopting microgrids can take the pressure off the central grid infrastructures. This change assists the transition to renewable energy sources, aligning with global sustainability goals aimed at reducing carbon emissions and promoting clean energy solutions. Microgrids also play a vital role in improving energy security by allowing communities to be self-contained during power grid failures or other difficult situations, thereby safeguarding key services and infrastructure (Uddin et al., 2023). The dynamic behavior of DERs and microgrids further positions microgrids as the key element in modern energy systems. As the share of intermittent sources such as wind and solar is increased, a comprehensive approach to controlling energy is required. Microgrids, with advanced optimization techniques at hand, can manage such variables optimally in order to offer a stable source of energy irrespective of changes in the level of generation. Current studies highlight the necessity for sophisticated energy management technologies in addressing such challenges, discovering that mixed-integer programming approaches have emerged as dominant methods in microgrid operation optimization. By leveraging the capabilities of DERs and advanced optimization techniques, microgrids enhance system reliability while minimizing operation costs, thereby enabling a sustainable energy environment with the ability to adapt to diverse community needs (Thirunavukkarasu et al., 2022). Besides, microgrids are also significant because of their long-term socio-economic contributions. They create job opportunities in the development and maintenance of renewable energy facilities through the facilitation of localized energy production and consumption. Further, microgrids promote community empowerment by allowing decision-making within the local community for energy production, leading to ownership and compliance with sustainable activity. The financing framework that provides for microgrids facilitates investment in renewable sources and efficiency technology that facilitates energy independence. To this extent, microgrids are not only a platform for integrating distributed energy resources but also a path towards a more equitable energy system. Their prospects in enabling resilience, efficiency, and sustainability make microgrids a cornerstone of modern energy solutions in the face of emerging global challenges (Uddin et al., 2023) (Figure 7.4).

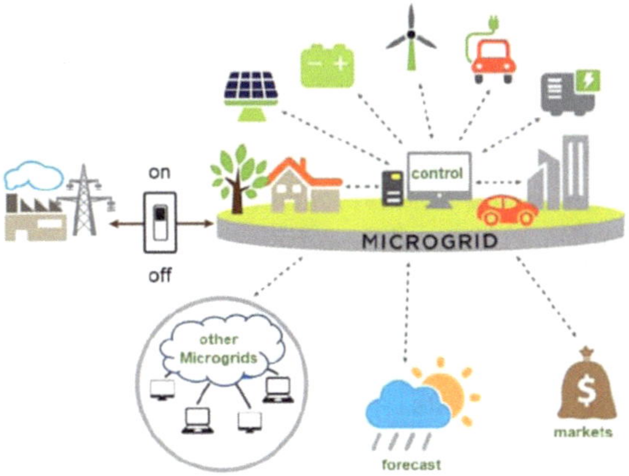

Figure 7.4. A microgrid is comprised of an interconnected network of load clusters and distributed energy resources (DERs). These nodes within the network have defined electrical boundaries. The microgrid can be connected to the grid or operate independently in island mode.

Source: https://eepower.com/uploads/articles/understanding-the-role-of-microgrids-in-the-energy-transition-fig1.jpg.

7.1. HISTORICAL CONTEXT

The evolution of microgrids and distributed energy resources (DERs) has its roots in the history of the evolution of the energy system over the last century. To start with, centralized power generation through fossil fuels was prevailing in the energy landscape, which was providing efficient and secure electricity supply to urban areas. However, with growing environmental awareness in the late 20th century, the frailties of traditional grids became apparent, particularly with regard to sustainability and resilience of the energy supply. Situations such as the 1973 oil crisis generated interest in alternative energy sources and stimulated investment in renewable technologies. Besides, international summits like Stockholm and Rio advocated for sustainable development, shaping policy towards decentralized systems. This set the stage for microgrids and DERs as realistic solutions to address energy efficiency and sustainability, shaping how energy is generated, distributed, and consumed globally. When the 21st century began, the imperative to address climate change accelerated the shift towards more sustainable methods of providing energy. The issue of greenhouse emissions from conventional electricity generation, which was a great source of global pollution, prompted stronger policy actions towards reducing the reliance on fossil fuels. The cumulative impacts of global climate agreements like the Paris Agreement emphasized that nations ought to build sustainable energy transition strategies that focus on renewable energy. The main components of this transition were the inclusion of microgrid systems, which aid in local energy management, optimized energy efficiency, and increased resilience during natural calamities. These configurations give power to communities as they allow for the integration of renewable energy sources alongside minimizing the environmental impacts associated with traditional energy systems (Kabeyi et al., 2022). Microgrids therefore became a significant feature in the global pursuit towards sustainable energy systems. The technological advancements have played the key role in the progression of microgrids and DERs through history. Restricted by technology and cost in the past, newer developments in energy storage, smart grid systems, and demand response infrastructure have made stronger and more resilient energy networks possible. Past energy outages, for instance, caused by Hurricane Sandy in 2012, exposed the vulnerability of centralized grids, driving communities to embrace microgrid technologies for greater energy self-sufficiency and sustainability. Research and development investments in microgrid technology exploded, leading to effective models for harnessing already available renewable resources while maintaining grid resilience. Besides, the increased prevalence of information technology has made possible a paradigm shift towards more decentralized, community-based energy systems, which is a general trend in history toward democratization of energy access and allowing global climate stabilization through sustainable approaches (Gad, 2022).

7.1.1. Evolution of Energy Systems

Evolution of energy systems is organically linked with an increasing demand for efficient and reliable power delivery, which has altered the face of conventional grids. Historically, centralized power production employed fossil fuels with their accompanying environmental concerns and heightened regulatory scrutiny. The introduction of distributed energy resources (DERs) came along with a paradigm change with localized power production employing renewable sources such as solar panels and windmills. Decentralization not only minimizes the green effects of fossil fuels but also enhances energy security and resilience. Consequently, when grid operators integrate DERs into the core of operations, they face new challenges, particularly managing the dynamic relationships between demand

and supply. This shift illustrates the main evolution of power systems, where microgrids are a strategic way to capitalize on DERs as well as reliability and performance. With the growing need for renewable energy, microgrids enhance the resilience of energy systems on multiple fronts. Microgrids encourage local energy generation and consumption, which reduces the burden on conventional energy grids that are inefficient and vulnerable to outages.

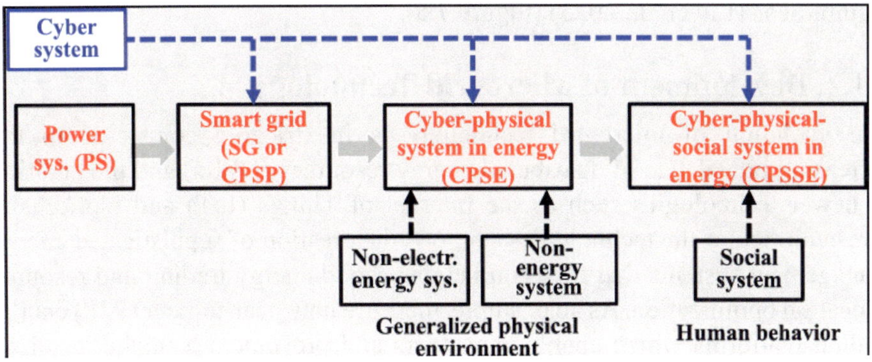

Figure 7.5. Evolution of energy systems.

Source:https://www.researchgate.net/profile/Yusheng-Xue/publication/321436277/figure/fig5/AS:581988169465856@1515768200121/Evolution-of-energy-systems.png.

Microgrids allow communities to tap renewable energy sources, which reduces transmission losses by half and enhances energy autonomy. But with the employment of advanced technologies and remote monitoring functionalities, there is also a concern for cybersecurity. The application of remotely controlled grid infrastructure enhances operational efficiency but also introduces potential vulnerabilities that would be prone to hacking by saboteurs. The mitigation of these challenges requires the establishment of robust cybersecurity measures and employing measures that facilitate the continuity of electricity supply even during potential cyberattacks, emphasizing the significance of energy system resilience in the transition(Topallaj et al., 2025). The significance of novel optimization techniques is at the core of facilitating the transition of microgrids and DERs towards sustainable solutions. Recent studies have also considered hybrid optimization methods for addressing complex scheduling issues inherent in microgrid operation, such as the DE-HHO algorithm, that offer significant cost savings and system stability because of their efficiency-focused approach. The approach not only addresses operational costs but also long-term environmental impacts, therefore making economic viability in line with ecological stewardship. With huge simulations, the performance of such algorithms demonstrates their capability to optimize diverse sources of energy including wind, solar, and micro-gas turbines, confirming the flexibility of the systems. With the ongoing transition of energy systems, what will be gleaned from microgrid optimization is invaluable, guaranteeing the future where distribution of energy will not only be efficient but sustainable, robust, and capable of addressing the worlds ever-growing demanding energy requirements(Liu et al., 2025) (Figure 7.5).

7.1.2. Development of Microgrid Technology

The evolution of microgrid technology is inextricably connected to the increasing prevalence of distributed energy resources (DERs) and the addition of newer technologies such as the Internet of Things (IoT) and blockchain. The evolution of the technologies supports the creation of sophisticated energy management systems that focus on neighborhood energy trading and resource allocation optimization. As an example, there are new peer-to-peer (P2P) energy trading platforms which enable consumers and prosumers to make localized exchanges independently. These platforms tap the use of IoT devices to collect detailed energy consumption information and utilize blockchain frameworks for ensuring secure, transparent trading experiences. Development of such systems is a testament to the potential of microgrids in facilitating efficient distribution of energy, boosting resilience, and accruing benefits to all players in the energy sector, particularly amid crisis or peak hours (Condon et al., 2023). Besides innovation in technology, the effectiveness of microgrid operation also relies on sophisticated demand-side management techniques. These methods aim to enhance the performance of microgrids with heterogeneous DERs by employing cost-based strategies and price variability models. Aggregators are central to this because they must take into account the inherent nature and levelized cost of energy (LCOE) of each type of DER, such as diesel generators, solar panels, and wind turbines. By triggering local electricity markets with re-priced electricity, aggregators are able to encourage energy usage patterns that align with the operational capacity of each resource, creating more efficient energy usage and the transition toward non-traditional, decentralized energy systems. This optimization is key to the real-world implementation of microgrids and their ability to operate in autonomous or parallel modes (Denysiuk et al., 2023). The successful rollout of microgrid technology also requires the tackling of infrastructural and regulatory challenges that occur with the integration of heterogeneous DERs. Policymakers, utilities, and technology developers need to collaborate and create enabling regulatory frameworks that facilitate the rollout and scaling of microgrid systems. Moreover, the greater reliance of the world on renewable sources of energy, the greater becomes the concern for resilience against natural disasters and disruptions. Combining smart grid technologies and battery storage systems into microgrid design enhances not only reliability but also local energy independence levels. Ultimately, with further development of microgrid technologies, they can transform the energy system into a decentralized, efficient, and sustainable model, challenging the sources of climate change and energy security threats directly (Condon et al., 2023; Denysiuk et al., 2023).

7.1.3. Key Milestones in Distributed Energy Resource (DER) Adoption

The transition towards DERs has been characterized by several key milestones that have informed their adoption. The initial years witnessed the enactment of policy support environments, which established the framework for the integration of renewable energy sources into conventional grids. Particularly, the innovation of advanced technologies such as photovoltaic solar cells and wind turbines deliberately fueled this transition, lowering costs and enhancing availability of clean energy. These technologies have compelled consumers and utilities to view microgrid systems as viable substitutes for traditional power production. By providing local energy solutions, microgrids enhance grid resilience against outages and provide economic benefits through reduced transmission losses and optimized energy efficiency. These developments underscore the necessity for regulatory support in developing a framework that is conducive to innovation and investment in clean technologies, leading to universal use of DERs in energy systems (Ibekwe et al., 2024). Another breakthrough in the use of DERs occurred with the focus on energy security and sustainability. With increasing global concern towards climate change, the majority of regions have come to value the environmental advantage of making a transition to decentralized energy systems. Policies for greenhouse gas abatement and diversification encouraged investment in DER technologies. A case in point is the United Arab Emirates, which has made significant strides in adopting this by revolutionizing its National Energy Strategy, doubling its goal for its renewable energy capacity by 2030 (Alnaqbi et al., 2023). Similarly, other nations have implemented similar objectives and launched schemes to promote renewable energy installations. These policy reforms not only enhance energy security but also enable the integration of various renewable sources in the grid. This has resulted in increased focus on the economic and environmental sustainability of energy systems, which has also helped establish DERs as a central aspect of modern-day energy policies. The culmination of these milestones is experienced in the growing complexity of microgrid and DER infrastructure, which currently utilizes sophisticated technologies and business models. As utilities and customers become increasingly aware of the benefits of decentralized energy production, technological advancements in energy storage, smart grid equipment, and demand response schemes have become imperative. Such technologies improve the reliability and efficiency of energy delivery and allow for more flexibility in managing energy. Moreover, the formation of partnerships between public and private organizations has accelerated innovation and best practices development for the implementation of DER. Collaborative projects that allow for learning between parties, knowledge transfer, and facilitation assist in dispelling existing barriers to wider application. Ultimately, the success of DERs hinges on ongoing investment, research, and the ability of stakeholders to innovate in accordance with the changing dynamics of energy, hence microgrids are a game-changing power in the power industry (Ibekwe et al., 2024).

7.2. COMPONENTS OF MICROGRIDS (MGS)

The design of MGs includes several key components to aid successful energy management and distribution. The core of such systems are distributed energy resources (DERs), or renewable energy technologies which encompass solar panels, wind turbines, and bioenergy systems. They are crucial for clean energy generation, reducing the consumption of fossil fuels, and promoting local power systems to be more sustainable. Aside from the generation capability, energy storage systems (ESSs) are also crucial in the functioning of a microgrid as they store excess energy generated during peak generation hours. With the inclusion of batteries or other storage solutions, microgrids can balance the energy supply and provide a constant power supply when there are variations in demand or in generation, thus allowing for uninterrupted supply and increased resilience against failure. Control strategies are also a vital microgrid component, ensuring proper operation and coordination between various energy sources and storage units. Advanced control methods enable real-time monitoring and management to facilitate efficient exchange of energy between DERs and the local loads or utility grid. Implementation of intelligent algorithms, such as those based on Fuzzy Logic Controllers (FLCs) and Genetic Algorithms (GAs), provides dynamic optimization against operating conditions and energy requirements. The control systems ensure efficient operation through minimizing energy loss and energy purchasing cost. Further, they improve system stability along with mitigating harmonic distortion and reactive power management challenges to facilitate optimal microgrid operation in standalone and connected modes, as discussed in Balikowa et al. (2024). For the effective implementation of microgrids, various challenges must be minimized, particularly in infrastructure, regulation, and technology. High initial capital costs are a barrier to wide penetration, typically necessitating innovative financing devices and enabling policies from regulatory bodies. Moreover, regulatory frameworks must change to accommodate the unique operating characteristics of microgrids and facilitate their entwinement with the totality of the energy ecosystem. To boot, ongoing research in microgrid design and engineering will assist in curbing present technical challenges, including enhancing system interoperability and offering reliable protection schemes. With new technologies emerging on a daily basis, the position of microgrids in providing decentralized energy solutions, enhancing efficiency, and promoting integration of renewable energy becomes clear, making them major players in the future of the energy system (Adegboyega et al., 2025) (Figure 7.6).

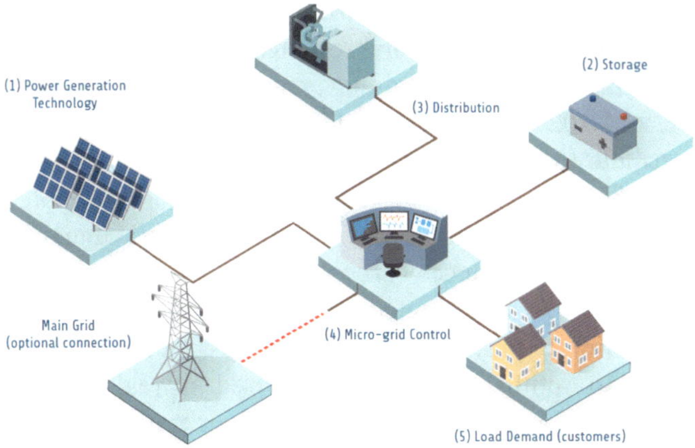

Figure 7.6. Micro-grid (MGs) components.

Source: https://www.researchgate.net/profile/Aran-Eales/publication/334277511/figure/fig1/AS:777682371485696@1562425334175/Micro-grid-components-Adapted-from-5.ppm.

7.2.1. Energy Generation Sources

The landscape of energy generation sources has been completely transformed by the integration of distributed energy resources (DERs) into microgrids. Microgrids are localized networks through which numerous energy generation technologies can be harnessed, like solar photovoltaic (PV) panels, wind turbines (WTs), and battery energy storage systems (BESSs). Microgrids utilize these technologies to formulate a decentralized approach towards power generation that enhances resilience and efficiency. Efficient functioning of such microgrids requires advanced optimization techniques to address the intricacies of various energy sources with reliability as well as cost-cutting. Continued advancements in smart grid technologies and control systems also further foster DER integration, resulting in better microgrid performance and reliability as a whole (Dr. Dixit et al., 2024). Microgrids are increasingly being praised for assuring energy security and sustainability, particularly in situations involving environmental threats and natural disasters. Traditional centralized power grids have long been vulnerable to extreme weather disturbances, leading to widespread outages and impacting community reliability. Microgrids address this by using prosumer models where consumers are also producers of energy on a local level, thus developing integrated networks that are more resilient in cases of crises. This model of operation allows communities to generate and supply electricity on their own, even from the national grid. Moreover, battery electric vehicles are among the on-board energy storage options that help in the augmentation of the flexibility and responsiveness of microgrid systems to meet changing demand while maintaining essential services in times of hardship (Thirumalai et al., 2024). The role of DERs is not only to produce energy; they also contribute to environmental sustainability and reducing greenhouse gas emissions. The transition of fossil fuel sources to renewable technology in microgrids is a milestone in cleaner energy strategies. Integration of solar and wind power not only reduces carbon footprints but also fosters a circular economy in which energy is generated close to the site of consumption, thereby reducing transmission losses. In addition, demand-side management strategies allow microgrids to optimize energy usage based on real-time data analysis. This forward-thinking approach not only makes operation more efficient but also enables communities to take ownership of their energy future, a step imperative in constructing a sustainable, resilient energy system (Dr. Dixit et al., 2024) (Figure 7.7).

Figure 7.7. Over 89m Kwh of electricity generated from renewable sources in a month.

Source: https://media.tehrantimes.com/d/t/2021/04/10/4/3738205.jpg.

7.2.2. Energy Storage Systems (ESS)

The advent of ESS has transformed the world of DERs and microgrids with the aspect of reliability and efficiency in power supply. These systems work as integral components that can accumulate excess energy generated from renewable resources such as solar panels and wind turbines and release it during peak hours. By facilitating energy management and mitigating supply and demand swings, ESS both enhances microgrid operational efficiency and stabilizes grids. Moreover, with the application of Battery Energy Storage Systems (BESS), there is a guarantee of local energy resilience, with communities having the capability to deliver power during outages or severe weather. This capability is significant since electricity grids are vulnerable to attack from different risks, making it important to have robust storage solutions during the shift towards renewable energy systems (Thirumalai et al., 2024).

Figure 7.8. Energy storage systems (ESS).

Source: https://assets.infineon.com/is/image/infineon/energy-storage-systems.png:asset-880-2x1?qlt=55,0&resMode=sharp2&op_usm=1.75,0.3,2,0.

The implementation of smart Energy Storage Management Systems (ESMS) also significantly enhances the efficiency level in applying energy storage in microgrids. These advanced systems leverage sophisticated algorithms to analyze energy consumption patterns and predict future demand, thereby maximizing the control of DERs. Through dynamic pricing schemes and enhanced load management, intelligent ESMS can enhance energy efficiency as demonstrated by research indicating an 11.7% enhancement in efficiency for solar generation management and reduced grid fluctuations by a significant margin (Guo et al., 2024). The capacity of such systems to handle variable renewable energy is crucial in ensuring stable power supply and reliability. Consequently, the incorporation of smart ESMS in microgrids has a crucial role in both enhanced energy management and the cost-effectiveness of DERs. The role of prosumers—individuals who generate and consume energy—brings added weight to the transformative role of ESS in microgrids. By generating electricity from decentralized sources and selling surplus power to and purchasing from their neighbors, prosumers maintain a decentralized energy system that enhances local resilience. Incorporating Electric Vehicle Batteries (BEVs) as movable energy storage units during crises demonstrates how multi-purpose energy storage systems can be employed to act on energy security requirements. Because prosumers are directly engaged in load management and demand response practices, they ensure operational continuity for critical services, particularly during unfavorable weather conditions. All these advancements of energy storage technology not only make integration of renewable sources of energy possible but also allow communities to become active stakeholders of their energy futures (Thirumalai et al., 2024) (Figure 7.8).

7.2.3. Control and Management Systems

Proper coordination of control and management systems within microgrids plays a vital role towards enhancing their functionality as well as their reliability. The center for this coordination resides in employing sophisticated controller architectures that facilitate smooth mode switching between operation. For instance, a local power-based droop controller for use in distributed energy resource (DER) converters facilitates power flow control during normal operation and bus voltage regulation in instances of upstream grid faults. This dual capability enables microgrid systems to switch automatically between power flow control and droop control without relying on complex detection schemes or communication with other components. This adaptability is crucial in maintaining the stability and efficiency of microgrids, particularly as they increasingly comprise a combination of renewable energy sources, therefore high-capacity control strategies are needed to support their long-term operation under dynamic conditions (Atherton et al., 1997). As there is greater reliance on renewable energy sources, the operational and scheduling patterns become more complicated. The coordination of interconnected microgrids managed by a Distribution System Operator (DSO) can go a long way in terms of improving operation efficiency as well as reliability. A stochastic method using the Unscented Transform technique addresses the inherent uncertainty of variables such as market prices, load demand, and renewable power generation. By application of these inexact parameters, the model enables optimization of resource scheduling, which is important in terms of scheduling the economic aspect of interdependent microgrids. The employment of the newly developed linearization techniques on the AC power flow equations enables one to develop a more manageable formulation of such intricate systems, which can be used as a point of departure to establish optimal operating decisions from.

Lastly, these control and management system advancements are essential to facilitate scalable and sustainable microgrid alternatives (Dabbaghjamanesh et al., 2018). Monitoring and optimizing the performance of microgrids through effective control systems also has significant effects on energy management. The implementation of these control strategies, particularly in situations of complete penetration of renewables, allows for improved sharing and allocation of power produced between microgrid units. Comparative analysis of proposals for power-based droop control compared to conventional droop control indicates noteworthy performance improvements, particularly under conditions of changing energy output from DERs. Not only does this shift towards employing power-based techniques enhance the reliability of isolated converters, but overall microgrid performance is also maximized, allowing it to be resilient against disturbances. By employing autonomous control systems that conform to dynamic real-time fluctuations of energy production and usage, microgrids have a better chance of maintaining operational integrity and service quality in the midst of the wild nature of renewable energy supply sources, thus forming a more stable energy climate (Atherton et al., 1997; Dabbaghjamanesh et al., 2018).

7.3. BENEFITS OF MICROGRIDS

Microgrids have many more benefits than the factor of mitigating load from old power systems; they also enhance robustness and operation efficiency. As natural disasters and weather extremes continue to threaten conventional power grids, the distributed nature of microgrids guarantees controlled and managed use of local energy resources. This is particularly the case in areas with past outages, where microgrids are able to operate independently, providing power even during extended grid failures. Utilizing solar and wind energy as renewable sources of power, microgrids not only reduce dependence on centralized power supply but also guarantee sustainability. The shift toward decentralized energy management is a reflection of a holistic approach towards energy production and consumption, promoting the involvement of people and cultivating a culture of collaboration and sharing of energy between local consumers (Babu et al., 2023). Apart from that, implementation of Peer-to-Peer (P2P) energy sharing between microgrids makes them highly efficient. By allowing users in a microgrid to share the excess energy, P2P systems ensure optimal resource utilization and zero energy wastage. This interoperability between microgrid actors is also reflected in tangible cost savings, making energy more affordable to consumers. The efficiency derived from P2P networks is particularly useful in areas of high density where energy demand fluctuates wildly, enabling users to tap into local energy reservoirs rather than being reliant on central grids. These dynamics not only maximize the utilization of resources but also maximize social connections by drawing communities into a shared participation in energy management, further endorsing the reality of microgrids being essential in the transformation towards a more sustainable and robust energy infrastructure (Babu et al., 2023). In addition to operational advantages, regulatory aspects of microgrids call for more concern, especially in deregulated energy markets of states. The case of the Pittsburgh energy district is a perfect example of the complexity associated with microgrid applications on a mass scale. The interaction between public utilities, government agencies, and private entities is a task in itself in regulating microgrid technology with prudence. During development, one must know about the impact of deregulation on microgrid projects. A well-defined framework can help spur innovation and facilitate microgrids to be seamlessly connected into the existing infrastructure, utilizing their ability to boost reliability and efficiency in the energy system. Addressing these regulatory issues in advance allows for microgrids to realize their maximum potential in enhancing resilience and keeping pace with shifting energy demands of communities (Figure 7.9).

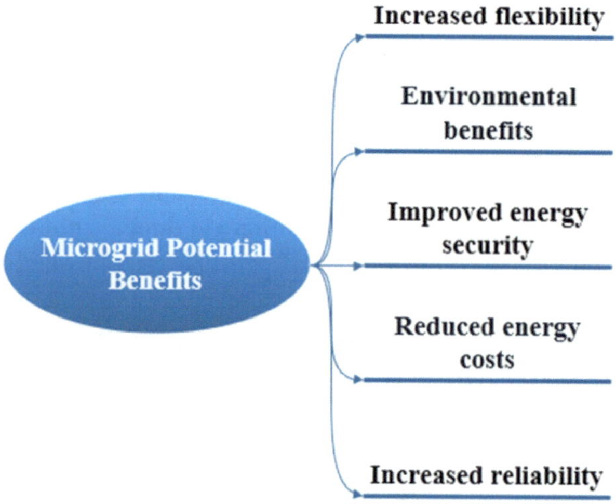

Figure 7.9. Microgrids' potential benefits.

Source: https://www.mdpi.com/sustainability/sustainability-15-06366/article_deploy/html/images/sustainability-15-06366-g002.png.

7.3.1. Enhanced Energy Resilience

The conversation of greater energy resilience more and more highlights the significance of microgrids in modern energy systems. Centralized energy infrastructure, based on its reliance on long-range transmission lines and behemoth power plants, is vulnerable to collapse by its nature. Microgrids, on the other hand, facilitated by Distributed Energy Resources (DERs), provide a solution to de-centralize that enhances resilience by allowing easier local management of energy.

Figure 7.10. Energy reliability and resilience.

Source: https://www.energy.gov/sites/default/files/2023-01/shutter-stock_570200689_0.jpg.

This strategic move allows communities to access renewable energy sources while lowering their dependence on fossil fuel, hence enhancing reliability and sustainability. Microgrids' flexibility can assist them to operate independently or in parallel with the primary grid, quickly reacting to outages or demand fluctuations, a key factor towards the availability of a reliable supply of power in the face of environmental stresses and infrastructure vulnerabilities (Kaal et al., 2025). Improvement of microgrid performance through state-of-the-art technologies such as Deep Reinforcement Learning (DRL) is a major factor in energy resilience. DRL enables intelligent management of various energy sources within a microgrid, with dynamic control of energy outputs and battery management. Through interaction with the energy infrastructure and learning to optimize energy distribution within microgrid components, DRL-based systems can maximize energy distribution efficiency at reduced costs and ensure stable power supply. Besides, the incorporation of Graph Neural Networks (GNNs) makes such smart systems responsive to varying network structures, with a robust framework that can adapt to complex microgrid operations under dynamic conditions. Such incorporation of DRL with GNNs is a giant step towards maximum energy resilience against rising energy demands and infrastructural challenges in the world's energy system (Kaal et al., 2025). With the technological innovations, addressing operational challenges related to inverter failure in microgrids is crucial in achieving enhanced energy resilience. DERs, which are typically interfaced through power electronic inverters, are essential for microgrid function but equally pose complications that need to be handled suitably. Recent advancements in data mining techniques have proven to be invaluable in the effective real-time detection and location of inverter faults, which are critical for maintaining microgrid operation integrity. By utilizing artificial intelligence methods, such as FaultNet-ML, facilitates a sophisticated method of distinguishing between system anomalies and real faults. This pre-fault management enhances the resilience of the microgrids, such that any perturbations can be corrected at once, avert broader grid failure, and enable long-term power availability in dynamic energy systems (Figure 7.10).

7.3.2. Environmental Impact and Sustainability

The embedding of Distributed Energy Resources (DERs) and microgrids in urban cities pays a high contribution to environmental sustainability through the enhancement of energy resilience as well as the reduction of greenhouse gas emissions. With towns in increasingly growing recognition of their pivotal position to fight climate change, microgrids create a state-of-the-art solution towards facilitating localized production and consumption of energy. Through this method not only is reliance on central fossil fuel-based power networks minimized, but also towns are able to incorporate their ambitious agendas of measures to fight climate change. Research has attested that combative urban strategies have the ability to decrease climate emissions by great amounts, hence proving the worth of local efforts towards the broader agenda of U.S. climate action plans. Improved energy infrastructure, like that facilitated through microgrid technology, is essential to sustaining and averting climate emissions that continue to be a challenge for many cities in the United States (Cleveland et al., 2019). The utilization of microgrids also addresses the imperative requirement of upgrading the energy infrastructure in the wake of increasing natural disasters and climate change complexities. Microgrids have the potential to provide enhanced reliability and efficiency in the supply of electricity, which is particularly significant as extreme weather conditions continue to increase in intensity and frequency. By establishing microgrid pilot projects in various states, including New York and California, cities such as Pittsburgh are leading the way to formulate advanced energy districts. These districts are designed to be implemented as strong infrastructure with the ability to address energy needs in a proficient manner and drive local systems sustainably. These broad implementations pose key regulatory concerns, particularly in deregulated energy environments, to highlight the potential of public and private partnerships toward advancing these innovative initiatives (Parks et al., 2019). Beyond the value of sustainability of providing energy, microgrids also possess economic value and social justice. By encouraging local generation of power, communities are able to access employment opportunities in the renewable energy sector, realizing economic sustainability. Besides, microgrids can empower disadvantaged communities by offering access to an assured source of energy, thus attaining energy equity. Urban areas that invest in microgrids and DERs not only enhance their own resilience against power outages but also contribute to social and economic well-being for all their residents. While cities are grappling with twin challenges of climate change and energy equity, the embracement of microgrid technologies takes center stage as a central strategy that acts towards augmenting environmental footprint and community resilience. As technology evolves, such systems will continue to be central in fostering sustainable future cities (Cleveland et al., 2019).

7.3.3. Economic Advantages for Communities

DER and microgrid integration has significant economic advantages to communities, particularly in reducing the energy cost and development of the local economy. Local generation, using such systems, minimizes transmission loss and reliance on the remote energy markets, thereby stabilizing energy price. Microgrid communities enjoy access to renewable sources such as wind and solar, whose operating cost is generally lower than the conventional fossil fuels. Additionally, the system is particularly useful during peak periods, providing residents lower rates and stable energy bills, which improve local business growth and job creation. The shift towards localized energy production also encourages investment in infrastructure benefitting the community directly, thereby directing financial resources back into local economies as well as contributing to general resistance against energy price volatility. Beyond saving money, communities are also benefited by the added energy independence derived from using microgrids and DERs. These infrastructure systems offer local control of the energy resources, which makes it easier to be resilient to external shocks, such as interruption of energy supply because of natural disasters. For instance, microgrids are capable of operating in isolation, or island, during grid outages, with essential services such as hospitals and emergency services still accessing electricity. Such increased resilience not only contributes to public health and safety but also enables economic stability in the community. Moreover, as communities become more energy independent, they become more attractive to investors and businesses looking for stable settings, which in turn creates windows of opportunity for economic development. By reducing exposure to unreliable global energy markets, microgrids enable short-term economic benefit along with long-term opportunities for growth. Maximizing the efficiency of operating microgrids is key to realizing the economic potential of microgrids. As identified in current studies, effective utilization of the microgrid resources allows for maximum efficiency and reliability and has a direct contribution towards economic returns for populations. Novel methods such as hybrid mode control and energy management systems that incorporate demand response capabilities offer a platform for dynamic optimization of energy generation and use based on local needs. These steps enhance the performance of the microgrid while aiding in cost-saving energy consumption, further reducing costs for community dwellers and businesses. By adopting such state-of-the-art control techniques, communities can achieve the highest economic benefits obtained from their on-site power resources and establish an environment in which sustainability and economic growth exist together, finally demonstrating the combined advantages of implementing microgrids and DERs in community networks (Haque et al., 2024; Yuvaraj et al., 2025).

7.4. CHALLENGES IN IMPLEMENTATION

With increasing emphasis on microgrid integration and DERs in energy policies, some challenges are apparent in their use. Uncertainty regarding regulations that involve these technologies, particularly in deregulated energy markets, is one of the prominent issues. In Pennsylvania, the lack of adequate statutory provisions for microgrids makes their implementation difficult, and stakeholders are unsure about how to classify, own, or assign roles as between public utility and private actor. This uncertainty has the potential to discourage investment and innovation, slowing down the development of microgrids that can enhance resilience and efficiency in local energy networks. When cities are seeking to implement microgrid projects, regulatory frameworks must be reformed so that they provide clarity and support for the implementation of these initiatives so that they flourish without obstruction from ineffectual or old legislative policies (Parks et al., 2019). Additionally, technological challenges pose a significant setback in the successful implementation of microgrids. With the utilization of complex data analytics in running and maintaining microgrids, complexities arise that must be addressed. The new field of big data analysis is critical in achieving the vast amounts of data generated by microgrid operations. However, stakeholders are typically faced with difficulties such as the need for sophisticated computational power and expertise to derive meaningful information from this data. The traditional analytical methods can fail due to the inability to handle the high complexity and variability inherent in new energy systems, thereby leaving space to integrate newer technologies such as artificial intelligence and machine learning. The ability of the technologies to improve efficiency of operations and decision-making in microgrids is essential to long-term performance and viability (Gharehpetian et al., 2019). Moreover, community acceptance and stakeholder engagement also have vital roles in successfully deploying microgrids. Initiatives can be confronted with opposition from the local population concerned with changes in their existing infrastructures or with fears of being intruded upon during implementation. Therefore, developing healthy relationships between various stakeholders, including government agencies, public utilities, and local community members, is a requirement. Public awareness and education of the benefits of microgrids, such as improved energy resilience and sustainability, can facilitate acceptance as well as collaboration. Additionally, addressing equity and access issues, especially in low-income communities, is essential to ensure equitable distribution of microgrid benefits. Unless these social problems are successfully resolved, even technically sound microgrid projects may have a hard time making progress among the people they are intended to benefit (Figure 7.11).

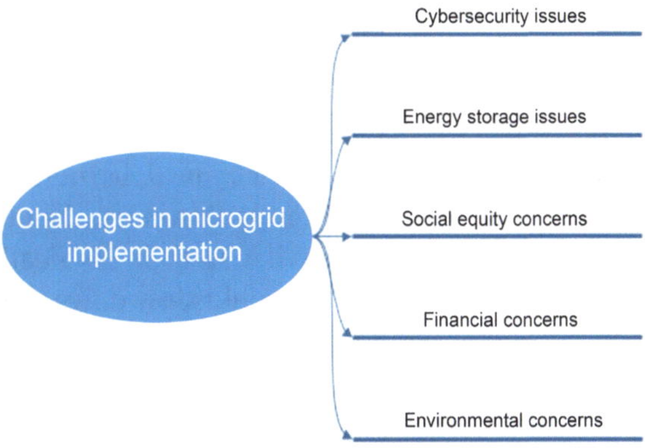

Figure 7.11. Challenges of microgrid development.

Source: https://www.mdpi.com/sustainability/sustainability-15-06366/article_deploy/html/images/sustainability-15-06366-g005.png.

7.4.1. Technical Barriers

Implementation of microgrids and distributed energy resources (DERs) has tremendous potential to enhance energy resilience and sustainability but is hampered by certain technical hurdles. A major challenge is the complexity of integrating such systems to the conventional power grids, where it can involve large-scale retrofitting of the decentralized generation and diverse technology interfaces common with DERs. Integration is not only a capital expenditure process but also necessitates advanced load flow modeling and system analysis in order to reach stable operation. In addition, as highlighted in current literature, community-level microgrids that incorporate wastewater resource recovery facilities (WRRFs) illustrate the technical difficulties in attaining technical feasibility without significant investment in infrastructure upgrades to allow for integration with renewable energy sources and the required communication technologies necessary for effective load management and continued provision of services in emergencies (Mohamed et al., 2023). Apart from infrastructure problems, the cybersecurity risks of smart inverters are a significant technical challenge to incorporating DERs in microgrids. While inverters are being designed to enhance grid-support functions and communication features, they are increasingly vulnerable to cyber-attacks, potentially jeopardizing the reliability and integrity of energy systems. The development of DERs introduces additional devices on the grid, presenting multiple access points to possible vulnerabilities. Hence, there has been increasing research targeting smart inverter cybersecurity with the goal of developing an efficient defense strategy and identifying key weaknesses. Addressing these cybersecurity challenges is essential to gaining confidence in microgrid technologies, thereby enabling more uses of DERs and ensuring that forecasted benefits are realized without compromising system security or consumer protection (Li et al., 2023). The microgrid and DER regulatory regime also creates additional technical barriers that complicate their installation and expansion. Legislation is slower to adapt to technological advancements, creating a lag between what is technologically feasible and legal. This piecemeal nature of regulations could slow down the deployment of cutting-edge technologies that would facilitate an increased number of DERs onto the existing grid. Also, different standards across different regions could add more complexity for anyone looking to streamline processes. There is a need to coordinate efforts between utility companies, government departments, and private institutions in order to overcome these challenges. As microgrid applications increase, it is important to promote harmonious regulations and standards to overcome the present technical barriers while facilitating smooth integration of renewable resources and generally improving the resilience of the entire grid (Mohamed et al., 2023).

7.4.2. Regulatory and Policy Issues

The integration of Distributed Energy Resources (DERs) and microgrids into the conventional energy grid has significant policy and regulatory implications. With various stakeholders, including policymakers, customers, and utility companies, vying to access the potential gains from these technologies, a need for effective regulatory frameworks is more and more obvious. Traditional energy markets lack adequate frameworks to facilitate the decentralized nature of DER generation, hence inhibiting the ability of small-scale producers to participate. In addition, regulations must develop to accommodate the integration of microgrids with major utility infrastructures in a safe, secure, and reliable manner. Though regulatory organizations think about evolution, it is critical to create policies that realize the heterogeneity of microgrid operations as well as the various roles various players have in facilitating the installation and integration of these assets within the broader energy landscape. Adding to the regulatory complexity is the necessity to deal with financial and economic implications of microgrids and DERs. Policymakers have the task of establishing profitable investment conditions while also dealing with high startup costs and maintenance fees that act as disincentives for adoption. Subsidies, tax credits, and other forms of fiscal incentives, as seen in the case of EV charging stations, play a crucial role in ensuring the adoption of these technologies at scale. In addition, ensuring balance between utilities' and DER operators' economic interests continues to remain pivotal to cultivating cooperative arrangements among players within a region. Such interest balance can be instrumental in defining the success of incorporating microgrids into national and regional power systems, necessitating thorough examination of long-term and short-term economic impacts on players in energy consumption and production. Lastly, addressing microgrid and DER regulatory issues calls for an overall understanding of multi-faceted challenges at stake. For instance, the geographical diversity of solar and wind sources presents operational uncertainties that must be dealt with by regulators when framing policies. The incorporation of DERs, especially in EV charging stations, is a desirable method to offset the utilization of local grids without putting existing infrastructure under pressure. But the need for ongoing research into user behavior, seasonal fluctuations in demand, and location of installation underscores the position of evidence-based decision-making in policy-making. As highlighted, future studies will need to investigate further on capital expenditure analysis and local impacts in formulating regulatory incentives towards innovative approaches that will spearhead the integration of DERs to create a stronger and more efficient urban power grid (Lee et al., 2024; Salvadori et al., 2024).

7.4.3. Financial Constraints

The financial limitations of microgrid and distributed energy resource (DER) implementation present significant hurdles for the stakeholders. These limitations can be attributed to a wide range of determinants, including the high upfront capital investment of putting in the microgrid infrastructure and innovations needed for effective energy management. More generally, using sources of renewable energy often requires complex systems in order to support reliability and efficacy. This becomes more crucial in the case of a microgrid scenario wherein reconciling disparate needs, such as social welfare and ecological aspects, contributes to financial expenses. Thus, possessing a strong financial model capable of raising revenues to pay for such initial expenses while maintaining long-term viability and affordability becomes instrumental for the effective deployment of microgrids (Liu et al., 2024). Additionally, the economic climate in which these microgrids operate is one of volatile energy prices and several regulatory regimes. In some regions, DERs, particularly those from renewable energy, will not be an economically viable choice due to unstable market conditions and subsidies. For instance, conventional sources of energy may be capable of generating cheaper short-run solutions, leading investors and utilities to keep at bay when considering performing the switch to microgrid systems. Besides, the capacity of such systems to efficiently manage demand and also harmonize the intricacies of the energy market holds a thin line upon which fiscal constraints must be keenly managed. The finding of optimal pricing principles for electricity and gas purchases can enhance microgrid viability but is heavily reliant on local market realities (Dzyuba et al., 2024). Finally, economic constraints to the use of microgrids require creative finance structures and risk-sharing arrangements. Public-private partnerships can provide the potential to enable passing on the cost of finance to small business access to microgrid technology. Shared pool of resources allows companies to overcome high capital expenses while everyone sharing pooled operating efficiencies and lower bills for energy. Besides, the development of tailored financing solutions—e.g., performance-based incentives or adaptive loan arrangements—can encourage investments in such systems. Mutualism between financial sustainability and operational performance is critical; therefore, defining a framework under which economic constraints are regulated in microgrid development policy can foster larger-scale applications of DER technologies in the long run (Liu et al., 2024).

7.5. CASE STUDIES

In the assessment of microgrid and Distributed Energy Resources (DER) efficacy, case studies come in handy as they present real-world applications and challenges. The Harmonic Stability Assessment (HSA) of power systems with large penetrations of Converter-Interfaced Distributed Energy Resources (CIDERs) is one study in focus. With a Harmonic State-Space (HSS) model, researchers demonstrated how such systems can be analyzed for harmonic stability, observing the importance of understanding the interplay among different harmonic components. The study revealed characteristic traits of system eigenvalues and sensitivity to control parameters, highlighting the requirement for accurate modeling in systems that utilize CIDERs. This study is a significant step towards alleviating harmonic instability in microgrids, as it aids in affirming the significance of sophisticated analytical techniques while the reliance upon DERs also increases with modern energy systems (Becker et al., 2024). Another robust case study considers the resilience of electrical microgrids during instances of severe weather, and it aids in emphasizing the significance of prosumer-centric designs. This research utilized case studies with varied damage levels to analyze microgrid performance under disaster conditions. Through simulations, the study demonstrated how the integration of distributed energy resources such as solar photovoltaic panels, wind turbines, and battery energy storage systems could be used to increase resilience during a blackout. In addition, the study highlighted the significance of electric vehicles (EVs) as rolling energy storage devices that can carry loads when conventional grid networks collapse. In the observation of these microgrids' performance, the case study demonstrated their capacity to perform in difficult weather conditions, and for the first time, defined possible load management and demand response options during emergency situations, thus affirming the feasibility of microgrids and DERs in real applications (Thirumalai et al., 2024). The outcome of these case studies illuminates the multilateral benefits microgrids and DERs accord to modern energy systems. They not only reveal the technical challenges with the integration of these resources but also the pragmatic measures that can maximize energy resilience and stability. For instance, the HSA study specifies the necessity for advanced modeling tools to accurately reflect the dynamics of CIDERs, whereas the prosumer microgrid analysis emphasizes the necessity for resilient communication systems for response to emergencies. Together, these findings extend the knowledge of how microgrids can evolve to meet increasing energy demands while minimizing risk in regard to grid instability and environmental risk. Thus, these papers offer sound foundations for subsequent research and implementation strategies in the emerging field of distributed energy resources (Becker et al., 2024; Thirumalai et al., 2024).

7.5.1. Successful Microgrid Projects

Microgrid projects growth accentuates the opportunity for microgrids to enhance energy sustainability and resilience at the community level. Successful applications have been initiated by supportive government policies that facilitate funding support as well as technological assistance. Microgrid deployments at community centers and schools, for instance, demonstrate an effort to move away from traditional sources of power. Projects powered by renewable energy, where both wind and solar energies are blended with energy storage systems, have been most effective. Not only do such blends improve energy autonomy, but also cater to national sustainability and emission reduction targets. By comparing various federal, state, and utility-level policies, one can appreciate the multi-faceted approach behind the success of microgrid projects all over the United States, as highlighted in present reviews of the microgrid business (Bao et al., 2018). The ability to accommodate a range of Distributed Energy Resources (DERs) is a feature of successful microgrid projects. These microgrid initiatives provide seamless integration of energy generation from local, renewable resources, effectively addressing the problem of energy distribution. Microgrids are also a great illustration of smart grid technologies since they can accommodate higher levels of incorporation of renewable energy without putting a strain on existing infrastructure. Microgrids store energy during peak usage times to ensure balance of supply and demand, making the overall system more reliable. Notably, the research has shown that prioritizing some loads in microgrid arrangements enhances the indices of reliability, particularly with consideration given to the stochasticity of renewable energy sources (Ansari et al., 2017). As communities accept microgrid alternatives more, their collective benefit of higher reliability and reduced cost of energy continues to be noticeable, and more investment and technological advancement in the industry becomes unavoidable. When evaluating the success of microgrid projects, careful performance analysis and lessons learned are significant inputs to future implementations. Positive outcomes from demonstration projects in different environments serve as benchmarks for organizations seeking to install similar systems in other places or countries. Analysis indicates that effective deployment of microgrids can delay or even entirely prevent major distribution network expansions, and thus reduce costs for utility companies. Additionally, improving the resilience of the grid during outages proves the strategic importance of microgrids in the broader energy landscape. With increasing numbers of stakeholders working on microgrid design and optimization, the expertise accumulated will undoubtedly drive the world towards more efficient, sustainable, and resilient energy systems so that microgrids become a core part of future infrastructure planning (Bao et al., 2018).

7.5.2. Lessons Learned from Failed Initiatives

It is important to investigate lessons learned from failed projects within the field of microgrids and distributed energy resources (DERs) in order to proceed with future initiatives. Numerous community energy initiatives have collapsed due to inadequate regulatory frameworks, inability to engage with stakeholders, and insufficient adequate understanding of local energy patterns. For instance, while Germany has achieved a great deal in community energy, the slow and piecemeal transposition of the Renewable Energy Directive (RED II) has hindered some projects from realizing their full potential. Germany's problems are indicative of the problems that other nations experience when attempting to introduce similar models of community energy, and this is an indication of a universal call for policy frameworks that are comprehensive enough to support seamless deployment and integration. This uneven development is used to demonstrate that policymakers must be concerned with developing robust infrastructure for promoting the growth of DERs and microgrids so that innovation does not implode behind administrative barriers (Krug et al., 2022). In contrast, Italy's relatively dynamic response to the RED II framework provides a valuable lesson in the importance of responsiveness and flexibility when facing regulative cues. The anticipatory transposition of the directive by the country has begun a proliferation of renewable energy communities (RECs), providing a tangible demonstration of how models of legislation can drive bottom-up action. However, this experience serves to show the necessity of ongoing structural reform within the system of governance to meet emerging challenges. The importance of multi-level governance (MLG) is uncovered in that effective coordination among varying levels of power is imperative in achieving maximum exploitation of resources as well as mobilization of the populace. Through Italy's experiences, other areas are in a position to learn the importance of policy-making adaptability while valuing the fact that achievement is not solely dependent on strong frameworks but also their responsiveness and elasticity to real problems (Krug et al., 2022). Moreover, digital technology-powered projects such as blockchain in the circular economy provide insights into how future microgrid projects can be successful. Failed implementations of blockchain-driven circular supply chains (CSCs) in most cases would primarily be due to a lack of understanding of the key success factors (CSFs) needed for successful implementation. Recognition of such determinants, including the significance of collaborative networks, illuminates the way partnerships can make microgrids more flexible and operational. As evidence on blockchains use in circular economies has indicated, successful programs tend to be characterized by collaborative methodologies enhancing participation and resource sharing. So, as one develops microgrid infrastructures, emphasizing collaboration among stakeholders—local communities, private businesses, and government institutions—may be pivotal in weathering challenges that have thwarted previous initiatives.

7.5.3. Comparative Analysis of Different Regions

The evolving dynamics of microgrids and distributed energy resources (DERs) indicate regional intensity in deployment and effect. For instance, where there are robust policy interventions and fiscal incentives, such as in certain regions of Europe, DERs are increasingly being incorporated into the energy mix, enabling consumers to generate their own electricity and ensure grid stability. This effective adoption reshapes traditional power dynamics, allowing a shift towards decentralized energy systems that foster greater consumer engagement. Conversely, regions lacking supportive regulations face barriers that hinder similar advancements. The comparative analysis highlights the need to identify best practices from regions successfully integrating DERs, as those experiences can inform policy adjustments elsewhere. Effective strategies may include the establishment of full-scale pricing mechanisms and the removal of barriers for participation in the market, as outlined in the analysis of ongoing attempts to utilize DERs for enhancing energy supply (Gómez Román et al., 2017). The technological frameworks of microgrids and DERs vary widely across locations, influencing the efficiency and reliability of energy distribution networks. For example, places like California have invested heavily in smart grid technologies that enable bi-directional energy flow and enhance communication between service providers and users. These Smart Grids employ advanced monitoring, analysis, and control systems, wherein the Internet of Things (IoT) is depended upon for data acquisition and management. At the other extreme are places with little technological investment, which are infested with archaic systems and which suffer from inefficiencies and losses of energy. The variance in the technological developments underscores the necessity for the incorporation of IoT and other smart technologies into the grid systems through an integrated approach. Comparing results of regions that have implemented such innovations to regions that have not will give stakeholders a glimpse of how to optimize energy management and enhance system resilience (Copeland et al., 2019). Furthermore, the sociocultural conditions of different regions play a crucial role in deciding the viability and acceptability of microgrids and DERs. If local drive and community involvement drive energy projects, as seen in Scandinavian regions, then the benefit of DERs is optimized. Cooperative energy solutions enjoy more trust among such communities, which results in more local investment and local community involvement in renewable energy projects. On the other hand, regions with economic constraints or lacking community involvement may not find it easy to mobilize support for such projects. This indicates the significance of tailored interventions that take into account unique sociocultural settings influencing the adoption of energy. Understanding these factors can guide policymakers and stakeholders in creating policies that inspire local environments' imaginations, effectively optimizing the effectiveness of DERs and microgrid development in various regions.

7.6. FUTURE TRENDS

The landscape in energy production and transmission is evolving extremely rapidly, and therefore the necessity for microgrids and distributed energy resources (DERs). Among the most prominent trends is application of advanced control methodologies, with the economics model predictive control (EMPC) making a cut in the top class. This is aimed at enhancing the efficiency of microgrid operation. EMPC applies real-time data and prediction models to enhance decision-making strategies aimed at reducing costs and maximizing benefits. With increasing sophistication of microgrids involving heterogeneous DERs, the necessity for sophisticated management systems becomes essential. This paradigm shift accords primacy not only to the operational benefits but also to the necessity of gaining a master level of understanding of emerging control practices, essential for enabling increased usage of microgrids in energy grids (Hu et al., 2024). Another salient trend is the realization of multi-microgrids (MMGs), which make more comprehensive integration of DERs possible through cooperative power trading between individual microgrids and the main grid. The impetus for this shift arises from the capability of MMGs to enhance the overall system stability, reliability, and efficiency with cost-effectiveness in mind. The operational feasibility of MMGs relies on energy management systems (EMS) efficiently planned, which are an imperative for maximum energy flow and load balance. This entails issues of uncertainty associated with intermittent generation and volatile market conditions. Optimal sizing strategies, EMS implementation, and uncertainty quantification are required here, emulating the need for a robust framework to address the future energy environment's complexity (Aeggegn et al., 2024). Furthermore, intelligent predictive modeling can significantly improve control and operation of DERs and microgrids by incorporating energy storage systems (ESSs) as a means for mitigating impacts of variability of generation and demand. With the advancements in energy storage technologies continuing, incorporation of these in microgrid infrastructures not only improves grid resilience but also enhances energy independence through the ability to create localized energy generation and consumption. The coupling of smart systems with renewable energy sources enables more responsive and adaptive grid solutions. Microgrids and DERs development in the future must, therefore, couple technological innovation with strategic deployment of ESSs within conformity to evolving regulatory landscapes and market directions so that the energy transition would be efficient and sustainable (Figure 7.12).

Figure 7.12. Distributed energy resources (DERs): Shaping the future energy grid.

Source: https://cdn.prod.website-files.com/640ad43a98dc7b4fca909775/654d3a3a5d59cdf60079a718_joule-case-vpp-blog-feature-image.png.

7.6.1. Technological Innovations in Microgrids

Microgrid technology history is a landmark milestone in the search for sustainable energy uses, particularly with the move towards decentralized networks worldwide. Microgrids employ distributed generation systems such as solar panels and wind turbines to produce and supply energy more efficiently through localized grids. This evolutionary approach not only reduces the reliance on traditional power sources but also enhances energy resilience and security during grid failure. As an integral component of modern smart grid initiatives, microgrids leverage advanced information technology to automate power distribution to ensure optimal performance and reliability. In addition, this convergence enables real-time management and monitoring of energy resources, touting the microgrids status as a game-changing innovation for the energy systems of the future (Uddin et al., 2023). Including technology advancements in microgrid systems also makes it possible for the integration of Distributed Energy Resources (DERs) effortlessly. These resources refer to energy generation or storage technologies placed near the point of consumption, providing flexibility and efficiency as the energy demands vary. Internet of Things (IoT) plays a significant role in enhancing microgrid functionalities through interconnecting of devices and systems such as smart meters and smart sensors, which provide real-time energy consumption monitoring. This interconnectivity supports better load management, predictive maintenance, and improved decision-making for energy operators. Besides, technological advancements in cloud computing allow scalable data processing and storage solutions to analyze vast amounts of data collected from DERs, thus allowing the provision of actionable insights to improve energy management strategies (Muthumanickam et al., 2022). Lastly, the transition to smart microgrids requires visionary policies and investment in infrastructure for the absorption of technological innovations. In this vein, policymakers and researchers have to focus on overcoming many microgrid deployment challenges, such as regulatory contexts, interoperability standards, and cybersecurity concerns. A thorough understanding of microgrids will facilitate stakeholders in establishing key strategies of priority that promote investment and collaboration within the energy sector. In overcoming such complexities, continuous investment in research and development is needed for unlocking the full potential of microgrids. This commitment will not only guarantee economic growth but also pave the way to a cleaner and stronger energy future, thereby meeting growing demands for cleaner sources of energy (Uddin et al., 2023) (Figure 7.13).

Figure 7.13. Technological innovations in microgrids.

Source: https://www.siemensgovt.com/assets/img/insights/articles/microgrid-1.jpg.

7.6.2. Integration with Smart Grids

The marriage of microgrids with smart grids introduces a revolutionary approach to the operation and control of electric power distribution, especially with respect to Distributed Energy Resources (DERs). This marriage ensures that renewable energy systems that are generally variable and intermittent in nature can be effectively integrated with conventional sources of power. Smart grids are accompanied by advanced information and communication technologies that augment bidirectional energy transfer, enable real-time monitoring, and ensure higher reliability. Through connecting microgrids to smart grids, the ability of operators is enhanced to coordinate demand-side management and optimize dispatching of distributed generation resources to meet load demands in an efficient manner. This integration also facilitates overcoming challenges related to energy pricing and resource allocation, leading to increased utilization of sustainable energy solutions facilitating grid stability. The operational dynamics of the interconnected microgrids are significantly improved by integrating them with smart grids, leading to improved cost-effectiveness and reliability. Current studies indicate that collaborating interconnected microgrids can reduce operating expenses with greater reliability.

Figure 7.14. RE integration and smart grids.

Source: https://www.eprmagazine.com/wp-content/uploads/2024/06/10-1200x800.jpg.

With a stochastic framework based on methods like the Unscented Transform, these systems can effectively predict uncertainties in generation output and load demand. Effective operation of interconnected microgrids involves optimizing energy management techniques that ultimately minimize the cost function while satisfying the load demand. These advances pinpoints the potential of smart grid technologies that provide the necessary infrastructure for microgrids to operate both independently and integrated with the grid, enabling resilience in a dynamic energy landscape (Dabbaghjamanesh et al., 2018). Along with this, novel control methods, such as power-based droop controllers, are also necessary components in integrating the DERs in smart grids. They facilitate seamless transitions between modes of operation and assist microgrids in handling power flow during normal and faulty conditions effectively. In regular operation, grid-interface converters control voltages and during disturbances, droop control assists with stability and power sharing among the DERs with no communication latency. This standalone capability not only enhances microgrid system resilience but also maximizes resource-sharing efficiency, offering microgrids as a reliable complementary option for central grid infrastructure. Experimental validation attests to the efficacy of such control approaches, which exhibit improved performance relative to traditional methodologies. These technological innovations underscore the stringent interdependence of microgrids and smart grids, which maximally drive their potential within the renewable energy sector (Atherton et al., 1997) (Figure 7.14).

7.6.3. Role of Policy in Shaping the Future

The shifting energy supply landscape, particularly through microgrids and Distributed Energy Resources (DERs), is highly reliant on policy directives that establish operating measures and generate innovation. These policies are critical in determining not just the technology rollout but also the socioeconomic viability of microgrid projects. If supportive legislation is passed, it draws in public and private investment, creating an environment friendly to technological innovation and bulk rollout. As witnessed in the United States, federal and state policies actively incentivize the development of microgrids, which offer more energy resilience amid natural disasters and climate change. This parliamentary backing is central to fueling the creation of microgrids, enabling them to meet energy demands in the future in a sustainable manner and provide critical services during power outages, thus reconfiguring how communities consume and generate energy (Bao et al., 2018). The integration of microgrids also raises significant regulatory matters that need wise facilitation. The implementation of huge microgrid projects, such as those being considered in Pittsburgh, brings out the complexities involved in regulating energy in a deregulated market. Government agencies, public utilities, and private enterprise must all collaborate in establishing operational standards and guidelines that will lead to smooth functioning in such systems. Concerns linger regarding whether microgrids must be treated as generation assets or storage facilities and how they will be regulated. As there are regulatory loopholes currently in states like Pennsylvania, defining explicit policies will be vital to making sure that these technologies have the opportunity to thrive while providing public safety and energy reliability (Parks et al., 2019). Besides, the effectiveness of microgrid and DER installations relies on policy evolution keeping pace with technological progress. Policymakers must consider the dynamic character of energy technology, bringing flexibility into rules which can be adaptive to future innovations. With a move of microgrids, embracing an array of renewable supplies and intelligent grid management systems, legislative frameworks must evolve in parallel. Regular evaluation and refinement of energy policy are necessary so that they support innovation and ensure protection of public interest while ensuring fair access to energy resources. Hence, the policy contribution to shaping the future of DERs and microgrids cannot be overstated; it is both the impetus and the basis for creating equitable and sustainable energy platforms for communities across the country (Bao et al., 2018).

SUMMARY

In conclusion, microgrid and DER integration is a pivotal step towards alleviating modern energy challenges. By taking advantage of local consumption and generation, microgrids facilitate greater utility grid disruption resilience, which ensures energy sustainability and security. ICT and sophisticated signal processing technologies used in the systems provide improved efficiency in electricity distribution and utilization, ushering in a new age in managing energy. It is important that as the cities continue to grow, they adopt such technologies so they not only meet current requirements but also lay the groundwork for the energy needs of the future. Such bold strategies reflect a long-term commitment to sustainability in the environment and society, key ingredients of 21st-century urbanization. Moreover, the integration of renewable energy sources such as sunlight in microgrid systems presents challenges and opportunities that are very different. Solar generation is intermittent and calls for advanced planning and management methods to counter grid instability possibilities. Microgrids are useful local solutions that expertly manage the natural uncertainties relating to renewable energy generation (Bahramirad et al., 2017). With the application of robust scheduling models that account for such fluctuations, communities can maximize their energy independence while simultaneously supporting grid stability at large. Such strategic community planning for local solutions to energy not only supports higher penetration levels of solar generation but also engages consumers proactively in energy management, fomenting the culture of conservation and efficiency. Lastly, what microgrids promise to deliver transcends the mere provision of energy management; they incorporate wider socioeconomic advantages that promote social cohesion and resilience. Microgrid- and DER-enabled communities are capable of being included in demand response schemes whereby domestic residences can be leveraged to enable load balancing and peak shaving initiatives (Bonetto et al., 2017). This collective endeavor both empowers local actors and improves the system's overall reliability. With cities increasingly faced with the reality of climate change and limited resources, the implementation of microgrids and DERs is an important strategy towards sustainable energy grids. Therefore, the future urban energy infrastructure will have to be centered around the development of these distributed systems in a way that they are properly integrated, efficient, and responsive to the emerging needs of an evolving society as well as the surrounding environment.

Microgrid and DER evolution is a significant shift in modern energy systems with an emphasis on sustainability and resilience. As the world continues to shift in changing energy requirements and the drive towards addressing climate change on a global scale, microgrids offer itself as a crucial solution through the encouragement of distributed energy generation and consumption. The integration of DERs into microgrrids offers the assurance of grid reliability, especially during emergencies and power outages. Specifically, mobile emergency generators (MEGs) and application of EVs for energy storage are dominant strategies for temporary restoration of critical loads. Optimization strategies that govern these resources do not just emphasize efficiency but also on-time provision of priority loads with the power, emphasizing the importance of strategic resource management in energy crises (Ayşe Kübra Erenoğlu et al., 2023). Those challenges of integrating DERs into microgrids must not be downplayed, particularly when there are fault management and system reliability issues. As DERs expand into energy networks, making them increasingly complex, traditional grid management techniques show their inadequacy. That's where newer, more innovative technological advancements such as AI and the IoT can introduce new ways of addressing these newly arisen issues. Adaptive overcurrent protection algorithms provide a great model of how modern solutions can bring the performance level of microgrids significantly higher with improved selectivity, reliability, sensitivity, and speed. The comparison between advanced algorithms and conventional methods proves that there are remarkable improvement in performance, which exhibit the critical role of technological integration in guaranteeing safe and stable operation of microgrids (Chandraratne et al., 2020). In general, the integration of DERs and microgrids is a transformative step toward constructing more sustainable and robust energy systems. Proper management of such resources provides greater self-sufficiency and reliability in power supply, particularly in situations of crisis. As the power system is evolving, the integration of advanced technologies like AI and IoT will play a key role in overcoming the particular challenges presented by DERs. This integrated approach not only improves operational effectiveness but also aids in achieving the overall ends of minimizing environmental effects and enhancing energy access. The ongoing research and implementation of real-time optimization methods and adaptive protection measures bring into focus the imperative need for innovation in energy management systems so that microgrid systems have the ability to manage existing and future energy needs efficiently (Ayşe Kübra Erenoğlu et al., 2023; Chandraratne et al., 2020).

REFERENCES

1. Adegboyega, A. W., Sepasi, S., Howlader, H., Griswold, B., Matsuura, M., & Roose, L. R. (2025). DC microgrid deployments and challenges: A comprehensive review of academic and corporate implementations. *Energies.* https://www.semanticscholar.org/paper/800c30b0a64fd75cf97b0635da10c5ca936e93b5 (accessed on 30th September 2025).
2. Aeggegn, D. B., Nyakoe, G., & Wekesa, C. (2024). A state-of-the-art review on energy management techniques and optimal sizing of DERs in grid-connected multi-microgrids. *Cogent Engineering, 11.* https://www.semanticscholar.org/paper/a6a1b7bfba0e7d91567cf6ac35f3af21572983da (accessed on 30th September 2025).
3. Alnaqbi, S. A., & Alami, A. H. (2023). Sustainability and renewable energy in the UAE: A case study of Sharjah. *Energies, 16*, 7034–7034. https://doi.org/10.3390/en16207034 (accessed on 30th September 2025).
4. Angjelichinoski, M., Popovski, P., Scaglione, A., et al. (2018). Decentralized DC microgrid monitoring and optimization via primary control perturbations. http://arxiv.org/abs/1703.10467 (accessed on 30th September 2025).
5. Ansari, O. A., Chung, C. Y., & Safar, N. (2017). Reliability assessment of microgrid with renewable generation and prioritized loads. http://arxiv.org/abs/1709.07970 (accessed on 30th September 2025).
6. Atherton, C., Bakwin, P., Brasseur, G., et al. (1997). Power-based droop control in DC microgrids enabling seamless disconnection from upstream grids. https://core.ac.uk/download/201723855.pdf (accessed on 30th September 2025).
7. Babu, K. V. S. M., Chakraborty, P., Dwivedi, D., et al. (2023). A resilient power distribution system using P2P energy sharing. http://arxiv.org/abs/2303.09471 (accessed on 30th September 2025).
8. Bahramirad, S., Khodaei, A., & Majzoobi, A. (2017). Capturing distribution grid-integrated solar variability and uncertainty using microgrids. http://arxiv.org/abs/1703.00120 (accessed on 30th September 2025).
9. Balikowa, S., Abdulkarim, A., Mundu, M. M., & Udoka, E. V. H. (2024). Enhancing the resilience and efficiency of microgrids through optimal integration of renewable energy sources and intelligent control systems: A review. *KIU Journal of Science Engineering and Technology.* https://www.semanticscholar.org/paper/f53833ca86db2e31991a59aef327832941c50885 (accessed on 30th September 2025).

10. Bao, Y., Feng, W., Jin, M., et al. (2018). A review of microgrid development in the United States: A decade of progress on policies, demonstrations, controls, and software tools. https://core.ac.uk/download/323310623.pdf (accessed on 30th September 2025).
11. Becker, J., Kettner, A. M., & Paolone, M. (2024). Harmonic stability analysis of microgrids with converter-interfaced distributed energy resources, Part II: Case studies. *IEEE Transactions on Smart Grid, 16*, 2327–2337. https://www.semanticscholar.org/paper/b630abc7bc94cd0742c73dc593facef6b07cd68c (accessed on 30th September 2025).
12. Bonetto, R., & Rossi, M. (2017). Smart grid for the smart city. https://core.ac.uk/download/84922332.pdf (accessed on 30th September 2025).
13. Caldognetto, T., Khodamoradi, A., Liu, G., et al. (2018). Plug and play DC-DC converters for smart DC nanogrids with advanced control ancillary services. https://core.ac.uk/download/186448304.pdf (accessed on 30th September 2025).
14. Chandraratne, A., Ramasamy, T. N., Logenthiran, T., & Panda, G. (2020). Adaptive protection for microgrid with distributed energy resources. *Electronics.* https://www.semanticscholar.org/paper/4997e9b73c1959439d49558229da42154fbf6833 (accessed on 30th September 2025).
15. Cleveland, J., Fox-Penner, P., Hatch, J., et al. (2019). Cities and climate change: Strategic options for philanthropic support. https://open.bu.edu/bitstream/2144/39182/1/CitiesStrategicOptions_Hatch_Cleveland-et-al-_Final_3.12.19b.pdf (accessed on 30th September 2025).
16. Condon, F., Franco, P., Martínez, J. M., Eltamaly, A., Kim, Y. C., & Ahmed, M. A. (2023). EnergyAuction: IoT-blockchain architecture for local peer-to-peer energy trading in a microgrid. *Sustainability.* https://www.semanticscholar.org/paper/62db1dcc98215fa56b2946ed499b3a090fcfcb3a (accessed on 30th September 2025).
17. Copeland, R., Crespi, N., Rehmani, M. H., et al. (2019). Internet of Things-aided smart grid: Technologies, architectures, applications, prototypes, and future research directions. http://arxiv.org/abs/1704.08977 (accessed on 30th September 2025).
18. Dabbaghjamanesh, M., Fard, A. K., Ferdowsi, F., & Mehraeen, M. (2018). A new efficient stochastic energy management technique for interconnected AC microgrids. http://arxiv.org/abs/1803.03320 (accessed on 30th September 2025).
19. Denysiuk, S., Derevianko, D., Bielokha, H., & Zaichenko, S. (2023). Aggregation price models for microgrids with

distributed energy resources. *Power Engineering: Economics, Technique, Ecology.* https://www.semanticscholar.org/paper/ba9b2ed7ba6d05ea7e7b76ad7235da5a6d423fd5 (accessed on 30th September 2025).

20. Dixit, S., Mehla, S., Vellanki, S., Latha, M., Bhatt, C., & Gandhi, A. (2024). Micro grid integration of distributed power generation optimal operating methods. *Journal of Electrical Systems.* https://www.semanticscholar.org/paper/7c1889ab07e35f5257fd431d023157e8b15455af (accessed on 30th September 2025).
21. Dzyuba, A., & Semikolenov, A. (2024). Industrial microgrids as tools for managing the energy efficiency in industrial regions. *Strategic Decisions and Risk Management.* https://www.semanticscholar.org/paper/25f9175afa98e6548c1ab67d8843f4ae0f0a9ea7 (accessed on 30th September 2025)..
22. Erenoğlu, A. K., & Erdinç, O. (2023). Real-time allocation of multi-mobile resources in integrated distribution and transportation systems for resilient electrical grid. *IEEE Transactions on Power Delivery, 38,* 1108–1119. https://www.semanticscholar.org/paper/c32f9ffe6272e4772837ca8686e9065298b4c6b9 (accessed on 30th September 2025).
23. Gad, A. G. (2022). Particle swarm optimization algorithm and its applications: A systematic review. *Archives of Computational Methods in Engineering, 29,* 2531–2561. https://doi.org/10.1007/s11831-021-09694-4 (accessed on 30th September 2025).
24. Gharehpetian, G. B., Marzband, M., Moradi, J., & Nafisi, M. (2019). Attributes of big data analytics for data-driven decision making in cyber-physical power systems. https://core.ac.uk/download/305121394.pdf (accessed on 30th September 2025).
25. Gómez San Román, T. (2017). Integration of DERs on power systems: Challenges and opportunities. https://core.ac.uk/download/83830035.pdf (accessed on 30th September 2025).
26. Guo, Q., Zhang, X., Wu, X., Miao, Y., Shen, J., Liu, F., & Wu, Y. (2024). Design and implementation of an intelligent energy storage management system for distributed energy resources. *Proceedings of the 2024 International Symposium on Integrated Circuit Design and Integrated Systems.* https://www.semanticscholar.org/paper/f7a1e07a713b7f394b73c7c20c4f25a13f045523 (accessed on 30th September 2025).

27. Haque, A., Kholilullah, I., Sharma, A., Mohammad, A., & Khan, S. I. (2024). Analysis of different control approaches for a local microgrid: A comparative study. *Control Systems and Optimization Letters.* https://www.semanticscholar.org/paper/37942701ff5e986293d3ab9be621c5cc65453041 (accessed on 30th September 2025).

28. Hu, J., Shan, Y., Yang, Y., Parisio, A., Li, Y., Amjady, N., Islam, S., et al. (2024). Economic model predictive control for microgrid optimization: A review. *IEEE Transactions on Smart Grid, 15*, 472–484. https://www.semanticscholar.org/paper/3dd39bbacfd7b1adc147c95a2bed572c480f3474 (accessed on 30th September 2025).

29. Ibekwe, K. I., Ohenhen, P. E., Chidolue, O., Umoh, A. A., Ngozichukwu, B., Ilojianya, V. I., & Fafure, A. V. (2024). Microgrid systems in U.S. energy infrastructure: A comprehensive review: Exploring decentralized energy solutions, their benefits, and challenges in regional implementation. *World Journal of Advanced Research and Reviews, 21*, 973–987. https://doi.org/10.30574/wjarr.2024.21.1.0112 (accessed on 30th September 2025).

30. Jansen, P., Schoor, G. V., & Uren, K. (2025). Load-profile based sizing method for distributed energy resources in industrial microgrids. *2025 33rd Southern African Universities Power Engineering Conference (SAUPEC).* https://www.semanticscholar.org/paper/418eea9839be096e37bab8b51a94d2d91e7a7a55 (accessed on 30th September 2025).

31. Kaal, T., & Rafiee, A. (2025). Deep reinforcement learning–graph neural networks–dynamic clustering triplet for adaptive multi energy microgrid optimization. *ISPRS Annals of the Photogrammetry, Remote Sensing and Spatial Information Sciences.* https://www.semanticscholar.org/paper/1b59b2892a6dbf4784fcacbb2dc9792afbb7313e (accessed on 30th September 2025).

32. Kabeyi, M. J. B., & Olanrewaju, O. A. (2022). Sustainable energy transition for renewable and low carbon grid electricity generation and supply. *Frontiers in Energy Research, 9.* https://doi.org/10.3389/fenrg.2021.743114 (accessed on 30th September 2025).

33. Kasimalla, S. R., Park, K., Hong, J., Kim, Y. J., & Lee, H. (2024). AI-enhanced inverter fault and anomaly detection system for distributed energy resources in microgrids. *ArXiv, abs/2411.08761.* https://www.semanticscholar.org/paper/35fa8bda2488d12a42593b11aba6e86535623b8f (accessed on 30th September 2025).

34. Kayıkçı, Y., Gozacan-Chase, N., Rejeb, A., & Mathiyazhagan, K. (2022). Critical success factors for implementing blockchain-based circular supply chain. *Business Strategy and the Environment, 31*, 3595–3615. https://doi.org/10.1002/bse.3110 (accessed on 30th September 2025).

35. Krug, M., Di Nucci, M. R., Caldera, M., & De Luca, E. (2022). Mainstreaming community energy: Is the Renewable Energy Directive a driver for renewable energy communities in Germany and Italy? *Sustainability, 14*, 7181–7181. https://doi.org/10.3390/su14127181 (accessed on 30th September 2025).

36. Lee, T., Yoon, G., Kang, B., Choi, M., Park, S., Park, J., & Park, S. (2024). Enhancing electric vehicle charging infrastructure: A techno-economic analysis of distributed energy resources and local grid integration. *Buildings*. https://www.semanticscholar.org/paper/0c6f32b6ce574f1953b640b5b1e1f2363e2588a5 (accessed on 30th September 2025).

37. Li, Y., & Yan, J. (2023). Cybersecurity of smart inverters in the smart grid: A survey. *IEEE Transactions on Power Electronics, 38*, 2364–2383. https://www.semanticscholar.org/paper/c6874a0d173de76c1e79df43324c4d58d30a30bc (accessed on 30th September 2025).

38. Liu, J., Hou, Z., Wang, B., & Yin, T. (2025). Optimizing microgrid energy management via DE-HHO hybrid metaheuristics. *Computers, Materials & Continuum*. https://www.semanticscholar.org/paper/eba2a1ff5a37e13984fb509d981147a5da4e0bfb (accessed on 30th September 2025).

39. Liu, L., Yang, G., & Wasly, S. (2024). Distributed event-triggered economic environmental resource management for islanded microgrids under DoS attacks. *IEEE Transactions on Automation Science and Engineering, 21*, 7158–7169. https://www.semanticscholar.org/paper/319ebfda155377cb4d2aa7a388fb515db1c10dae (accessed on 30th September 2025).

40. Mohamed, A., Zafar, K., Vaidya, D., Salmeron, L., Kanwhen, O., Esa, Y., & Kamaludeen, M. K. (2023). Grid impact of wastewater resource recovery facilities-based community microgrids. *Smart Cities*. https://www.semanticscholar.org/paper/55e1cfe9aed84ade3e720423362c28e1d367e0ae (accessed on 30th September 2025).

41. Muthumanickam, D., Poongodi, C., Kumaraperumal, R., Pazhanivelan, S., & Ragunath, K. P. (2022). Smart farming: Internet of Things (IoT)-based sustainable agriculture. *Agriculture, 12*, 1745–1745. https://doi.org/10.3390/agriculture12101745 (accessed on 30th September 2025).

42. Parks, R. (2019). Microgrids: Legal and regulatory hurdles for a more resilient energy infrastructure. https://core.ac.uk/download/212892728.pdf (accessed on 30th September 2025).
43. Qashqaie, F., Saghafi, H., Sadeghi, R., & Fani, B. (2025). Impedance-based adaptive droop control for islanded AC microgrids and overview. *International Transactions on Electrical Energy Systems.* https://www.semanticscholar.org/paper/c8284b2340094e8db3f242ccde69fd2bb3b79cd0 (accessed on 30th September 2025).
44. Salvadori, F., Leandro, P. G. M., Soares, M. B., de Freitas, I. S., & dos Santos, E. C. (2024). Hardware-in-the-loop (HIL) simulation: Modeling, characterization and analysis of a distributed energy resource in a microgrid network. *IECON 2024–50th Annual Conference of the IEEE Industrial Electronics Society*. https://www.semanticscholar.org/paper/510f61754164cdde2a90bce0e6582dd616470c8f (accessed on 30th September 2025).
45. Thirumalai, M., Hariharan, R., Yuvaraj, T., & Prabaharan, N. (2024). Optimizing distribution system resilience in extreme weather using prosumer-centric microgrids with integrated distributed energy resources and battery electric vehicles. *Sustainability*. https://www.semanticscholar.org/paper/159d70bd9694f71bfe51cbd05645724b5d4365bb (accessed on 30th September 2025).
46. Thirunavukkarasu, G. S., Seyedmahmoudian, M., Jamei, E., Horan, B., Mekhilef, S., & Stojcevski, A. (2022). Role of optimization techniques in microgrid energy management systems—A review. *Energy Strategy Reviews, 43*, 100899–100899. https://doi.org/10.1016/j.esr.2022.100899 (accessed on 30th September 2025).
47. Topallaj, K., McKerrell, C., Ramanathan, S., & Zografopoulos, I. (2025). Impact assessment of cyberattacks in inverter-based microgrids. *ArXiv, abs/2504.05592*. https://www.semanticscholar.org/paper/76dda0bdb040b23d19443adf42d2d9581b6ab5ab (accessed on 30th September 2025).
48. Uddin, M., Mo, H., Dong, D., Elsawah, S., Zhu, J., & Guerrero, J. M. (2023). Microgrids: A review, outstanding issues and future trends. *Energy Strategy Reviews, 49*, 101127–101127. https://doi.org/10.1016/j.esr.2023.101127 (accessed on 30th September 2025).

CHAPTER 8

ENERGY STORAGE TECHNOLOGIES

Contents

INTRODUCTION

The ever-increasing need for energy, coupled with the necessity of a transition toward renewable energy, has put energy storage technologies at the forefront of the debate in modern energy systems. With the world redrawing how energy is generated and consumed, the integration of good storage solutions becomes essential. Energy storage technologies not only help counter the randomness of the renewable energy sources such as solar and wind but also make the grid more reliable and resilient. By enabling greater flexibility in balancing energy supply and demand, these technologies are vital to making the transition to sustainable systems a reality. Improved understanding of energy storage technologies, how they operate, and their uses is critical for stakeholders and policymakers who aim to shape and transform the future of energy management. Significantly, energy storage systems are the foundation of many applications, ranging from home energy management to grid stabilization at the scale of large grids.

Figure 8.1. Energy storage technologies.

Source: https://www.azonano.com/image-handler/ts/20230719120057/ri/950/src/images/Article_Images/ImageForArticle_6513_16897824560828570.jpg.

In residential settings, emerging technologies such as DG allow residents to harvest and store green energy and reduce their reliance on conventional electric utilities. For instance, Meridian Energy Ltd has developed a decision-making tool that helps measure the feasibility and potential benefits of employing DG technologies in residential markets, also pointing out the revolutionary contribution that these technologies play in consumers' control of energy (Redman et al., 2002). As such, energy storage technologies are a vital stepping stone towards having an integrated energy system that enables individuals and communities to play an active role in managing their energy needs with a view to contributing towards an overall shift towards sustainable energy systems. Further research into the production of innovative materials for energy storage is also creating potential for more efficient and environmentally-friendly solutions. Developments in battery technologies, such as lithium-ion and sodium-ion batteries, are driving innovation for new materials that enhance performance without damaging the environment. Support towards sustainability is demonstrated through increasing interest in carbon-based electrodes from biomass, indicative of dual purposes to enhance the efficiency of energy storage as well as alleviate ecological footprints (Gómez Martín et al., 2021). With the evolution of the technology of energy storage, there is a need to consider beyond the technology itself and what technological innovations are being achieved, but also consider the larger environmental implications and how these technologies will define the sustainable energy of the future (Figure 8.1).

Definition of Energy Storage Technologies

Energy storage technologies are a part of today's energy systems as they enable the efficient management and use of energy resources. Fundamentally, these technologies consist of devices that capture energy when there is excess generation or off-peak demand and spew it out during deficiency in generation or peak demand. Such a function is particularly vital towards balancing the renewable power sources such as wind and sunshine, which by their nature are intermittent and non-dispatchable. By balancing day-to-day fluctuations in energy demand and supply, various forms of energy storage, such as batteries, pumped hydro storage, and thermal energy systems, balance the energy grid. Beyond making the grid more reliable, storage promotes the integration of renewable resources, which is in line with sustainable ends and catalyzes the transition toward cleaner energy systems (Villanueva et al., 2019). Among the emerging energy storage technologies, thermal energy storage systems are gaining popularity due to their unique advantages. These systems store heat generated through off-peak energy production and release it when needed, thus meeting electrical as well as thermal energy demands efficiently. One such emerging process involves the use of transcritical carbon dioxide charge and discharge cycles along with thermal energy storage systems. This technique provides an acceptable substitute for the traditional storage systems, such as pumped hydro and compressed air energy storage. Solar energy, in combination with thermal systems, has been affordable, and more efficiency is achieved when solar input is incorporated. With over 60% round trip efficiencies in some instances, thermal energy storage proves capable of meeting energy storage demands through sustainable means while contributing to the worldwide shift towards consuming renewable energy (Villanueva et al., 2019). Furthermore, the idea of microgrids has now become the symbol of an integral part of energy storage technologies, particularly in terms of maximizing local energy stability and resilience. Microgrids are self-sustaining energy systems that can generate, store, and distribute energy on-site independently of the general grid. Microgrids have gained much attention due to their capacity to integrate diverse renewable energy sources and other storage technologies and provide flexibility to the electricity market. With favorable state and federal policy, microgrids enable multiple technologies to be integrated and provide power reliability in the event of outages or fluctuating demand. As evident in a series of successful projects in the United States, microgrid system development is a crucial innovation in energy storage technologies, not only facilitating greater integration of renewables but also allowing communities to have better control over energy matters (Figure 8.2).

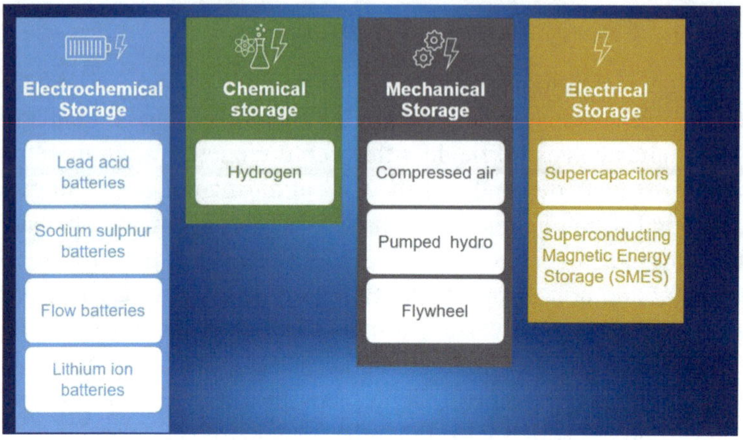

Figure 8.2. Overview of energy storage technologies.

Source: https://www.ricardo.com/media/cc2pvzub/overview-energy-storage-techs.jpg?rmode=max&width=801&height=583.

Importance of Energy Storage in Modern Energy Systems

With shifting world trends in energy use, the importance of storing energy in alternative energy systems has become more pronounced. Traditional energy grids, designed primarily for steady and controllable energy supply, are no match for the uncertainty of intermittent renewable sources such as wind and solar. Energy storage technology, ranging from batteries to pumped hydroelectric plants, enables the incorporation of such variable resources into the grid, leveling supply and demand. By storing extra energy at periods of maximum output and releasing it when demand is greatest, these systems play a critical role in grid efficiency and reliability. The shift towards a smart grid with the help of advanced Information and Communication Technologies (ICT) suggests the necessity of energy storage in upgrading aged infrastructure to meet the demands of the current energy situation and the environmental laws, thus enabling a stronger energy system, as argued in Borghetti et al. (2013). In addition to enhancing grid reliability, energy storage systems support the increased application of renewable energy resources and thus play an important role in curbing climate change. Energy storage technologies mitigate fossil fuel dependence carbon intensity by facilitating increased penetration of renewables, which are pivotal to international decarbonization. For instance, hydrogen has emerged as a viable energy carrier for large-scale introduction of variable renewable energy sources. While the hydrogen economy has been making promises for years, developments in storage methods of hydrogen, particularly those from cryogenic technologies, work to promote its introduction (Glowacki et al., 2016). Collectively, these energy storage innovations not only enhance energy security but also environmental sustainability, thus establishing the crucial position of storage solutions in contemporary energy systems. Moreover, the ability of energy storage systems to provide ancillary services such as frequency regulation and voltage support cannot be overestimated. These functions are essential to maintaining grid stability and preventing outages, particularly as increasing numbers of distributed energy resources enter the grid. Energy storage under smart grid technology can dynamically engage supply and demand signals to optimize performance and efficiency. As energy technology develops, energy storage will probably become more important, not just for its ability to hold energy, but also because it may interact synergistically with future technologies. To this end, the integration process of such technologies into our power systems will make them unavoidable components in achieving an efficient, sustainable, and robust power system, affirming observations found in Borghetti et al. (2013) and Glowacki et al. (2016).

8.1. TYPES OF ENERGY STORAGE TECHNOLOGIES

An examination of the energy storage technologies reveals a broad range of methods, each suitable for application in various uses and requirements. Traditional mechanical storage systems such as pumped hydroelectric storage utilize gravitational potential energy by pumping water to higher levels during times of low demand and releasing it during peak demand. The procedure has been a safe way of storing energy, accounting for the majority of the worlds vast storage capacity. On the other hand, new technologies such as flywheels have fast response times and high power density, and are well suited to grid stability and frequency regulation. With the future of the energy systems, these mechanical solutions exemplify the intrinsic significance of energy storage in optimizing supply and demand and enhancing grid resiliency. With the application of various types of storage technologies, the energy sector can enhance the management of variability inherent in renewable energy generation. Electrochemical energy storage systems, particularly batteries, have also gained prominence due to their multitasking capability and efficiency in static and mobile applications. Lithium-ion batteries are the most prevalent now, with high energy density and efficiency, making them the best for consumer electronics and electric vehicles. However, research persists on new chemistries, such as solid-state batteries and flow batteries, which are likely to break through current limitations, such as safety and capacity degradation over time. Battery development using nanotechnology is highly promising in that it enhances material properties in order to drive overall efficiency and sustainability.

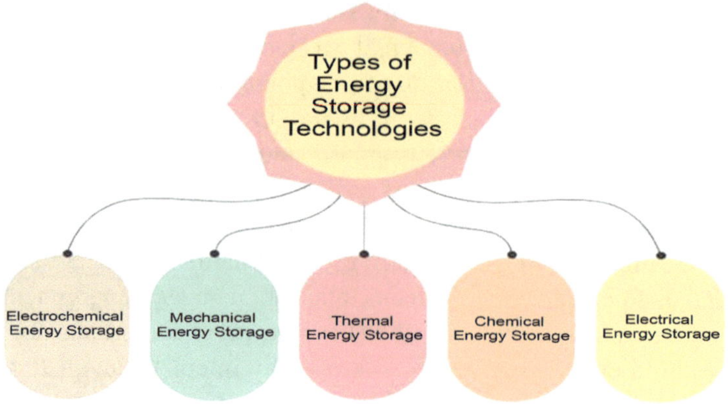

Figure 8.3. Types of energy storage technologies.

Source: https://b1879915.smushcdn.com/1879915/wp-content/uploads/2025/04/Types-of-Energy-Storage-Technologies.jpg?lossy=2&strip=1&webp=1.

These developments look promising not only for the future of electric vehicles but also for large-scale energy storage devices in renewable energy systems. As the sector makes efforts to obtain a balance between energy consumption and production, electrochemical storage advances as part of a sustainable energy system (Beena et al., 2025). Thermal and chemical storage technologies are other ways in which energy is made more efficient and consistent. Thermal energy storage devices that store energy as heat provide valuable benefits to integrate renewable resources like solar energy. For example, molten salt facilities can store solar energy received during the day and release it in the form of electricity at night, making way for a continuous flow of energy. At the same time, chemical storage devices such as hydrogen fuel cells convert excess electricity into hydrogen, which can be utilized in fuel cells to generate electricity. Both chemical and heat storage solutions not only mitigate the challenges associated with the intermittent renewable sources but also have ancillary benefits of grid stability as well as energy security. As the energy storage landscape continues to transform, it will be important to review the economic and technical viability of the technologies in determining their role in the future energy landscape (Figure 8.3) (Dubey et al., 2024).

8.1.1. Mechanical Energy Storage

The relevance of mechanical energy storage in modern energy systems cannot be overstated, especially as the demand for reliable energy storage systems grows. Mechanical energy storage systems utilize conventional physics principles, with a primary focus being energy conversion and storage in physical forms such as kinetic or potential energy. Examples of this include pumped hydroelectric storage, where water is pumped into a reservoir during off-peak demand and released to generate electricity when demand is on. Such a technique not only stabilizes demand and supply but also has maximum energy efficiency and scalability.

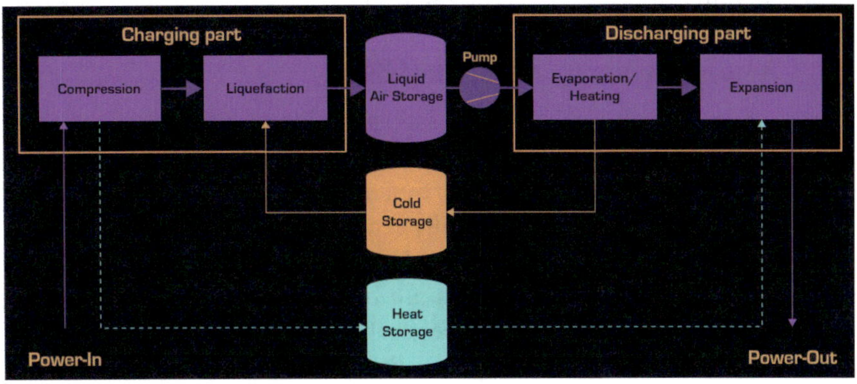

Figure 8.4. Functioning principle of a liquid air energy storage (LAES) system.

Source: https://planetdecarb.com/wp-content/uploads/2021/11/7-2.png.

These types of systems are particularly beneficial towards stabilizing grids with intermittent renewables such as solar and wind by providing the backup power required and enhancing grid resilience overall. Among the innovative technology in mechanical energy storage, systems like inertial energy storage have gained attention on how they can benefit various applications. One of the key concepts is the Mechanical Capacitor which can be utilized in spacecraft power systems with composite rotors and magnetic suspension to store energy lighter in weight and longer life. The system not only has improved voltage regulation and efficiency but also reduced lifetime costs over traditional electrochemical storage devices like NiCd and NiH2 batteries (Baer et al., 2025). As the demand for energy solutions in space increases, such innovation highlights the universality of mechanical energy storage technologies across a diverse set of industries from earth-based to space exploration. A closer examination of mechanical energy storage reveals several new methods that will change the way grid-scale storage is managed. A few recent studies identify a few new technologies, including solid mass gravity energy storage (GES) and liquid air energy storage (LAES), both with comparable cost and scalability (Mäkelä et al., 2024). Such a system functions through storing and releasing electricity through gravitational potential or heat energy, offering fascinating alternatives to conventional ways. New technologies, such as floating liquid-piston accumulators and buoyancy energy storage, are likewise creating prospects for offshore energy storage applications to address the growing need for efficient and sustainable solutions for a new energy environment. With the ongoing improvement of mechanical energy storage technology, its utilization in energy systems will be instrumental in enhancing the efficiency and dependability of future grids (Figure 8.4).

8.1.2. Chemical Energy Storage (CES)

Energy storage technologies remain an indispensable area for solving problems related to the integration of renewable energy. Among numerous technologies, CES stands out due to its ability to store energy as chemical bonds, which can be released when needed. This is typically accomplished through the conversion of electricity into chemical energy through reactions such as occur in fuel cells and batteries. Curiously, CES systems are best positioned to smooth out the intermittent nature of renewable sources that has a tendency to destabilize the grid. As highlighted by current research, the assessment of such technologies involves an integrated life cycle analysis encompassing technical performance, economic costs, safety evaluation, and environmental impact, making it imperative to guarantee CES solutions align with the evolving energy paradigm based on sustainability and reliability (Yi et al., 2024). In addition to efficiency in energy management, environmental and safety issues related to chemical energy storage systems also necessitate serious deliberation. Safety concerns, especially related to batteries, encompass thermal runaway and chemical leakage risks with horrendous implications for end users and surrounding societies. These need careful resolution, as advanced battery management systems and other mitigation strategies are being investigated to enhance system reliability. As well, the environmental footprint of CES cannot be overlooked, given resource extraction intricacies and life cycle emissions. Sustainable habits such as responsible sourcing and recycling of materials should be employed to minimize environmental traces and ensure that the use of CES technologies has a positive effect on the massive utilization of renewable energy options (Simpa et al., 2024). The complicatedness of chemical energy storage systems demands innovative solutions to their optimization. For instance, the development of second-life applications for depleted batteries not only extends the life of these materials but also avoids waste issues. Additionally, ongoing research is investigating the other chemical storage media such as hydrogen and ammonia, which are of great potential for large-scale energy storage and transfer. Looking at various applications for CES, from home energy management to grid stability, reflects the variety of the technologies. Thus, as the world evolves toward decarbonized energy infrastructures, chemical energy storage becomes pivotal, delivering solutions to facilitate the objectives of security, efficiency, and green sustainability in the world's energy mix (Yi et al., 2024) (Figure 8.5).

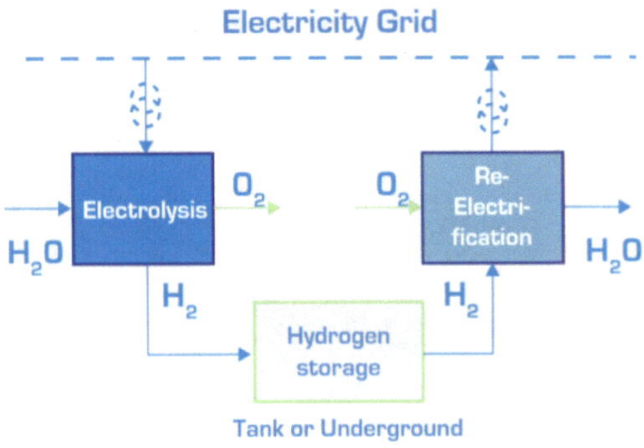

Figure 8.5. Hydrogen chemical energy storage (CES) principle.

Source: https://planetdecarb.com/wp-content/uploads/2021/12/34.png.

8.1.3. Electrical Energy Storage

The advent of electrical energy storage facilities has changed the energy management landscape, offering solutions to counter the uncertainty and randomness of renewable energy sources. As the world energy production grows more aligned with wind and solar technologies, grid stability is now the goal. Battery Energy Storage Systems (BESS) are at the forefront of this integration, and in doing so, provide essential services such as frequency stabilization and load balancing. The role of BESS is particularly significant as it enables the integration of renewable sources in the grid in an uninterrupted fashion, addressing the issue of intermittency by energy arbitrage and ancillary services. With grid enhancement in terms of performance and reliability, BESS not only contributes to the provision of stable electricity supply, but also advances the way towards future sustainable energy, noting the necessity of regulatory policies that facilitate their deployment and operation (Akinsooto et al., 2024). Technical advancements in energy storage technologies have introduced various systems aligned to meet various needs for energy. These include electrochemical energy storage in batteries, which have gained popularity for their flexibility and efficiency. This category includes not just lithium-ion batteries but also future technologies such as solid-state batteries and flow batteries, with expected increased performance and safety. These systems are all designed to supply a variety of grid services from peak shaving to excellent integration with renewable sources of energy.

Figure 8.6. Electrical energy storage systems.

Source: https://www.renhotecev.com/wp-content/uploads/2023/07/Electrical-Energy-Storage-System.webp.

In addition, the application of power electronic converters plays a significant role in grid integration of BESS to support enhanced functions that enhance the capabilities of the grid support. Increased interest in deploying these state-of-the-art solutions points towards ongoing research and development and the establishment of standards to promote interoperability across multiple energy storage systems (Farivar et al., 2023). Furthermore, regulatory frameworks have an important contribution to make in supporting the deployment and uptake of electrical energy storage technologies. Policymakers must contend with a changing landscape fueled by burgeoning technological development and the growing contribution of renewables to the energy mix. Establishing standards for the acceptance of BESS can go a long way towards helping support the operation of the grid, fostering certainty for actors within the energy space. Strong standards of framework and economic incentives act to overcome entry barriers that are perceived to new technologies, ultimately leading to mass deployment of BESS. Impacted positively by regulatory activities, projects have been successful in case studies with better grid performance resulting from the partnership between grid operators and energy storage providers. These collaborative initiatives are required to realize the complete potential of electrical energy storage, delivering a stable and robust energy system for the future (Akinsooto et al., 2024) (Figure 8.6).

8.2. MECHANICAL ENERGY STORAGE SYSTEMS (MESS)

MESS are of great importance in the broader context of energy storage technologies, particularly as the necessity for secure and sustainable energy options becomes more and more amplified. They utilize the mechanism of energy conversion and storage in mechanical forms, such as gravitational potential energy or kinetic energy. Both the most widely used mechanical energy storage forms are compressed air energy storage (CAES) and pumped hydro energy storage (PHES). PHES stores energy using gravity as water is pumped to the top when there is a low demand for energy, and is released to generate power when there is a high demand. CAES stores compressed air within underground caverns or tanks, which are then heated and expanded to drive turbines.

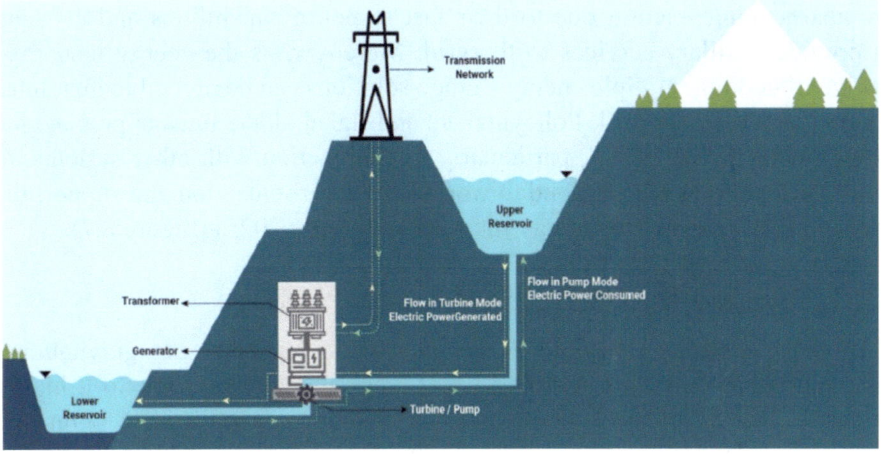

Figure 8.7. A pumped-hydro energy storage (PHES) system.

*Source:*https://www.researchgate.net/profile/Ammu-Jacob/publication/357900492/figure/fig1/AS:1113526802350080@1642496886358/A-pumped-hydro-energy-storage-PHES-system.png.

Both systems play key roles in power grid stabilization and integration of renewable energy sources, particularly solar and wind, which are prone to variability of production. In considering economic viability with respect to mechanical energy storage systems, consideration must be given both to capital expenditure as well as operational efficiency. Mechanical systems like PHES, as per recent findings, have long life cycles and relatively minimal maintenance expenses, which further contribute to their utility-scale benefits. While the initial capital investment can be costly due to location-specific geographical and constructional requirements, their ability to provide mass-scale energy storage solutions and grid stability makes them invaluable. LCOS for PHES is comparable with other technologies, especially in terms of long-term applications where the dispatchability of energy is a requirement. The results of the analysis of such systems indicate that while they are geographically limited, the capability to effectively utilize renewable power positions MESS at the vanguard of a transition towards sustainable power systems (Aleksandrov et al., 2025). Developments in the fields of technology and energy management in the recent past also enhance the effectiveness of mechanical energy storage systems. The integration of advanced energy management systems (AEMS) improves operation efficiency through improved energy dispatch forecasting and optimization. These technologies can potentially gain up to 10%–30% efficiency, thereby guaranteeing higher utilization of stored energy during peaking demand. Emerging technologies such as flywheel energy storage are also gaining traction due to their fast response capabilities and capacity to provide ancillary services to the grid. Moreover, as the energy landscape evolves, bundling multiple energy storage solutions can be pivotal to guarantee a strong and resilient grid. Policymakers and stakeholders must appreciate the growth potential of MESS, particularly in conjunction with other systems, to implement policies that respond to worldwide decarbonization and support the transition towards a clean energy future (Afreh et al., 2025) (Figure 8.7).

8.2.1. Pumped Hydroelectric Storage (PHS)

PHS is a predominant method of energy storage that harnesses gravitational potential energy through pumping water from a lower reservoir to a higher reservoir. During off-peak hours, surplus electricity is utilized to pump the water upwards, creating a reservoir with potential energy. When electricity demand is high, stored water is let down, flowing through turbines to generate electricity. This is an extremely effective process, often providing round-trip efficiencies of 80-90%. As grid stability and penetration of renewable energy increases, PHS plays a critical role in balancing supply and demand, especially with the higher penetration of intermittent energy sources such as solar and wind power. The significance of PHS in the overall context of energy storage is underscored by its ability to provide not just energy but also ancillary services such as frequency regulation and reserve supply. The operational flexibility of pumped hydroelectric storage plants is particularly useful in managing the intermittencies of renewable energy sources. According to recent studies, such integrated systems can react almost instantly to variations in energy consumption and generation, hence being important for maintaining grid stability. The integration of PHS in power grids enhances the chances of increased application of dispersed generation as it accommodates the dynamic nature of renewable energy. Furthermore, technology innovation is extending the applications of PHS so that systems can be designed to suit various geography and power demands. By storing energy effectively when it is available and releasing it where it is needed, pumped hydroelectric storage has the ability to transcend the limitation of renewable resources and help create a strong energy infrastructure, hence making it a vital part of the next generation of energy scenarios (Ault et al., 2005). Despite its advantage, development and installation of pumped hydroelectric plants are not without issues. Choosing the site is critical because suitable geographical locations where there is a substantial elevation difference are rare, and permitting can be lengthy and complex. Additionally, while PHS is an economically sound choice in most cases, it entails high up-front capital expenses, which are bound to be a disincentive to development. But as noted in recent research, the growing demand for energy storage technology due to integration of renewable energy sources promises to improve the viability of PHS facilities. Innovative developments in hydropower and enabling regulatory policies can make the environment conducive to investment in PHS schemes. In conclusion, the history of pumped hydroelectric storage indicates its significance in the transition to a sustainable and resilient energy future (Nikola, 2018) (Figure 8.8).

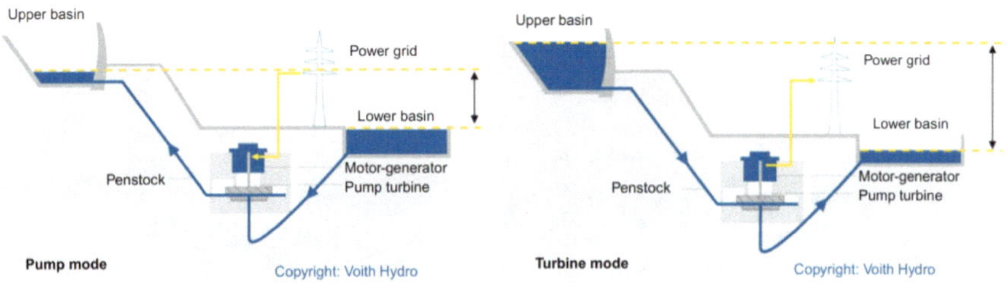

Figure 8.8. Pumped hydro-storage (PHS) working principle.

Source: https://planetdecarb.com/wp-content/uploads/2021/11/1-1.png.

8.2.2. Compressed Air Energy Storage (CAES)

CAES is a focal point of modern-day energy storage technologies, offering a feasible solution to the problems caused by intermittent renewable energy sources. The concept behind CAES is to store excess electrical energy as compressed air, which will be released to drive turbines and electricity production when needed. This is done most ideally for renewable power like wind power and solar power, whose generation is fluctuating with external environmental factors. As seen in recent studies, CAES systems are competitive in terms of economics compared to traditional energy storage systems, becoming a building block for the future energy system towards decarbonization and sustainability (Zhang et al., 2025). Moreover, new developments in CAES technologies have led to interest in various configurations, such as diabatic, adiabatic, and isothermal systems, all with purposes of boosting efficiency and operational performance. CAES system efficiency and scalability have been enhanced through ongoing improvements in technology and progression towards larger installations. CAES has traditionally progressed from simple application in mechanical propulsion to sophisticated energy storage complexes capable of helping provide grid stability as well as meet peak period demand. Current deployments are typically in gas storage facilities, where the stored air is compressed and prepared to utilize energy rapidly and revert to electricity.

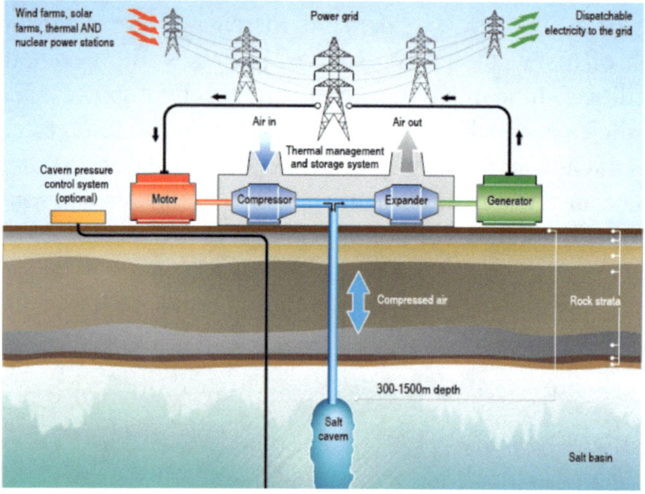

Figure 8.9. Compressed air energy storage (CAES).

Source: https://www.oilfree-air.eu/wp-content/uploads/CAES-schematic.png.

A comprehensive study demonstrates that LCOS of large-scale CAES is barely higher than that of established pumped hydro storage, demonstrating its economic viability and feasibility for its mass application (Jankowski et al., 2024). Given the evolving energy scenario, such advanced CAES projects are likely to expand further, dealing with increased energy capacities and efficiency suitable for the needs of the modern energy situation. Secondarily, the future for CAES is towards better utilization of renewable sources of energy. The current requirements in terms of sustainability of the environment demand technologies that not only optimize the efficiency of compressed air systems but also keep their carbon footprint low. New concepts are being investigated, such as coupling CAES with thermal storage technology to offer improved energy management and performance enhancement. Development review of the CAES technologies indicates tremendous potential for the systems to be utilized as fully fuel-free energy storage systems. Focusing on key areas such as compressors, expanders, and storage chambers, researchers are optimistic that they can identify means of utilizing even greater performance improvement and commercialization. With the advent of such technologies, they are likely to be pivotal in the overall goals of energy reliability and resilience in electricity supply and assisting in facilitating the transition towards a low-carbon economy (Zhang et al., 2025) (Figure 8.9).

8.2.3. Flywheel Energy Storage

Flywheel energy storage is one of the most important aspects of the evolving face of energy storage technology, particularly for maximizing grid stability and integrating renewable energy. The fundamental operation of flywheel technology is founded on angular momentum preservation, wherein energy is stored as kinetic energy in a spinning mass. The method has some benefits such as high power density, long life, and minimal maintenance. With the increasing requirement for efficient and adaptable energy storage, flywheels are also being targeted as they are able to offer immediate bursts of power, which makes them particularly well-suited to address the problem of intermittency of renewable energy production. In addition, the continuous progression in material technology enhances the capability for performance of flywheel systems to suggest that advancements in this field can further solidify their position in the future energy system (EPSRC (Funder) et al., 2008). The technical design of the flywheel system is also vital because the selection of electric machines plays a significant role in ensuring optimum performance. Most importantly, Permanent Magnet Synchronous Machines (PMSM) and Induction Machines (IM) are two common technologies employed in flywheel energy storage systems (FESS). The operation characteristics of these machines have direct effects on charge-discharge cycle efficiency, since electrical drive losses may impact the performance of the overall system. Importantly, squirrel-cage induction machines possess less standby loss and hence are more desirable for power applications involving constant energy regulation. Research in this area has demonstrated that good modeling and control strategies can optimize system responsiveness in various operating phases, enhancing the effectiveness of flywheel systems in practical applications (Breeze et al., 2019). With the availability of an endless supply of renewable energy sources, the absolute significance of robust high-energy storage solutions is becoming more evident. The integration of flywheel energy storage technology is not only promising on the technical merit but also in that it can synergize with other modalities of energy storage. With the combination of flywheel technology with such systems as electrochemical batteries or supercapacitors, the hybrid systems formed possess better performance characteristics, such as better energy density and safety of operation. This synergy in energy storage is paramount because no technology alone will be able to respond to requirements placed on contemporary energy systems. Overcoming challenges at hand with a polymeric approach in research and development, particularly in materials science and engineering in the field of flywheels, will be central to achieving the ambitious vision for future energy storage as planned within the next few decades (EPSRC (Funder) et al., 2008) (Figure 8.10).

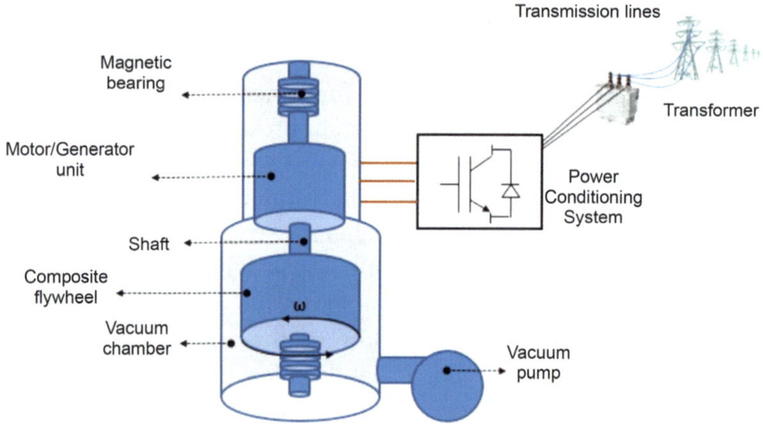

Figure 8.10. Schematic diagram of a flywheel energy storage system.

Source: https://planetdecarb.com/wp-content/uploads/2021/11/5-2.png.

8.3. CHEMICAL ENERGY STORAGE SYSTEMS

Chemical energy storage systems are a significant sector in the field of energy storage technologies, offering unique advantages for successful energy storage and release. Chemical energy storage systems apply chemical reactions to store energy in chemical bonds, offering high energy densities that surpass traditional mechanical and electrochemical means of energy storage. The versatility of chemical energy storage systems is observed in application ranging from renewable energy integration to portable power and electric cars. Increased demand for clean energy alternatives has led to a rise in research activity in order to fine-tune and optimize chemical storage methods. For instance, the ongoing development of thermochemical energy storage, where heat is stored by reversible chemical reactions, illustrates the energy efficiency potential of solar thermal systems' chemical systems, which merits their inclusion in the future energy dynamics (Copeland et al., 2025).

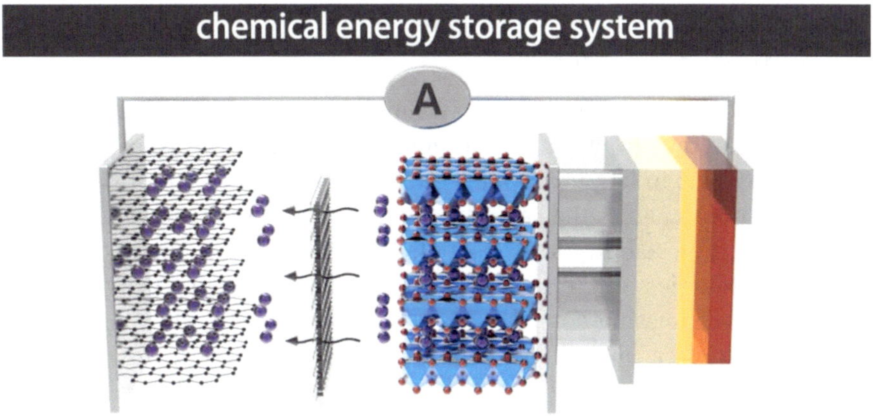

Figure 8.11. Chemical energy storage system.

*Source:*https://www.huntkeyenergystorage.com/wp-content/uploads/2023/08/chemical-energy-storage-system.webp.

The dynamic nature of chemical energy storage is also illustrated by hybrid energy storage systems that incorporate various energy storage systems. For instance, electrochemical batteries, together with supercapacitors and fuel cells, form a hybrid energy storage system with balanced energy density and performance efficiency. These arrangements are particularly useful in electric vehicle markets, where range and power output are extremely critical. Current studies indicate that hybrid systems are capable of reducing system weight significantly and enhancing vehicle range, and thus eliminating energy and performance limitations characteristic of stand-alone electrochemical batteries (Bingham et al., 2005). The feasibility of such hybrid systems offers the innovative promise of chemical energy storage for transforming transportation technologies and mitigating fossil fuel reliance. Besides, ongoing research into advanced chemical storage technologies continues to drive the advancement of energy storage systems. Direct contact latent heat storage and thermochemical energy storage technologies are demonstrating promising promise for increasing efficiency and reducing cost in a variety of applications. The rigorous assessment and ranking methods set in place to rank the performance metrics and economic viability of thermal storage facilities indicate the scientific community's resolve to see improved chemical energy storage technologies (Copeland et al.,

2025). Lastly, due to the evolving demands of society towards cleaner energy approaches, the role of chemical energy storage systems remains significant in facilitating more sustainable energy systems, underscoring their importance in modern energy systems (Figure 8.11).

8.3.1. Battery Technologies

Under the scope of energy storage technologies, battery technologies are prominent solutions to a range of applications, varying from consumer electronic products to large renewable energy systems. The use of batteries in these applications enables the smooth operation of electronic devices, electric vehicles, and grid stability when coupled with renewable energy systems. Lithium-ion batteries differ most notably in having high energy density, efficiency, and decreasing cost, that have stimulated their application in many industries. Moreover, alternatives such as lead-acid batteries, flow batteries, and sodium-ion batteries offer unique advantages and constraints particular to specific applications. A critical evaluation of these technologies has distinctive characteristics, including energy yield, charge cycles, and environmental factors, that play critical roles in stakeholder decision-making when it comes to energy storage technology for use in modern applications. A comparison of these kinds of batteries can be instrumental in optimizing them in the application in energy systems to enhance overall efficiency and sustainability (Abdumannob et al., 2024). As battery technology improves, development and design advancements in them are increasingly affecting the energy storage capacity. Advancements such as solid-state batteries and new lithium-sulfur batteries are set to enhance performance parameters by offering more capacity and lifespan with reduced thermal overheating and degradation risks. Such new technologies are being actively researched and developed to overcome shortcomings of existing lithium-ion systems, such as cycle life and rare material reliance. The quest for next-generation battery technology is not just about chemical makeup advances but also new ways of making them that reduce cost and enhance scalability. The pace of these innovations is part of a broader movement toward sustainable energy habits and with an acute demand for reliable energy storage technology that can keep pace with the mounting addition of renewable energy sources to the grid. These developments are crucial in that they address both near-term requirements and future issues of energy storage (Abdumannob et al., 2024). The application of battery technologies intersects with current demands for more energy resilience, particularly in situations impacted by renewable energy sources and electric vehicles. With California pushing towards high levels of renewable energy penetration, battery integration implications become increasingly more important. Their connection into the grid not only assists in voltage stability but also addresses potential problems caused by intermittent energy inputs from wind and solar sources. Volt/VAR Optimization and Control (VVOC) practices existing procedures must attend to these challenges; however, existing methodologies must transform to address growing complexities of decentralized energy systems. Whereas research highlights such imbalances, the exploration of emerging technology and its application in the development of VVOC is a turning point for development. Such interactive dynamics between battery technologies and grid management are symptomatic of the central significance of smart energy storage solutions in realizing a sustainable energy future (Figure 8.12).

Figure 8.12. Advances in battery technology for energy storage.

Source: https://beasfa.com/wp-content/uploads/2024/08/Picture1-1.png.

8.3.2. Hydrogen Storage

Hydrogen storage has emerged as a front-stage technology in the energy storage solution ecosystem. With the increased emphasis on renewable sources of energy such as solar and wind power, it has created the pressure of maintaining efficient mechanisms for energy storage to provide a steady power supply. Hydrogen storage technologies are a means to bridge the energy production and utilization gap by storing excess energy produced during periods of peak production. This hydrogen can then be utilized for producing electricity, eliminating the intermittency problem of renewable energy. As the world moves towards decarbonization, technological innovation in hydrogen storage is a significant milestone in achieving sustainable energy systems. Besides, the integration of power generation and hydrogen storage can optimize efficiency in the system and contribute towards an equilibrium energy supply, emphasizing its contribution as a solution to energy demand in a developing energy economy. Hydrogen storage technologies boast one of the most visible applications as Hydrogen-Electricity Coupling Energy Storage Systems (HECESSs).

Figure 8.13. Hydrogen storage.

*Source:*https://media.geeksforgeeks.org/wp-content/uploads/20230103100704/Hydrogen-Storage.jpg.

These systems offer an integrated framework in which hydrogen serves as a go-between between producers and users of various energy sources. Through this, HECESSs enable maximum efficiency in the consumption of energy through the storage of excess energy in the form of hydrogen during times of low demand. When there is high demand, this excess hydrogen can be converted to electricity or utilized in industrial applications. As pointed out, some of the current applications of hydrogen storage technologies are power generation, storage, and hydrogen production, showcasing their diversity and potential for energy security contributions. It is important to comprehend HECESSs and the operational modes in which they operate because they have the potential to make significant changes in how we handle energy resources and respond to rising demands of an energy-based modern society (Zheng et al., 2024). Hydrogen storage research and development play a critical role in closing the gap of issues like efficiency, cost, and scalability. Material science breakthroughs like the application of biochar as an electrode material in hydrogen storage systems show promising directions for increased performance. This renewable material not only reflects the dynamics of circular economy but also illustrates desirable physicochemical characteristics, such as large surface area and chemical stability, that are feasible for energy storage applications (Prabakar et al., 2024). In addition, application of deep reinforcement learning algorithms can improve processes within complex energy models based on hydrogen storage. With the advancement in technology, technological advancements and environmental impacts must be taken into account so that hydrogen storage is a contributing factor to global energy systems and decreases environmental problems. Therefore, continued research and development of hydrogen storage technologies will be essential to creating a sustainable energy future (Figure 8.13).

8.3.3. Thermal Energy Storage (TES)

TES can contribute a lot in making energy consumption more efficient and in raising the use of alternative sources of energy. Through the utilization of surplus thermal energy during low demand periods and releasing it during high consumption periods, TES facilities can effectively balance demand and supply. This aspect is particularly useful in renewable energy, where supply may not be consistent. Integration of thermal energy storage within the building structure presents a unique opportunity to maximize the utilization of electricity with the least demand for costly infrastructural adaptations. Integration can result in substantial cost saving and improved energy efficiency without compromising on available space in a building, thus paving the way for more environmentally friendly architectural designs (Allison et al., 2017). The convergence of renewable energy and TES represents a very promising path for the further evolution of energy storage technology that is economically viable and environmentally friendly. The breakthrough in direct thermal energy storage systems also demonstrates the revolutionary potential of TES in the generation of electricity. Such systems are often employed in concentrating solar power (CSP) stations to effectively harness solar power. Comparative analysis of various direct TES technologies has illustrated the advantages of modifying classic plant configurations to enhance performance and cost savings. Single-tank storage technologies, for example, are capital-saving compared to dual-tank systems but provide similar thermal storage performance. Further, optimization of heat transfer fluid management in such systems can significantly minimize temperature degradation during discharge hours, thereby enhancing overall system efficiency (Manenti et al., 2013). These innovations not only ensure solar energy projects become economically feasible but also ensure more consistent energy generation, and thus thermal energy storage forms a part of renewable energy systems in the future. In addition to their application in utilities, thermally integrated storage solutions have significant implications for energy policy and management as well. With energy grids increasingly able to handle larger percentages of renewable sources, the need for systems that allow for active load management increases. Active thermal storage systems can actively adjust to the swings of energy availability and market price, allowing buildings to be active players within the energy market. This active participation reverses the old passive energy storage ways of doing things, making for more responsive, more nimble energy systems. By creating a taxonomy of the current thermal energy storage solutions, researchers have the ability to recognize gaps within the contemporary environment and suggest design improvements to suit particular building applications (Allison et al., 2017). This strategic intervention not only highlights the status of innovation within TES but also its role in creating a sustainable future for energy (Figure 8.14).

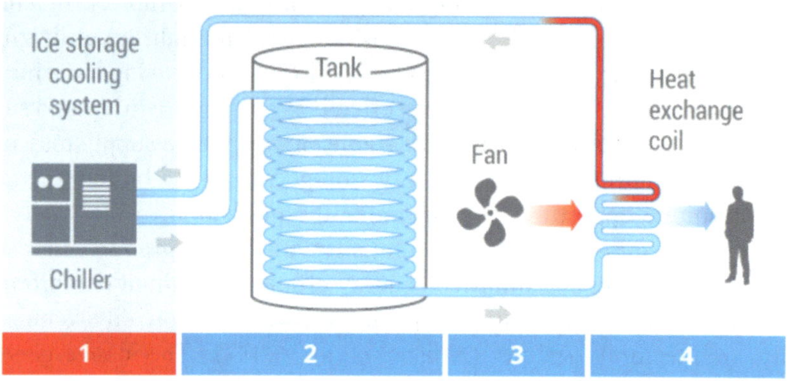

Figure 8.14. Thermal energy storage (TES).

Source: https://www.calmac.com/stuff/contentmgr/files/0/5052bccbfda20e9b18f57f773dfa5170/images/imagemap1_1.png.

8.4. ELECTRICAL ENERGY STORAGE SYSTEMS (EESS)

The development of EESS is set to address the intermittence issues in renewable energy supplies. With growing energy demands globally and the use of renewable sources on the rise, having the ability to store energy efficiently matters. EESS stabilizes not only the power grid but also adds reliability to the supply of energy during peak hours. Technologies such as lithium-ion batteries and redox flow batteries (RFBs) are promising access to scaleable solutions that can service large and small markets. While current technologies such as pumped hydro and compressed air still reign supreme, newer ones are being seriously considered on the basis of their particular strength in efficiency and deployability (Villanueva et al., 2019). One of the most recent advances in electrical energy storage is transcritical carbon dioxide systems combined with thermal energy storage. This innovative technology has been discovered to pose economic potential because it can function well with solar thermal resources, thus providing energy storage needs while allowing for renewable integration. The efficiency improvement while integrating solar power is evidence of the ground-breaking potential of the technology. With more than 60% round trip efficiencies being reported when employing solar input, transcritical carbon dioxide systems present a competitive alternative to existing options (Villanueva et al., 2019). With better configurations and performance parameter analysis, researchers are finding ways to optimize energy output while remaining within various scales of energy. The increasing interdependence of electrical energy storage and renewable technology increases the need for further research in the field of material science, and in this case, with the development of efficient electrodes and separators in battery systems. The investigation of carbon nanotube/polymer composites indicates promising avenues for performance enhancement in electrochemical storage technology. The composites provide twin benefits of greater conductivity and mechanical stability, which are critical to the optimization of battery longevity and efficiency. As the need for energy solutions in portable forms grows, the development of small-scale systems is also equally essential and provokes research in fields such as consumer electronics and electric vehicles (Jia et al., 2016). Overall, the combination of novel materials and advanced energy storage technology is a revolutionary era that will remain at the core of next-generation energy systems, with interdisciplinary thinking and research investment being the decisive factors.

8.4.1. Supercapacitors

Among all the energy storage devices in the process of being researched today, supercapacitors stand out because they possess distinctive attributes that enable them to discharge and charge at a higher rate compared to traditional batteries. Their superior performance is best explained by their electrochemical double-layer capacitance, which allows supercapacitors to store energy in the process of charge carriers separation. In contrast to batteries, which rely on chemical reactions to generate electricity, supercapacitors tap into electric fields, which are highly efficient for the rapid release of energy. That they also do not lose capacity with thousands of charge-discharge cycles makes them an ideal candidate for utilization in energy management systems in various industries, including renewable energy integration and electric vehicles. The use of supercapacitors in hybrid power systems emphasizes their potential for enhancing energy efficiency in various technological applications. For instance, in Cube Satellites, supercapacitors are being used in electronic power systems (EPS) to provide reliable energy storage devices. According to research work carried out by research institutions, the utilization of supercapacitors allows for the integration of energy storage technologies consisting of both supercapacitors and conventional batteries. This hybrid method ensures power availability during the time of eclipses when the opportunity for solar generation does not exist. Based on this research, supercapacitors can be effectively used to complement batteries in energy storage systems and lead to increased system performance and longevity as seen through projects such as those conducted by the Robotic Systems Lab of Santa Clara University (Lynch et al., 2014). In aviation, supercapacitors are increasingly being utilized to aid in meeting the stringent needs of emerging aircraft propulsion systems. As electric propulsion systems increase power, electric energy density and low mass storage is becoming a preeminent need. Research for NASA's Subsonic Fixed Wing Project indicates that advancements in energy-storage technology, including supercapacitors, will be the key to allowing substantial increases in motor efficiency and power density over the next few decades. Future energy density growth in supercapacitors would potentially render application in hybrid propulsion systems more viable, thereby accelerating emissions and fuel burn reduction. The incorporation of supercapacitors in such systems has the potential to do away with the weight penalties of traditional turbine engines, thus paving the way for a cleaner aviation future (Choi et al., 2025) (Figure 8.15).

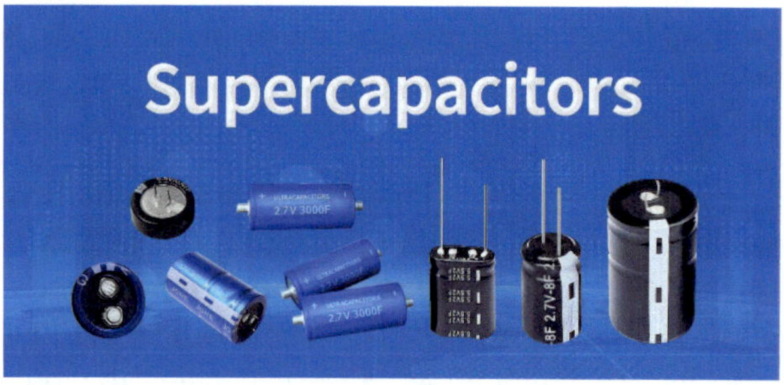

Figure 8.15. Supercapacitors.

Source: https://www.jeccapacitor.com/data/watermark/20250530/6839776045e01.jpg.

8.4.2. Grid-Scale Batteries

Grid-scale batteries represent an innovation leap for energy storage technologies to facilitate the imperative transition to sustainable energy systems. By storing reserve energy generated from renewable sources such as solar and wind, these batteries improve the efficiency and reliability of the grid, especially since electricity demand fluctuates during the day. Their ability to discharge stored energy during peak periods reduces pressure on traditional power generation, decreasing the risk of brownouts and aiding grid reliability. With the globe moving towards cleaner energy, the use of grid-scale batteries is necessary for harnessing the uncertainty of intermittent power sources, thereby minimizing their uncertainty challenges and making them the foundation of the next generation's energy infrastructure (Proser et al., 2011). The economics in the situation of grid-scale batteries are rapidly evolving, accelerated by technology and market demands.

Figure 8.16. Grid-scale battery storage.

*Source:*https://www.alsym.com/wp-content/uploads/In-Depth-Look-at-Grid-Scale-ESS.jpg.webp.

New battery technologies like lithium-ion and zinc-bromine flow batteries are generating interest because they can potentially bring about low-cost solutions for energy storage on the grid. Analysis estimates that while wholesale load shifting may not currently be financially viable, other applications of energy storage will at least break even when appropriately applied (Kluza et al., 2009). This evolving context increases the call to find certain use cases and operational contexts where grid-scale batteries can deliver maximum value, thereby enhancing overall system efficiency and reducing operational costs for utilities and grid operators. Despite the promising developments in grid-scale battery technologies, the regulatory environment presents formidable challenges that must be addressed to enable them to reach their full potential. Current regulatory frameworks often fail to adequately remunerate the numerous benefits of these storage facilities. That no less than the present concerns of the industry attest, making regulations responsive to the innovative potential of energy storage technology is critical to precipitating widespread usage. Policymakers need to recognize potential for grid-scale batteries to impact generation, transmission, and distribution, thus building a stronger and more dynamic energy climate. Therefore, framing a collaborative conversation between stakeholders—regulators, utilities, and technology innovators—will be key to driving the integration of these technologies forward and releasing their potential to transform our energy markets (Kluza et al., 2009) (Figure 8.16).

8.4.3. Flywheel Energy Storage (FES)

Implementation of flywheel energy storage (FES) systems is progressively picking up pace as a resilient solution in the domain of energy storage technology. By employing a rotating mass to store energy in the form of kinetic energy, flywheels provide a highly efficient and quick release means of storing energy. FES systems are particularly well adapted to uses requiring quick bursts of energy, such as grid stabilization and frequency regulation, primarily due to their high power-to-weight and long cycle life durability. This technology can fulfill the need for continuous and stable power supply, especially with the growing utilization of renewable energy sources. Due to this, the integration of flywheels with other storage technologies, such as lithium-ion batteries, could introduce more efficient operational performance and longevity in energy systems, emphasizing the necessity of enhancing disparate storage solutions (Maroufi et al., 2025). With renewable energy sources (RES) expanding to support increasing energy demands, the issue of how to address their stochastic character is at the forefront. Flywheel energy storage is such a technology that is perfect in its capacity to smooth out the fluctuations that come with intermittent energy resources like wind and solar. Where massive renewable energy targets are being set for countries like India, utilization of flywheel systems can be a determining component in terms of achieving grid stability and reliability. The ability of FES to dynamically inject or absorb power makes it very suitable for countering the inherent variability of renewable energy sources. Moreover, flywheels can serve as a more eco-friendly supplement to traditional energy storage technology, e.g., lead-acid batteries, thanks to the high specific energy and peak power capability of flywheels, which reinforces their competitive edge in current energy application (Kumari et al., 2024). In addition, the future of flywheel energy storage technology is one of ongoing advancement in materials science as well as system design. The advancements of SoC correction technologies, together with more sophisticated power management techniques, are critical to the optimization of the operational effectiveness of flywheel systems. The introduction of fuzzy logic controllers and moving average algorithms are some of the innovative approaches being developed to further improve system performance. With the utilization of such methodologies, FES is capable of facilitating effective energy requirements and improving power redistribution, thereby mitigating performance differences among storage technologies. Such technologies indicate the potential of flywheel systems not only as key contenders in energy storage but also to facilitate smoother transition towards a renewable energy infrastructure that is able to deal with the demands of an increasingly evolving energy landscape (Maroufi et al., 2025).

8.5. APPLICATIONS OF ENERGY STORAGE TECHNOLOGIES

The function of energy storage technologies has grown increasingly vital in furthering the efficiency and effectiveness of modern power systems. One of the most important applications is balancing demand and supply in real-time. As intermittent power sources such as solar and wind become more prominent, storage technologies can be employed to reverse energy generation fluctuations. Storing excess power produced during peak-generation periods and releasing it when demand is high creates a smoother supply of energy. Besides, energy storage systems allow the strategic management of electricity flows within microgrids, hence ensuring outage resilience and overall grid performance optimization. The capability to control energy availability is central to the highlighting of the need for energy storage in transitioning towards sustainable energy systems. Energy storage technologies are also imperative in facilitating the growth of electric vehicles (EVs). With the automotive industry shifting towards electrification to combat climate change, efficient energy storage holds the key to optimum range and performance of EVs. Lithium-ion batteries, for example, have gained market share due to low energy density and declining costs, becoming the preferred choice for the industry. Another push for the electrification of transport is the development of fast-charging stations with energy storage devices. By facilitating fast recharging as a way of cushioning pressure on the power grid, these solutions become more accessible to consumers while also facilitating the increased adoption of cleaner means of transport. The interoperability of energy storage technologies with electric vehicles is a mutually beneficial relationship that inspires various sectors to grow sustainably. In power infrastructure development with resilience, energy storage systems are relied upon for matching distributed generation systems with the grids. Decentralized generation has proven to be a trend with the requirement for systems to control localized energy sources without jeopardizing the grid's integrity. Technologies such as battery storage, pumped hydro storage, and thermal energy storage possess the ability to optimize the energy extracted from these distributed sources and channel it towards peak load management or ancillary services. Referral to existing storage technologies explores primary uses that can be employed in various situations. For instance, thermal storage systems can be capable of providing cheap solutions in thermal energy utilization, while electrification and integration of renewables demonstrate the need for top-level management strategies to optimize energy storage potential (Ault et al., 2005; Vecchiola et al., 2009). Therefore, innovative energy storage technology's contribution is essential in transforming our approach towards energy use and supply.

8.5.1. Renewable Energy Integration

The integration of renewable energy resources into existing energy systems is both an opportunity and challenge for modern energy systems. Energy from sources such as solar and wind has been increasingly popular over the last few years with growing concern over climate change and the need for sustainable energy solutions. However, their natural intermittency necessitates a secure energy storage solution to maintain supply and demand stability. Because the available capacity for renewable generation is usually larger than consumption, particularly at peak production, energy storage technology, such as various ways of geological reservoir methods, is critical to reduce supply volatility, enhance grid management, and provide security of energy. Effective storage of energy will not only facilitate the transition to a low-carbon economy but also ensure more stable energy infrastructure, healing both environmental and economic problems in power production (Carneiro et al., 2016). Different energy storage technologies emerge as viable solutions to facilitate larger-scale integration of renewable energy and ensure grid stability. Of these, Underground Gas Storage (UGS), Compressed Air Energy Storage (CAES), and Thermal Energy Storage (TES) stand out because they have the capacity to store a large amount of energy with minimum environmental impact. All these technologies utilize some form of geological structures—e.g., depleted hydrocarbon reservoirs, aquifers—to effectively store energy. The utility of such rock formations is governed by factors that are site specific, and these play a critical role in the selection of an appropriate storage technology for specific applications. With energy demand continuing to rise and grid management becoming more complex, accessing geological energy storage potential has the ability to significantly boost the efficiency of renewable energy systems as a whole by ensuring that power produced is made more accessible when it's required and reducing greenhouse gas emissions in the process (Carneiro et al., 2016). Innovative concepts in terms of energy storage, such as next-generation thermal energy systems, have huge potential in addressing the challenges associated with renewable energy non-dispatchability. By implementing transcritical carbon dioxide cycles in conjunction with thermal energy storage systems, these technologies seek to maximize the production and utilization of energy. The efficiency of such systems is heightened when integrated with solar energy, raising the energy efficiency of the systems and lowering the cost of operation. For example, the integration of solar thermal energy can boost the system's efficiency, proving that merging renewable sources with energy storage technology is not only possible but desirable. The cost-effectiveness of such measures again aligns with greater objectives of energy conversion, providing a path forward for the augmentation of renewable energy penetration worldwide without sacrificing increasing energy requirements.

8.5.2. Grid Stability and Reliability

Grid reliability and stability are critical parameters that make possible the efficient functioning of contemporary energy systems. As renewable energy sources like solar and wind energy take center stage, the very fact that these sources are intermittent ones presents enormous management issues to the grid. Volatility of energy production requires a robust system in order to achieve a balance between demand and supply. Energy storage technologies are the ultimate solution to assist in bridging the challenges to be overcome, to store excess power generated during periods of maximum production. The integration of storage solutions not only reinforces the resilience of the grid but also aids in the provision of improved voltage conditions and reduced power losses, ultimately resulting in the provision of more reliable energy to consumers. The complementarity between advanced energy storage and renewable energy sources can thus be viewed as a cornerstone for the achievement of greater grid stability and reliability in a transforming energy regime. The application of energy storage technologies, such as lithium-ion batteries and flow batteries, plays a fundamental part in ensuring grid resilience. Such technologies can provide services such as frequency regulation, peak shaving, and energy arbitrage. Energy storage can counterbalance supply variability caused by intermittent renewable sources through storing energy during high availability and releasing it during peak demand. In addition, intelligent grid systems, such as high-level communication and control systems, improve the effectiveness of energy storage in managing the operation of the grids. Using predictive analysis and real-time data, grid operators can enhance their response to shifting energy conditions, ultimately resulting in a more secure and stable electrical grid. The power quality management and disturbance reduction performance analysis of various energy storage implementations has been shown to be significant improvement. The integration of photovoltaic systems into electrical power grids also depicts the extensive impact of energy storage on reliability and stability. In studies on the comparison of the effect of photovoltaic power plants on the performance of the grid, it has been established that with the addition of energy storage devices, these systems effectively insulate the fluctuation in solar power generation. The results forecast the potential of renewable installations not only to complement power generation but also to enhance the grid's reliability as a whole by reducing the frequency and duration of outages. By enabling better management of active and reactive power currents, energy storage systems enhance the seamless integration of renewables, maintaining stability throughout the grid. With advancements in energy storage technologies, they have the potential to craft a more stable electric grid, ultimately leading to broader sustainability objectives and energy security as an answer to increased renewable energy adoption (Kumar et al., 2025; Zeljković et al., 2025).

8.5.3. Electric Vehicles (EVs) and Transportation

The transition to EVs has emerged as a key element in the general energy storage technologies agenda. As urbanization accelerates and global transport demand keeps rising, conventional fossil fuel consumption contributes a large share to greenhouse gas (GHG) emissions. There is a need for a paradigm shift to clean technologies that address not only the challenges of energy supply but also minimize local pollution. The advent of EVs, facilitated by advanced energy storage technologies such as lithium-ion batteries, is a critical solution to these problems. Powered by clean energy sources and energy-efficient technology, EVs are at the forefront of reducing the carbon footprint of the transport sector, an imperative link between energy storage technology and sustainable transit options. In addition to their environmental benefits, EVs are rapidly acquiring value for their economic prospects in an era of rising oil prices and economic competition. As world oil reserves decline, the price of continuing to rely on traditional fuels is unaffordable. The integration of hydrogen power, addressed in the context of alternative technologies, highlights the future prospects of EVs as being powered not only by traditional electricity but also by hydrogen fuel cells. This diversification can enable energy security through a stable efficient substitute for oil-based fuels. Additionally, economic advantages of EVs extend beyond the operational cost to include reduced health and environmental cleanup costs associated with vehicle emissions. Energy storage technologies thus play a double role in maximizing transportation dynamics while enabling economic feasibility. Electric vehicle implications extend into societal and infrastructural realms beyond transportation. Mass adoption and infrastructure readiness are essential for mass EV adoption, involving a considerable outlay on charging points and renewable energy generation. As cities embark on this new future, mechanisms will need to be put in place to manage peaking of energy demand resulting from increased EV usage. Furthermore, integration of mobile energy storage technologies—such as vehicle-to-grid technology—can contribute to a more stable and robust grid. This broad strategy is an overarching rethinking of transportation infrastructures, with EVs playing not the role of alternatives to traditional cars but of cornerstones of an advanced transport infrastructure. By embracing innovation in power storage and different propulsion modes, society can drive itself towards a cleaner and more efficient transport landscape, ultimately building a sustainable future for mobility.

8.6. CHALLENGES AND LIMITATIONS

The technology of energy storage is beset by several challenges that make them unpracticable and not widely used. The most significant challenge is the inherent limitations in the storage material. For instance, hydrogen storage means such as cryogenic, compressed, and liquid hydrogen are plagued with enormous technical issues in terms of volumetric density and efficiency. The low density requires using huge storage systems, which accordingly causes the space problem in applications, particularly in transportation. Also, there is a problem with cooling and boil-off, where the hydrogen that is stored at cryogenic conditions tends to evaporate with time, causing possible efficiency losses as well as safety risks. Research identifies that these are challenges that require intense investigation into better insulation and pressure vessel materials in order to increase storage capacity, showing the need for material science innovation to break through obstacles in energy storage systems (Reyes et al., 2007). Additionally, economic viability is another significant limitation to energy storage technologies. The huge capital required in the development and deployment of advanced storage systems, such as lithium-ion and flow batteries, is too costly, particularly for developing nations. Capital-hungry investments tend to be made in such technologies, and without proper financial incentives or complementing policies, their penetration in existing energy systems is inactive. Furthermore, lifecycle impacts of manufacturing and end-of-life battery disposal pose environmental concerns, complicating the cost-benefit calculation. As evident from various studies, all stakeholders must consider not only the economic factor but also sustainability factors so that technological development does not compromise environmental objectives (Miskam et al., 2013). Finally, technological sophistication and operational limitations of energy storage devices render them a serious challenge. The majority of storage technologies, including battery systems, incur degradation of performance with time, impacting their efficiency and reliability. Additionally, installation of such systems into existing energy grids is technically cumbersome, requiring sophisticated management and control systems for optimal operation. Factors like losses associated with conversion processes and varying discharge rates make their use in operational applications difficult, thus continuous research efforts are inevitable for creating more flexible and resilient solutions. As the energy landscape shifts, these challenges will be critical to overcome in creating a secure and sustainable energy future, again emphasizing the need for continued innovation and investment in energy storage systems (Reyes et al., 2007).

8.6.1. Cost and Economic Feasibility

The economic feasibility of energy storage technologies is central to the transition toward sustainable energy systems. With the increasing energy demand, driven by increasing vehicular populations and use of energy, the cost of current sources of energy is no longer viable. Reddy et al. (2025) identifies four closely related reasons to shift towards alternative fuels and technologies: shrinking oil resources, environmental issues, competitiveness, and energy security. All these factors emphasize the need for the development of energy storage technologies that not just supply the energy where and when it is required but also do so in an economical manner. In placing economic feasibility ahead in the development of new storage technologies, one can create solutions not only sustainable but also accessible, minimizing the potential of energy crises and fostering public acceptance. Further, a comparison of different energy storage systems exhibits significant variation in their economic efficiency for different uses. The comparison of existing technologies like lithium-ion battery, Redox Flow Batteries (RFB), and Adiabatic Compressed Air Energy Storage (A-CAES) reveals the variation. According to Adetunji et al. (2025), while lithium-ion batteries are being extensively used, systems like A-CAES exhibit good performance in the context of levelized cost of storage. The mixed-methods approach employed in this analysis emphasizes the importance of economic indicators—such as net present value and internal rate of return—in ascertaining the viability of energy storage alternatives. By targeting these quantifiers, decision-makers are able to make preferential investments in technologies that are not only superior under some operating conditions but also financially sound in the long term. Furthermore, the integration of renewable energy involves a redefinition of traditional economic principles. The majority of renewable integration cases foresee inefficiencies under which high price volatility is necessary to achieve profitability, particularly in day-ahead arbitrage. This aligns with observations presented in Adetunji et al. (2025) that all the technologies are proved efficient in intraday arbitrage but reliance on some methods can be financially vulnerable under more stable situations. To enhance the viability of energy storage technologies, innovation in reducing cost across different platforms, especially for those like Liquid Air Energy Storage (LAES), which persistently exhibit worse financial performances, is paramount.

With the world moving steadily towards an energy future, embracing a mix of storage technologies from smart batteries to other platforms will be crucial in terms of minimizing the cost and making it easy to integrate renewable energy into the grid.

8.6.2. Environmental Impact and Sustainability

The nexus between energy storage technologies and the environment is increasingly being recognized as most important in the quest for sustainability. Effective energy storage systems enable the achievement of integration of renewable energy sources, reducing the use of fossil fuels. As the systems improve, they can reduce greenhouses emissions by significant measures. The environmental impacts of the installation of such systems must be assessed with caution, however. For instance, certain energy storage technologies such as lithium-ion batteries have green issues surrounding mining of raw materials, which causes loss of habitat and contamination. Hence, even though energy storage can facilitate a transition to greener energy systems, sustainability goals need to be assured without exacerbating other environmental issues with proper knowledge of the environmental expenses. Additionally, the deployment of energy storage technologies must be aligned with energy management practices that are sustainable in orientation. The integration of concepts such as cleaner production and lifecycle analysis has the capacity to make these technologies more effective. For instance, anaerobic digestion is not only used to store and generate energy but also to utilize waste materials, and therefore aligns energy production with waste management practice to make industries, such as food processing, more sustainable in their processes. By maximizing the use of energy, industries can reduce their carbon footprint while increasing efficiency. It is apparent that in the absence of an energy storage strategy complemented by sustainability, the beneficial impacts of environmental footprint might be compromised by less obvious operational inefficiencies or harmful byproducts resulting from technology deployment (Benoit et al., 2012). Finally, the role of novel technologies like carbon capture and storage (CCS) towards making energy storage options more sustainable cannot be overstated. As economies evolve towards less-emitting industries, CCS technologies have the potential to play a vital support role in energy security while meeting environmental requirements. While CCS is yet to emerge and has many challenges confronting it—technical, economic, and safety-related—it has a big role to play as a bridging technology. Effective integration of CCS with energy storage would have the ability to decrease marginal costs of achieving climate objectives while encouraging a rich energy mix required for energy sector longevity and sustainability. This kind of synergism between different options for mitigation highlights the imperative of integrated strategies with respect to environmental sustainability on the one hand and technological innovation acceptance on the other (Praetorius et al., 2025).

8.6.3. Technological Limitations and Advancements

The rapid advancement of energy storage technology holds both massive achievements and inherent limitations that characterize their usage. Lithium-Ion (Li-Ion) batteries rank among the front-running technologies due to their high energy density, efficiency, and falling prices that have made them the pillars of modern energy storage systems. Despite their skyrocketing demand, material scarcity and ecological footprint concerns have come to the forefront. Specifically, the reliance on strategic minerals like cobalt is where supply chain sustainability and geographical concentration concerns come into play. Ultimately, even as innovation has positioned Li-Ion batteries at the pinnacle, such restrictions necessitate a two-pronged plan that is simultaneously innovative and prudent when it comes to being a good steward of resources to achieve long-term energy storage technology sustainability (Dey et al., 2025). In exploring the provision of alternative means of energy storage, one observes that technologies are appealing but also present special challenges. Such technologies as pumped hydroelectric and thermal energy storage hold immense potential for balancing supply and demand in grids. However, geographical limitations of such systems—in terms of requiring specific shapes of the topography—limit their applicability. Besides, the diversification of renewable sources necessitates the development of homeostatic energy management systems that will be able to optimize more than one storage technology simultaneously. This is a level of complexity that raises engineering challenges in terms of scalability and deployment timetables for emerging technologies. Overcoming these limits involves more than technological innovation but innovation in how energy storage systems are organized, deployed, and operated in varied landscapes (Kumar et al., 2024). As the global energy landscape shifts, the relationship between development and constraint is increasingly significant. Never have there been so many calls for efficient, cost-effective energy storage technology, particularly with renewables like solar and wind taking center stage in projections of power generation in the future. The development of newer battery types, including solid-state batteries and flow batteries, is gaining pace on the back of efforts to overcome issues of energy density, safety, and longevity. Moreover, materials science advancements are essential to develop storage systems to support demanding operating conditions. However, effective translation from the laboratory to real-world applications relies on technical breakthroughs as well as on economic feasibility. Thus, the potential for energy storage technology relies on continual innovation combined with a comprehensive understanding of the intricate challenges present in today's energy system (Dey et al., 2025; Kumar et al., 2024).

SUMMARY

Briefly, energy storage technologies are the breaking point in the pursuit for sustainable energy systems. Besides optimizing the utilization of renewable resources such as solar and wind power, the technologies enhance grid reliability and robustness. The monetary implications are even more critical, as evidenced in recent research that indicates integrating solar photovoltaic systems with battery storage is able to save tremendous amounts of money by minimizing electricity costs for various stakeholders. Perhaps most importantly, in the affordable housing market of California, solar installation tied to battery storage can give homeowners essentially zero electric bills, thus rendering renewable energy more viable for low-income communities (Airth et al., 2016). Such outcomes indicate the potential of policies and industry players championing energy storage technologies as a clean path to pursuing economic and environmental objectives. The technological review also mirrors how they can reverse intermittency issues in renewable energy generation. The availability of new storage systems, such as steam accumulators in biomass combined heat and power plants, provides a more flexible power supply, with the capacity to consider changes in energy demand without compromising thermal capacity (Abdessamad et al., 2019). Such power generation decoupled from traditional means of energy production is a prime example of how energy storage can facilitate higher operational efficiency and responsiveness in the energy sector. As energy systems globally continue to transform, the use of such technologies will be crucial in addressing peak demand as well as variability of renewable generation in the long run, thereby creating a more resilient and responsive energy grid. Lastly, strategic deployment of energy storage technology is central to propelling the transition to a sustainable future in terms of energy. With both economic analysis and technological progress showcasing the conformity of energy storage with renewable generation, compatibility promises a range of benefits spanning from reduced costs and improved reliability. Further research and development will not only reinforce empowerment of communities but also solidify global action towards reducing greenhouse gas emissions. As energy policies adapt to incorporate these advances, the concerted action of stakeholders—industry, government, and education—will bring to bear an atmosphere in which energy storage technologies are not only tolerated but celebrated as essential components of a fair, robust, and sustainable energy system. The science of energy storage technologies is at the core of the grid's electrification in today's world, particularly with societies moving toward renewable sources of energy such as solar and wind. These power systems are inherently changeable, generating challenges in delivering a consistent power supply. As a response to such problems, energy storage solutions play a critical role in ensuring power grids are not vulnerable to disturbance. Recent vulnerability testing shows that effective storage systems can be able to enhance the reliability of electric grids, especially when renewable energy generation exceeds demand. As much as the traditional sources of energy have been in the lead, the integration of smart technology with energy storage systems can significantly decrease the risks that come with the transition, thus opening up a way for a more resilient energy system in the future (Millis et al., 2019). Aside from the technical benefits, energy storage technologies are significant in enabling the move toward low-carbon policy. As outlined in recent discussions on future energy strategies, there is growing recognition of the need for coordinated policies favoring carbon pricing and promoting low-carbon technologies. The creation of energy storage technology not only assists in increasing renewable energy reliance but also aids in support for integration with the overall climate goals defined for 2050. Through investment in energy storage technology, policymakers have the ability to bridge the gap between current infrastructure and a green future, bring about smooth transition towards cleaner energy, and assist in bringing down greenhouse gases at the global level (N/A, 2010). In addition, there should be the adoption of energy storage solutions and then rigorous evaluation methods to determine their effectiveness under various operating scenarios. In particular, multi-period optimal power flow models have shown that storage significantly enhances resilience in systems with surplus generation while reducing as an advantage in constrained systems with insufficient generation capacity. This evolved understanding of the function of energy storages in grid operations is critical to the path ahead for investments and innovations in the industry. With threats to grid stability on the rise, the prudent use of storage technology will undoubtedly be a corner stone of electricity infrastructure planning, supporting the vision of the country for a secure and efficient energy environment (Millis et al., 2019).

REFERENCES

1. Abdessamad, S., Conti, F., Greenough, R. M., Stark, R., et al. (2019). Steam storage systems for flexible biomass CHP plants–Evaluation and initial model-based calculation. https://core.ac.uk/download/228199241.pdf (accessed on 30th September 2025).
2. Abdumannob, I., Fozilbek, J., Diyorbek, K., Bahrom, O., Ergashali, R., & Shuxrat, S. (2024). Battery technologies: Exploring different types of batteries for energy storage. https://core.ac.uk/download/599170302.pdf (accessed on 30th September 2025).
3. Adeleye, A., Koponen, K., Pongracz, E., Taponen, H., et al. (2025). Economic feasibility of medium-term energy storage for intermittent power producer. https://core.ac.uk/download/653751238.pdf (accessed on 30th September 2025).
4. Afreh, P., Gao, L., Passi, B. J., & Onwuagbu, C. C. (2025). Future energy storage: Technologies, management systems, and pathways for sustainable integration. *Academia Green Energy*. https://www.semanticscholar.org/paper/182f53b33a05144904990d06ff77ab78e4daa081 (accessed on 30th September 2025).
5. Airth, B., Milford, L., Schwartz, M., Constantine, S., Mullendore, S., & Waite, W. (2016). Closing the California clean energy divide: Reducing electric bills in affordable multifamily rental housing with solar+storage. https://core.ac.uk/download/75785738.pdf (accessed on 30th September 2025).
6. Akinsooto, O., Ogundipe, O. B., & Ikemba, S. (2024). Regulatory policies for enhancing grid stability through the integration of renewable energy and battery energy storage systems (BESS). *International Journal of Frontline Research and Reviews*. https://www.semanticscholar.org/paper/dc6fea377f5ee7bacb80574249275462e9c9716d (accessed on 30th September 2025).
7. Aleksandrov, A., Chekichev, A., Parushev, A., Shehova, D., & Popov, R. (2025). Analysis of energy storage technologies. *2025 19th Conference on Electrical Machines, Drives and Power Systems (ELMA)*. https://www.semanticscholar.org/paper/f93811abedd11f907fa15e086bd0c348377e3fdf (accessed on 30th September 2025).
8. Allison, J., Kelly, N., Marinho de Castro, M. M., Sharpe, T., et al. (2017). A taxonomy of fabric integrated thermal energy storage: A review of storage types and building locations. https://core.ac.uk/download/146489477.pdf (accessed on 30th September 2025).

9. Arghandeh, R., Brown, M., Cibulka, L., Miller, S., et al. (2013). Distribution system voltage management and optimization for integration of renewables and electric vehicles: Research gap analysis. https://core.ac.uk/download/288432751.pdf (accessed on 30th September 2025).
10. Ault, G. W., Currie, R. A. F., Foote, C. E. T., McDonald, J. R., et al. (2005). Ubiquitous energy storage. https://core.ac.uk/download/9034038.pdf (accessed on 30th September 2025).
11. Baer, D. A., Rodriguez, G. E., & Studer, P. A. (2025). Assessment of flywheel energy storage for spacecraft power systems. https://core.ac.uk/download/pdf/42851423.pdf (accessed on 30th September 2025).
12. Bao, Y., Feng, W., Jin, M., Liu, Y., et al. (2018). A review of microgrid development in the United States–A decade of progress on policies, demonstrations, controls, and software tools. https://core.ac.uk/download/323310623.pdf (accessed on 30th September 2025).
13. Becerra Villanueva, J. A., Calderón, B., Carvalho, M., Chacartegui, R., et al. (2019). Trans critical carbon dioxide charge-discharge energy storage with integration of solar energy. https://core.ac.uk/download/322845447.pdf (accessed on 30th September 2025).
14. Beena, T., Kannan, N. S., Reddy, G. B., Kathiravan, T., Ramaswamy, R., & Logasundari, T. (2025). Nanotechnology applications in battery energy storage systems for next generation. *E3S Web of Conferences*. https://www.semanticscholar.org/paper/480fe580e835b3e0fad3fc4d6cc5d5f1e1cabd00 (accessed on 30th September 2025).
15. Benoit, G., Boustead, I., Callaghan, D., Finkbeiner, M., Hall, G. M., et al. (2012). Energy from waste and the food processing industry. https://core.ac.uk/download/6112944.pdf (accessed on 30th September 2025).
16. Bingham, C., Schofield, N., & Yap, H. T. (2005). Hybrid energy sources for electric and fuel cell vehicle propulsion. https://core.ac.uk/download/54446.pdf (accessed on 30th September 2025).
17. Borghetti, G., Menga, G., Procaccianti, G., Kok, K., Ardito, L., Morisio, M., Mo, Y., et al. (2013). Smart grid technologies in Europe: An overview. https://core.ac.uk/download/pdf/11431923.pdf (accessed on 30th September 2025).
18. Breeze, J., Chapman, M., Diaz-Gonzalez, F., Faraji, B., Harris, R., Kim, J., Krack, M., et al. (2019). Performance and loss analysis of squirrel cage induction machine-based flywheel energy storage system. https://core.ac.uk/download/237199414.pdf (accessed on 30th September 2025).
19. Carneiro, J. F., Matos, C., & Silva, P. P. (2016). Large scale underground energy storage for renewables integration: General criteria for

reservoir identification and viable technologies. https://core.ac.uk/download/75982369.pdf (accessed on 30th September 2025).

20. Choi, B. B., Dever, T. P., Duffy, K. P., Lowe, R., et al. (2025). Assessment of technologies for noncryogenic hybrid electric propulsion. https://core.ac.uk/download/pdf/42720308.pdf (accessed on 30th September 2025).
21. Copeland, R. J., Wright, J. D., & Wyman, C. E. (2025). The SERI solar energy storage program. https://core.ac.uk/download/pdf/42865251.pdf (accessed on 30th September 2025).
22. Dey, A., Mondal, S., Chakraborty, B., Dalai, S., & Bhattacharya, K. (2025). Advancements in Li-Ion battery materials for battery energy storage systems. *2025 Devices for Integrated Circuit (DevIC)*. https://www.semanticscholar.org/paper/a12caed8cc948d914281af53ea4409ec4a18e962 (accessed on 30th September 2025).
23. Dubey, P. K., Singh, B., Singh, D., Narayan, P., & Verma, K. S. (2024). Energy storage technologies: Types, recent trends, and development. *2024 3rd International Conference on Power Electronics and IoT Applications in Renewable Energy and Its Control (PARC)*. https://www.semanticscholar.org/paper/e78932336fcdf85fbbdc2257017ce86d1d47dab5 (accessed on 30th September 2025).
24. EPSRC (Funder), & Hall, P. J. (2008). Energy-storage technologies and electricity generation. https://core.ac.uk/download/9023216.pdf (accessed on 30th September 2025).
25. Farivar, G., Manalastas, W. W., Tafti, H. D., Ceballos, S., Sanchez-Ruiz, A., Lovell, E., Konstantinou, G., et al. (2023). Grid-connected energy storage systems: State-of-the-art and emerging technologies. *Proceedings of the IEEE, 111*, 397–420. https://www.semanticscholar.org/paper/c6be36dd7d8fc6e4b0083e3c0ddb5012e8c0c9d9 (accessed on 30th September 2025).
26. Glowacki, B., Krishnamurthy, S., & Nuttall, W. (2016). Next steps for hydrogen – Physics, technology and the future. https://core.ac.uk/download/42537746.pdf (accessed on 30th September 2025).
27. Gómez Martín, A. (2021). Biomass-derived carbon materials for energy storage applications. https://core.ac.uk/download/225124073.pdf (accessed on 30th September 2025).
28. Jankowski, M., Pałac, A., Sornek, K., Goryl, W., Żołądek, M., Homa, M., & Filipowicz, M. (2024). Status and development perspectives of the compressed air energy storage (CAES) technologies—A literature review. *Energies*. https://www.semanticscholar.org/paper/

fcc924a1b1b49f1cda1e707876bfee271555da40 (accessed on 30th September 2025).

29. Jia, C., Yuan, D., & Zhu, Y. G. (2016). Carbon nanotube–polymer composites for energy storage applications. https://core.ac.uk/download/322426758.pdf (accessed on 30th September 2025).
30. Kluza, J. J. (2009). Status of grid scale energy storage and strategies for accelerating cost effective deployment. https://core.ac.uk/download/4419733.pdf (accessed on 30th September 2025).
31. Kumar, P., Bhavana, K., & Kumar, C. A. (2025). Optimizing renewable energy integration: A hybrid system architecture for sustainable energy production, storage, and grid stability. *International Research Journal of Innovations in Engineering and Technology*. https://www.semanticscholar.org/paper/116c58e77df26f1f0cd3e3e040a3a0d7f119012a (accessed on 30th September 2025).
32. Kumar, R., Lee, D., Ağbulut, Ü., Kumar, S., Thapa, S., Thakur, A., Jilte, R., et al. (2024). Different energy storage techniques: Recent advancements, applications, limitations, and efficient utilization of sustainable energy. *Journal of Thermal Analysis and Calorimetry*. https://www.semanticscholar.org/paper/857970a38536282bb02b94fdbcf530fb33fb2f1a (accessed on 30th September 2025).
33. Kumari, R., & Dubey, K. (2024). A critical analysis of flywheel energy storage systems' technologies, applications, and prospects. *2024 IEEE International Conference on Power Electronics, Drives and Energy Systems (PEDES)*. https://www.semanticscholar.org/paper/9a5115458b0b5e24543afca3b556bc8c1ca432bc (accessed on 30th September 2025).
34. Lynch, B., & Wallace, C. (2014). CubeSat electronic power system. https://core.ac.uk/download/72853039.pdf (accessed on 30th September 2025).
35. Mäkelä, N. (2024). Novel mechanical energy storage methods. https://core.ac.uk/download/630298166.pdf (accessed on 30th September 2025).
36. Manenti, F., Zuniga Linan, L., Lima, N. M. N., & Ravaghi-Ardebili, Z. (2013). Study of direct thermal energy storage technologies for effectiveness of concentrating solar power plants. https://core.ac.uk/download/55243664.pdf (accessed on 30th September 2025).
37. Maroufi, S., Karrari, S., Rajashekaraiah, K., & De Carne, G. (2025). Power management of hybrid flywheel-battery energy storage systems considering the state of charge and power ramp rate. *IEEE Transactions on Power Electronics, 40*, 9944–9956. https://www.semanticscholar.org/

paper/3139676ca9b03f6c538c0d5937e18e367c3c0bce (accessed on 30th September 2025).

38. Millis, M. (2019). Vulnerability analysis of modern electric grids: A mathematical optimization approach. https://core.ac.uk/download/215464251.pdf (accessed on 30th September 2025).

39. Miskam, N. (2013). *Transformasi Pekan Parit Raja: Kajian Terhadap Kesejahteraan Hidup Masyarakat*. https://core.ac.uk/download/12007108.pdf (accessed on 30th September 2025).

40. N/A. (2010). In brief: Update on the 10-50 solution: Progress toward a low-carbon future. https://core.ac.uk/download/71352424.pdf (accessed on 30th September 2025).

41. Nikola, M. (2018). Low carbon technology in the distribution network. https://core.ac.uk/download/186449305.pdf (accessed on 30th September 2025).

42. Prabakar, P., Mert, K. M., Muruganandam, L., & Sivagami, K. (2024). A comprehensive review on biochar for electrochemical energy storage applications: An emerging sustainable technology. *Frontiers in Energy Research*. https://www.semanticscholar.org/paper/a0254ab8093f5fedacd54d96d675b7d7113f2c3d (accessed on 30th September 2025).

43. Praetorius, B., & Schumacher, K. (2025). Greenhouse gas mitigation in a carbon constrained world: The role of carbon capture and storage. https://core.ac.uk/download/pdf/6566362.pdf (accessed on 30th September 2025).

44. Proser, N. (2011). Energy storage: Technology for a more efficient grid. https://core.ac.uk/download/70968321.pdf (accessed on 30th September 2025).

45. Reddy, A. S., & Balachandra, P. (2025). Hydrogen energy for Indian transport sector–A well-to-wheel techno-economic and environmental feasibility analysis. https://core.ac.uk/download/pdf/6469658.pdf (accessed on 30th September 2025).

46. Redman, G. (2002). *Residential Distributed Generation: Decision Support Software to Evaluate Opportunities in The Residential Market: A Thesis Submitted in Partial Fulfilment of The Requirement for The Degree of Masters of Engineering at Massey University, Palmerston North, New Zealand*. https://core.ac.uk/download/159449365.pdf (accessed on 30th September 2025).

47. Reyes, A. B. (2007). Cryogenic, compressed, and liquid hydrogen fuel storage in vehicles. https://core.ac.uk/download/4405315.pdf (accessed on 30th September 2025).
48. Simpa, P., Solomon, N. O., Adenekan, O. A., & Obasi, S. C. (2024). The safety and environmental impacts of battery storage systems in renewable energy. *World Journal of Advanced Research and Reviews*. https://www.semanticscholar.org/paper/fb71cdc87a6ff69c31f8f74b4ffafd66a6d0fc48 (accessed on 30th September 2025).
49. Vecchiola, A., de Assunção, D., Thain, D., Erwin, D. W., Foster, I., Broberg, J., Kim, K. H., et al. (2009). Cloudbus toolkit for market-oriented cloud computing. http://arxiv.org/abs/0910.1974 (accessed on 30th September 2025).
50. Yi, Y., Chang, L., Wu, B., Zhao, J., Peng, H., Li, L., & Wang, A. (2024). Life cycle assessment of energy storage technologies for new power systems under dual-carbon target: A review. *Energy Technology*. https://www.semanticscholar.org/paper/748bd94f6200cae4fbd13849bc5cdf04fd86820a (accessed on 30th September 2025).
51. Zeljković, S., Helać, V., & Hanjalić, S. (2025). Integration of photovoltaic systems into smart grids: Assessment of impact on grid stability and reliability. *2025 24th International Symposium INFOTEH-JAHORINA (INFOTEH)*. https://www.semanticscholar.org/paper/10612e201793c971b509021f57dc9393557bd84c (accessed on 30th September 2025).
52. Zhang, R., & Zhao, G. (2025). A comprehensive review of compressed air energy storage technologies: Current status and future trends. *Journal of Renewable and Sustainable Energy*. https://www.semanticscholar.org/paper/55ed1e2acad284946feaec3348f63b8f0e61df37 (accessed on 30th September 2025).
53. Zheng, J., Su, Y., Wang, W., Li, Z., & Wu, Q. (2024). Hydrogen-electricity coupling energy storage systems: Models, applications, and deep reinforcement learning algorithms. *Clean Energy Science and Technology*. https://www.semanticscholar.org/paper/0d83a6137f65092434377446b79057d0f3f72ded (accessed on 30th September 2025).

CHAPTER 9

ELECTRIC VEHICLES AND SMART CHARGING

Contents

INTRODUCTION

As cities struggle to cope with the impacts of rapid population growth and increased use of motor vehicles, the environmental impacts have been more significant. Higher concentrations of air pollutants and energy depletion tied to traditional internal combustion engine cars provide an imperative for a mode change in transport means. That is where electric cars (EVs) and smart charging systems come into the picture. The transition to battery electric vehicles (BEVs) is an established solution to address these adverse effects. It has been argued that the transport mode has largely contributed to urban air quality issues in the past, and BEVs provide a cleaner solution, which could change city landscapes and public health gains (Dalton et al., 2013). The availability of BEVs, made possible through smart charging technology, is a breakpoint towards sustainable city mobility to provide cleaner cities. Addressing the challenge of providing growing EV demand in urban areas entails addressing this issue from numerous aspects.

Figure 9.1. Smart charging of electric vehicles (EVs) could promote renewable energy adoption.

Source: https://environment.ec.europa.eu/sites/default/files/styles/oe_theme_medium_2x_no_crop/public/2022-10/NA1401_shutterstock_1668063934.jpg?itok=gUIsy4lo.

City planners and city authorities face the challenge of not just promoting the creation of BEV technology but also sustaining the infrastructure required to make this transition possible. The new methods could include the creation of charging points and intelligent grid technologies capable of providing efficient delivery of energy. Successful urban deployment of EVs is of high importance to city planners and electricity suppliers who aim to stay competitive in the evolving landscape of transportation, as per ongoing research (Fox-Penner et al., 2019). Such an association of stakeholders is the key to the establishment of a strong framework that will be capable of providing the growing demand for EVs on our roads and creating a sustainable environment that is good for the environment and the consumer alike. Generally, the transition to electric vehicles and the incorporation of smart charging systems are a major component in the reduction of the numerous challenges facing urban transportation. These developments not only seek to ease the pollution burden but also an environment that promotes sustainable development. Knowledge of the history and technological advancements that have enveloped BEVs gives us an idea of where the market stands today and where innovation is needed next. For instance, the chronology of BEV technology history is a promising sector with concomitant challenges, affirming the need for continued research and investment in policy and infrastructure strategies. As urban areas keep expanding, electric cars and smart charging systems will no doubt be at the center of the future of clean urban mobility (Figure 9.1).

Definition of Electric Vehicles (EVs)

EVs represent a new trajectory in transportation technology, distinct from traditional vehicles as much as they do not have internal combustion engines. They are powered by one or more electric motors, charged from recharging batteries, which reduces their dependence on fossil fuels to a very large extent. Thus, the environmental impact of EVs is very small, resulting in less greenhouse gas emission and improved air quality in cities. Evolving of the contemporary global EV market has led to several cities putting infrastructure support development high on their agenda as they struggle with elevated demand for the deployment of electric vehicles. This urban transformation highlights the necessity for city planners, authorities, and utilities to appreciate the complexities of integrating EVs into existing transportation infrastructure (Fox-Penner et al., 2019). Electric vehicles are referred to as a progression of various forms of vehicles, including BEVs, plug-in hybrid electric vehicles (PHEVs), and fuel cell electric vehicles (FCEVs). Each of these types owns a different operating system and benefits, fulfilling various consumer preferences and usage demands. BEVs operate using pure electric power from batteries, being the best example of emission-free cars. PHEVs supplement electric power with traditional combustion engines, being convenient for consumers who require longer driving ranges. Fuel cell electric vehicles operate with hydrogen to create electricity, enabling zero-emission mobility. As heterogeneity in EV technology progresses, an improved understanding of such distinctions becomes essential for their adoption and enhanced infrastructure development as per their unique characteristics (Dadlani et al., 2020). Amidst the raised awareness of electric vehicles, the relationship between their adoption and smart charging solutions cannot be overlooked. Smart charging enables the effective regulation of energy use, with EVs charged when electricity demand is low during off-peak times. This supports stabilization of the grid and promotes more environmentally friendly energy use. Smart charging solutions coupled with electric vehicles grow more pertinent to urban environments where infrastructure has to change to accommodate the unique challenges of electrification. By embracing these pioneering strategies, cities are capable of optimizing their networks for distributing energy and confirming the general efficiency of EV networks. Finally, the definition and development of electric vehicles are inextricably linked with the development of smart charging solutions since both elements play important roles in developing a sustainable, eco-friendly transportation future (Fox-Penner et al., 2019) (Figure 9.2).

Figure 9.2. Electric vehicles (EVs).

Source: https://www.aranca.com/assets/uploads/special-reports/Indias-Electric-Vehicle.jpg.

Importance of Smart Charging

The value of smart charging in the case of EVs cannot be overstated in the face of the environmental crisis that is triggered by traditional means of transport. Electric automobiles represent a paradigm shift toward cleaner practice. However, the integration of these automobiles into our existing power grid in a successful manner hinges on the innovation of clever charging solutions. These systems not only allow charging times to be maximized for energy supply but also contribute to enhancing the stability and efficiency of the grid. Through synchronization of energy consumption with renewable energy supply, such technologies can neutralize the negative impacts on the environment from conventional cars and push towards a cleaner transport future (Dalton et al., 2013). Smart charging eventually enhances the eco-friendliness of electric vehicles by fighting energy management and grid overload problems during peak usage hours. Smart charging also has a key role in fighting the severe threat of greenhouse gas emissions in the transport industry, which generates the highest proportion of United States emissions. As climate change advances further, as recognized by evidence found in documents such as the Fourth National Climate Assessment, the need for rapid and radical emissions cuts has never been greater (Arroyo et al., 2019). In addition to allowing electric vehicles to be charged during off-peak times when energy from cleaner sources is more viable, smart charging technology facilitates vehicle-to-grid (V2G) technologies. This enables EVs to pump back excess energy into the grid, allowing the system to be more reactive and dynamic. This implies that these technologies not only combat climate change but also enhance energy security and resilience for society. Further, strategic rollout of smart charging systems is poised to drive economic growth and innovation in the EV sector. With increasing EV demand, smart charging infrastructure investment has the potential to unlock jobs and innovation. Policy action that promotes EV take-up and innovation in smart charging technologies will catalyze a low-carbon transition. Partnerships between the public and private sectors will be essential to this, as they can aggregate the resources and knowledge needed to create new charging networks and innovations. By focusing on smart charging, societies are able to benefit from reduced air pollution, improved public health, and economic advantages and ultimately pave the way towards a more sustainable world (Arroyo et al., 2019; Dalton et al., 2013).

9.1. THE RISE OF ELECTRIC VEHICLES (EVS)

The rise of EVs is a revolution in the automobile industry driven largely by the urgency to combat climate change and growing awareness regarding ecological sustainability. As America's largest greenhouse gas emitter, the transportation sector is under immense pressure to limit its carbon footprint. Research such as the Fourth National Climate Assessment calls out the pressing impacts of climate change, saying that conventional use of fossil fuels exacerbates environmental degradation and public health. Transition to electric cars is not only a step needed to achieve emission reduction goals; it is also synergistic with efforts toward promoting innovation and economic growth through the utilization of cleaner technologies. As cities adapt to this electrification pattern, the creation of supporting infrastructure is needed in order to facilitate a mass transition to EVs and incorporate them in everyday life in a sustainable manner (Arroyo et al., 2019). Motivating forces beyond ecological benefits also drive electric vehicle growth. Technology developments in battery life and charging have raised EV performance levels, as well as making them more desirable to consumers and potential customers. Furthermore, public policies and incentives have played a significant role in establishing a supportive environment for the adoption of EVs, with interventions leading to increased investments in electrical mobility. The collaboration of private and public investment in EV charging infrastructure meets the requirement for interdependent relations in meeting increasing demand. Additionally, urban planners and city managers must manage the intricacies involved in developing adequate charging facilities, ensuring equal access and convenience to everybody. With this realization and deployment of novel concepts, cities can position themselves as the leaders of the next wave of electric mobility (Fox-Penner et al., 2019). Though there are multiple challenges associated with the advent of electric vehicles, there are enormous possibilities of revolutionary transformation. As cities and communities begin to prioritize smart charging solutions, development of effective electricity management systems will be the backbone of the EV infrastructure. Smart charging not only optimizes energy supply but also facilitates easier integration with renewable sources of energy, making the transport system more sustainable. Smartened grid features enable smooth connectivity of electric vehicles with power supplies and balancing demand while reducing peak loads. Cities can design a strong infrastructure to accommodate growing numbers of electric vehicles while enabling overall climate goals by investing in smart charging technologies. Overall, the transition to electric vehicles, coupled with smart charging technology, is a critical moment for society to embrace green habits while meeting the challenges of climate change (Arroyo et al., 2019; Fox-Penner et al., 2019) (Figure 9.3).

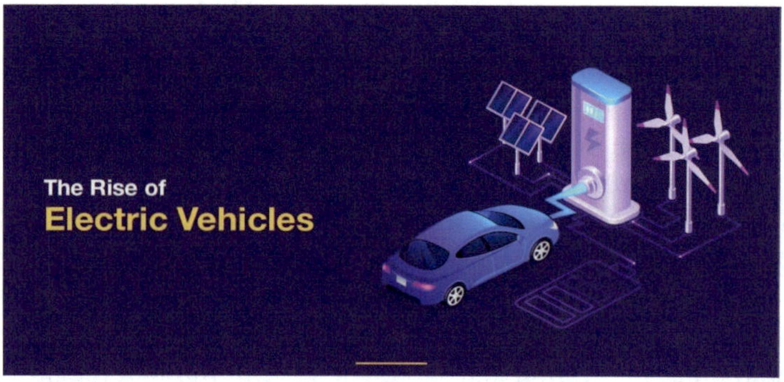

Figure 9.3. The rise of electric vehicles (EVs).

Source: https://trackobit.com/wp-content/uploads//04.jpg.

9.1.1. Historical Development of Electric Vehicle (EV) Technology

The historical development of EV technology has evolved significantly since the early days of the automobile. Initially, during the late 19th century, electric cars gained traction as a viable alternative to their gasoline counterparts, primarily due to their ease of use and lower pollution levels. However, the rise of internal combustion engines, supported by mass production techniques, led to a rapid decline in electric vehicle adoption. Notably, electric automobiles were effectively replaced by gasoline-powered automobiles in the early 20th century, which were more efficient and had longer range. This pivotal moment in the automobile industry's history laid the groundwork for an extended duration of time when electric automobile development was largely hindered until late 20th century environmental concerns spurred renewed interest in more sustainable transportation technology. In the second half of the 20th century, a renewed interest in electric car technology coincided with growing environmental awareness and technologies.

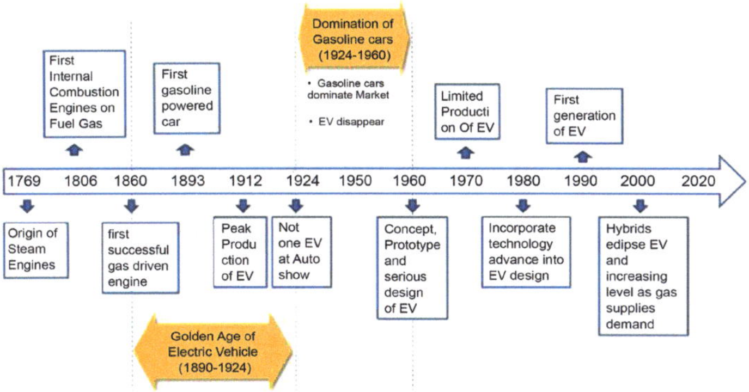

Figure 9.4. Historical development of automobile and development of interest and activity in the electric vehicle (EV) from 1890–2020. EV merged into hybrid electric vehicle.

Source: https://archive.nptel.ac.in/content/storage2/courses/108103009/module1/lec1/images/3.JPG.

Advances such as better battery chemistry and electric drives began to emerge in this period, enabling cars to travel greater distances on a single charge and improve the rate of charging. Government policies to promote clean energy alternatives also kicked in, with subsidies and economic incentives putting the spotlight once again on electric mobility. Large businesses such as Tesla came to dominate the industry, pushing the boundaries of battery technology and opposing public opinion against electric cars. This was a turning point because the interaction of new engineering innovation, increasing public demand for sustainability, and supportive policies began to make accessible a return to EV technology. With the progression of the 21st century, smart charging integration with electric vehicles became essential in bridging the challenges introduced by rising EV penetration into the energy supply system. Researchers began investigating the means by which smart grid technology could enhance the efficacy of electric vehicles while reducing their impact on existing power grids. Perhaps most significantly, research designed to develop methods for Volt/VAR Optimization and Control (VVOC) emphasized grid management enhancements that would enable electrification of transportation (Arghandeh et al., 2013). Moreover, electric cars emerged as an integrated energy network, working as load-on-wheels that could potentially respond dynamically to grid loads (García Caro et al., 2019). This history-changing moment of electric vehicle technology aligning with smart charging solutions is a milestone toward bringing sustainable mobility and energy efficiency to the forefront, emphasizing how EVs can help design a cleaner, safer energy future (Figure 9.4).

9.1.2. Current Market Trends and Statistics

With the ongoing evolution of the electric vehicle (EV) market, modern trends reflect a gigantic shift in consumer desires and production strategy. Over the past decade, the world has witnessed a spectacular increase in EV adoption, spearheaded by the need to address environmental concerns, technological advancements, and supportive government policies. One is that projections for EV penetration forecast a potential doubling of EV sales over the next two years. The increased demand sets challenges for urban planning as cities struggle to ramp up their infrastructure to accommodate the influx of such vehicles. The research compiled by leading experts emphasizes the fact that even though there are certain cities which are perfectly adapted to this shift, there remains a strong necessity for coordinated strategies to integrate EV facilities into present-day urban planning (Fox-Penner et al., 2019). This point of transition calls for the necessity of not only being aware of the current market trends but also anticipating future demands and the necessary alterations. The growing popularity of home charging solutions is another increasing trend in the EV market. Home power outlet charging is beginning to be an easy option for the majority of consumers due to its economic and environmental advantages. However, growing use of EVs leads to greater loading on local distribution networks, and that is a big challenge for DNO. Current studies point out that without coordination, even a minimal fraction—just 25%—of EVs on the road can impose severe overloads on distribution transformers. That is a situation which has necessitated the possibility of considering new technologies, such as coordinated charging systems, which will be able to redistribute energy loads on the grid optimally and minimize potential stress on the distribution infrastructure (Butans et al., 2016). This growth reflects the necessity of advanced planning strategies that consider both EV infrastructure development and its impact on electrical distribution networks. Furthermore, the intersection of market trends and emerging charging technologies presents an intriguing subject area in the broader context of electric mobility. With the demand for EVs increasing, stakeholders such as government bodies, car manufacturers, and utility companies are challenged to meet the challenges of infrastructure requirements and electricity usage. Current data show that as the market for EVs expands, investments in smart charging technology increasingly play a vital role. Smart charging not only increases grid efficiency but also maximizes user experiences, allowing drivers to exploit off-peak power prices and ultimately contributing to a greener energy model. In this transformed world, the use of smart charging technologies is vital in balancing supply and demand, ensuring that the path of EV market growth is sustainable and economically viable (Fox-Penner et al., 2019).

9.1.3. Government Policies Promoting Electric Vehicle (EV) Adoption

Government policies that bring about EV adoption contribute significantly to the shift away from traditional combustion engines towards cleaner technologies. Tax rebates for consumers and manufacturers, EV purchase subsidies, and investments in charging infrastructure are vital in creating an environment that will embrace the growth of EVs. These initiatives not only lower the cost burden for potential EV buyers but also drive market interest and production capacity of car makers. Hence, widespread public and private collaboration is crucial to facilitate the required ecosystems for a wholesome EV enterprise. Increasingly, cities now are looking for innovative policy templates to meet increasing demand, declaring the role of local leadership with the capacity to frame successful reforms in the city space in preparation for the future of transportation (Fox-Penner et al., 2019). In addition, the character of government policies promoting EV adoption extends beyond the area of environmental sustainability, where they adopt a pivotal role in stemming foul emissions. With the world striving to uphold international climate agreements, electric mobility policies feature prominently in the achievement of these targets by curbing greenhouse gas emissions by a significant amount. For instance, various policies are geared towards encouraging the production and use of two-wheel electric vehicles, especially among the Third World nations. Based on studies, electric two-wheelers are a more feasible option in such nations because they possess relatively lower acquisition costs than electric four-wheelers. Through incentives and educational initiatives focused on these more affordable options, policymakers can promote wider uptake, thus improving public access to sustainable transport modes and complementing economic development with environmental conservation (Albrecht et al., 2020). Aside from finance and ecological motives, policy incentives from the government also catalyze improvements in smart charging infrastructure. The incorporation of EVs into existing grid infrastructure works not in a vacuum but with concerted planning and investment. Smart charging can be promoted by governments through policy that supports grid resilience and renewable energy consumption during off-peak hours. This not only optimizes energy use but also reduces pressure on electrical grids when EV take-up is boosted. While urban centers make efforts to expand their EV network, they must do so with the adoption of smart technology solutions that offer seamless integration and compatibility with renewable energy sources. Thus, through the judicious mix of policy, technological advancement, and public engagement, governments can strongly establish the foundation for mass adoption of EVs with a secure energy future (Fox-Penner et al., 2019).

9.2. TYPES OF ELECTRIC VEHICLES (EVS)

The ground of EVs is diverse and comprises diverse kinds that meet different consumer needs and technical capabilities. Primarily, battery electric vehicles (BEVs) stand out since they operate on electricity from batteries only, with no emissions from the tailpipe and high energy efficiency. This vehicle is one great divergence towards environmentally friendly transport solutions, mitigating the adverse effects of traditional combustion engines. Moreover, plug-in hybrid electric vehicles (PHEVs) combine an electric drive system and an internal combustion engine, providing freedom for long-distance travel coupled with reduced emissions for the electric propulsion alone. The growing list of EVs shows a technological revolution to address the environmental challenges of conventional vehicles (García Caro et al., 2019). Identifying the characteristics of these different models is key to consumers and policymakers looking for cleaner forms of transportation. A second essential category of electric cars is the hybrid electric vehicle (HEV), blending electric power with gas motors, achieving maximum fuel economy without additional charging needs. HEVs work by switching between the two power sources automatically to realize optimal efficiency based on driving conditions, hence being attractive to consumers who are not ready to adopt full electric models due to range anxiety. Their design often enables urban drivers to experience significant emissions reductions when driving in stop-start traffic where electric running is most efficient. Apart from that, with the advancement of battery technology, HEVs have become more competitive by providing a seamless driving experience and lesser running costs than regular vehicles (García Caro et al., 2019). Blending hybrid technology gives a transition approach towards total electrification, and over time, cleaner mobility solutions become increasingly acceptable. As the world evolves, new technologies like electric buses and delivery trucks are changing urban transportation networks as well as logistics. These new electric cars are built to meet the demands of commercial and public transportation, thus decreasing air pollution in urban metropolises. Electric buses, for example, are increasingly becoming utilized in numerous city transit systems, decreasing greenhouse gas emissions and harmful air pollutants significantly compared to diesel counterparts. Besides, the adoption of electric freight vehicles is gaining momentum in logistics markets, where efficiency of operations and sustainability are increasingly important (Dalton et al., 2013). Such electric vehicle types are at the heart of demonstrating the flexibility and versatility of EV technology across different industries, eventually leading to an overall integration of more intelligent charging infrastructure that suits their respective operational requirements (Figure 9.5).

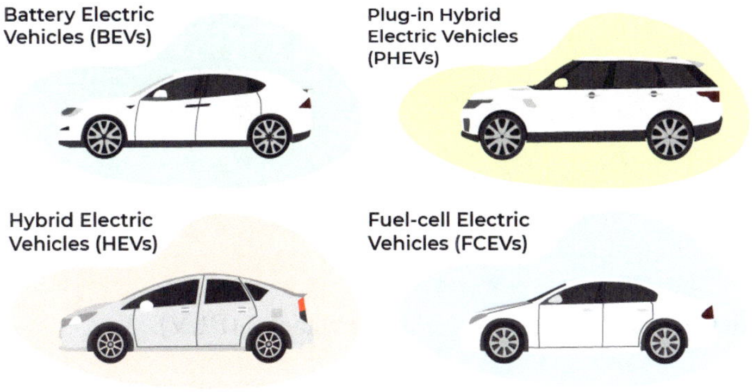

Figure 9.5. Types of electric vehicles (EVs).

Source: https://i0.wp.com/evnews.co.in/wp-content/uploads/2024/07/4-types-of-electric-vechile.jpeg?fit=960%2C480&ssl=1.

9.2.1. Battery Electric Vehicles (BEVs)

BEVs are a landmark in green transport, offering a cleaner, better alternative to traditional fossil fuel-based vehicles. By utilizing electric motor technology powered by rechargeable batteries, BEVs emit no tailpipe pollutants and have reduced greenhouse gas emissions, which contributes to improved air quality. The shift away from internal combustion engines is central to addressing the world climate crisis. Additionally, the adoption of BEVs accords with alternative policy frameworks for supporting energy efficiency and sustainability. As cities around the world implement stricter emissions standards, the momentum to transition towards electric mobility grows, driving R&D in battery technology and charging infrastructure, which are important drivers of achieving mass market penetration of BEVs. Furthermore, inclusion of smart charging technology is critical in enhancing the efficiency and utility of BEVs. Smart charging technology facilitates dynamic electricity use management by scheduling charging so that it aligns with low periods of electricity demand or higher availability of renewable energy. This optimization enhances the overall stability of the grid as well as lowering consumers' energy costs through off-peak charging.

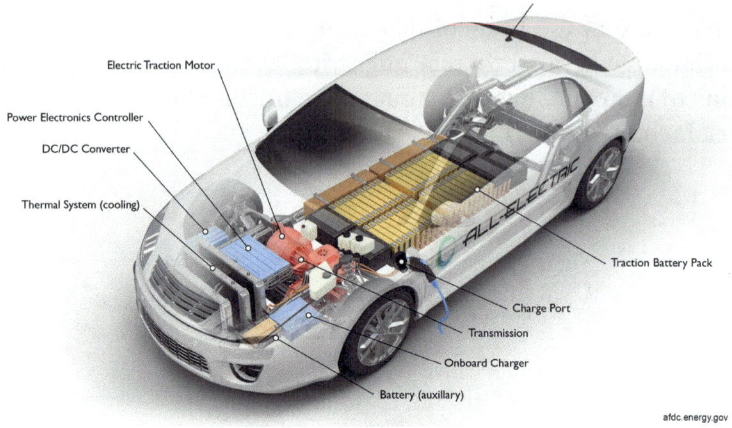

Figure 9.6. Components of battery electric vehicles (BEV).

Source: https://www.ev-resource.com/uploads/5/5/9/6/55967337/electric-high-res_orig.jpg.

In the context of intelligent electric vehicle infrastructure development, the combination of smart charging and BEVs can help improve the new smart grid technology, which aims to improve energy distribution as well as consumption habits. Through stimulation of innovation in such areas, policymakers and industry stakeholders can further stimulate the adoption of BEVs, hence promoting an integrated response to sustainable city transport (Alahakoon et al., 2023). Additionally, ongoing technological innovation in batteries is imperative to enhancing the performance and range of BEVs, alleviating many of the primary concerns that would-be adopters may have, such as charging time and battery life. Enhanced lithium-ion battery technology, as well as emerging solutions like solid-state batteries, have a great deal of promise to significantly increase energy density so that longer driving ranges and faster charging are feasible. These innovations come to fruition in such initiatives as the Technology Opportunities in Nordic Energy System Transitions (TOP-NEST), which explore efficient means of battery production and lifecycle management (Klitkou et al., 2015). The efficient integration of these innovations not only increases consumer confidence but also ensures the positioning of BEVs in the broader context of sustainable energy system transitions. Thus, the synergetic relationship between BEVs and battery technology plays a crucial role in defining the future of electric vehicles and their suitability for smart charging technologies (Figure 9.6).

9.2.2. Plug-in Hybrid Electric Vehicles (PHEVs)

Application of PHEVs represents the next leap in clean transportation that will make the path to cleaner car technology. PHEVs combine an internal combustion engine with an electric motor, providing drivers with the range of both electric and traditional sources of energy. This dual benefit not only decreases greenhouse gas emissions but also alleviates the range anxiety so typically associated with purely electric cars. With electric as well as gasoline power, PHEVs can excel at urban conditions where the electric range is utilized to the fullest without compromising on longer distances without recharging. Recent studies tell us that the convenience of charging infrastructure being available is essential for the uptake of PHEVs, emphasizing the need for an integrated strategy to make electric mobility possible (Bunch et al., 2020). The presence of robust charging infrastructure influences consumers' uptake of PHEVs directly, highlighting technology and usability intertwinement. PHEV driving experience is determined by cost, accessibility, and availability of charging points, with these factors playing important roles. Research indicates that PHEV drivers charge at home, work, or public refueling locations depending on economic incentives and convenience (Bunch et al., 2020). Additionally, placing charging points strategically is also crucial in urban planning so that these vehicles will be a viable option for the average buyer. Policymakers and stakeholders will need to cooperate and develop a system where PHEV use is encouraged by optimal infrastructure investment, which can reap greater numbers of PHEVs on the roads along with less carbon emission from the transport sector. Besides, PHEVs reflect the growing demand for AFVs and their potential to transform the automobile industry. As battery technology and renewable energy improve, the future is bright for PHEVs as they come closer to smart charging systems. Not only do these technologies enable efficient energy use but also enhance the use of PHEVs through charging from renewable sources, reducing emissions further. Strategically applying smart charging technologies, energy distribution can be optimized to achieve maximum energy distribution, facilitating peak demand management and enhancing grid stability. Engaging the public and private sectors to develop participating strategies for integrating PHEV will ultimately double their environmental benefits while leading society towards sustainable transportation habits (Graff, 2011) (Figure 9.7).

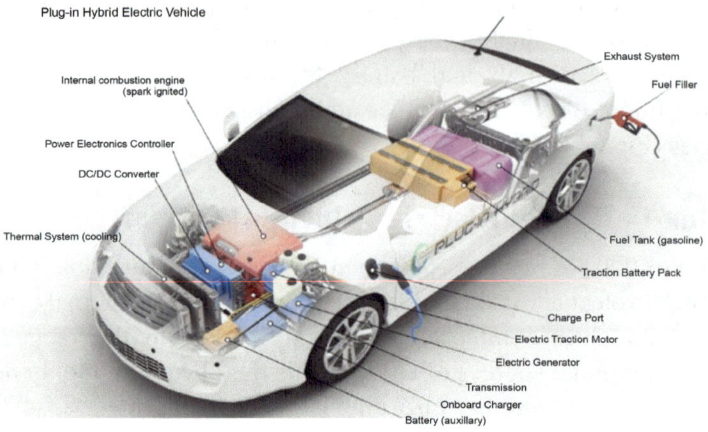

Figure 9.7. Components of plug-in hybrid electric vehicle (PHEV).

Source: https://evreporter.com/wp-content/uploads/2019/08/plugin-high-res-1024x588.jpg.

9.2.3. Fuel Cell Electric Vehicles (FCEVs)

The development of FCEVs is a significant advancement in the quest for sustainable transportation options. FCEVs differ from conventional electric vehicles that operate on electricity alone in that they generate electricity as a waste product of a chemical reaction between hydrogen and oxygen, with only water vapor being a byproduct. This unique aspect renders FCEVs extremely efficient vehicles, particularly for people seeking extensive range capability with no emissions. The development and deployment of such vehicles, however, depend on a highly advanced interdependence of technologies that enhance their efficiency, specifically in the realm of energy management. Smart charging networks form a key component of this by optimizing the refueling cycle and efficiently balancing electricity usage in relation to hydrogen production and transmission networks, thus making FCEVs viable to facilitate a fast-developing urban setting (Klitkou et al., 2015). While they hold promise, FCEVs are confronted by a number of challenges that need to be overcome in order to make the most out of their integration in sustainable transportation systems. Production expense is one key hurdle, with the production processes for fuel cells and supporting infrastructure presently economically unfeasible at mass scale. The sparse deployment of hydrogen refueling stations also limits their popular use, prompting thoughtful investment in infrastructure to support FCEVs optimally. Intelligent energy management is required in this case, since they can maximize the efficiency of hydrogen production and storage, as well as optimize supply and demand issues. Intelligent algorithms, control systems, and novel converter technologies research is required for the improvement of FCEVs capabilities and ensuring that they fit into a broader smart grid infrastructure (Ali et al., 2024). With constant R&D work, the emphasis can be increasingly observed to be laid on energy management strategies in a bid to enhance the overall performance and reliability of FCEVs. Fresh studies indicate the requirement of integrating sophisticated control algorithms and optimization techniques in an attempt to allow higher energy efficiency and operational performance. Such a diversified approach not only transcends current limitations but also attempts to bolster the resilience of the hydrogen supply chain, ultimately making mainstream FCEV deployment a reality. The insights gained from ongoing research into smart charging mechanisms, energy management plans, and infrastructure planning are imperative for automotive engineers and researchers alike to guarantee a cleaner and greener transport future. Mitigation of the complexities of FCEVs through targeted research and new technologies is important to overcome existing bottlenecks and achieve the full potential of hydrogen-powered automobiles in a sustainable environment (Ali et al., 2024) (Figure 9.8).

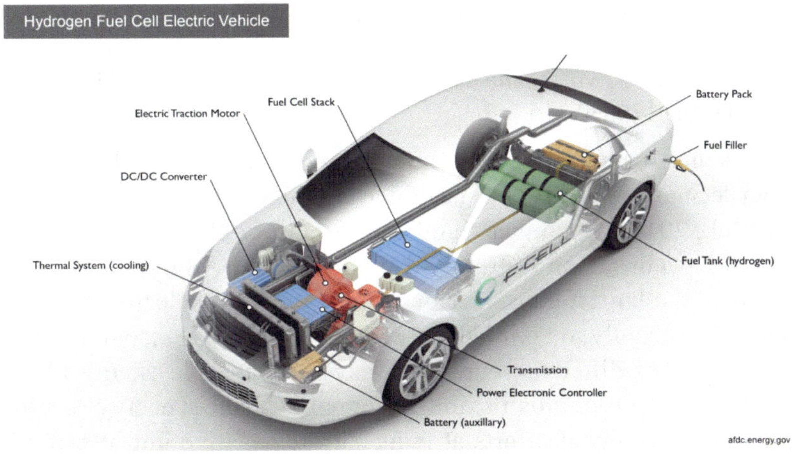

Figure 9.8. Components of fuel cell electric vehicle (FCEV).

Source: https://evreporter.com/wp-content/uploads/2019/08/hydrogen-high-res-1024x588.jpg.

9.3. BENEFITS OF ELECTRIC VEHICLES (EVS)

The transition to EVs is a significant step towards reducing the green footprint of transport. Traditional internal combustion engine vehicles are large sources of air pollution and climate change due to their reliance on fossil fuels and the exhaust emissions from their combustion. Electric vehicles, and particularly battery electric vehicles (BEVs), have no tailpipe emissions, so they improve urban air quality and prevent the health effects of air pollution. As per studies until 2015, the development of BEV has been marked by increasing technological innovation and attention in the sector for sustainability, and more recognition of electric vehicles as a sustainable mode of transport compared to conventional transport.

Figure 9.9. Benefits of electric vehicles (EVs).

Source: https://www.evmechanica.com/wp-content/uploads/2024/11/Untitled-design-2024-11-30T120050.315.jpg.

Thus, cities and manufacturers are urged to promote increased utilization of this clean technology to offer a healthier environment to future generations (Dalton et al., 2013). One other significant benefit of electric cars is that they can reduce dependence on finite fossil fuel resources. As the transport sector remains to be the largest consumer of energy, a move to electric vehicles can contribute significantly to taking the pressure off global oil resources. EVs can utilize a variety of different sources of energy, renewable energy like solar, wind, and hydroelectric. This flexibility not only enhances energy security but also contributes to global efforts at reducing the negative impacts of climate change. Further, the strategic utilization of smart charging systems maximizes the use of energy by enabling drivers to charge their electric vehicles off-peak or during times of abundant renewable energy. This synchronized effort can decrease pressure on the power grid, increase efficiency in energy use, and aid in the creation of a more environmentally friendly energy system (Fox-Penner et al., 2019). Besides energy and environmental benefits, electric vehicles can also bring about economic advantages. EV adoption can spur the creation of jobs in various industries, such as manufacturing, technology, and infrastructure development. As cities invest in developing charging infrastructure and associated services, there is a multiplier effect that stimulates the local economy and innovation. Additionally, the lower operational cost of electric vehicles, which tend to have less maintenance requirements compared to their gasoline counterpart, makes them compelling for buyers. That financial sustainability is also further encouraged through government subsidies to stimulate the purchase of EVs, such as tax rebates and rebates on charging infrastructure. Promoting electric vehicles, governments can foster economic development while ensuring the attainment of sustainability goals and improved public health measures (Dalton et al., 2013; Fox-Penner et al., 2019) (Figure 9.9).

9.3.1. Environmental Impact and Reduction of Emissions

Shifting to electric vehicles (EVs) presents a prominent window of opportunity for reducing greenhouse gas emissions and mitigating environmental impact. EVs release no tailpipe emissions, which is particularly advantageous in already polluted cities that have poor air quality. Implementing EVs over conventional internal combustion engine vehicles complements government and international efforts to lower climate change. The environmental advantage of EVs is contingent upon the power that is being utilized to charge them, however. As renewable energy resources such as sun and wind gain more popularity in the energy portfolio, emissions related to use of EV can be minimized further. Future electric mobility technologies therefore must incorporate means intended to enhance the sustainability of charging infrastructures and thereby further contribute to large-scale emissions mitigation (Bezruchonak et al., 2019). Public transportation systems, particularly electric buses, are another significant sector for environmental protection in cities. Not only do electric buses minimize air pollution, they also limit noise pollution in the city, leading to a healthier urban life. According to a report on electric bus technology, there are markets like the UK and the Netherlands that are leading fleet adoption, supported by good legislative foundations and mutual encouragement between regional governments (Bezruchonak et al., 2019). Moreover, proper planning must be done in order to integrate electric buses into existing transport systems, with adequate charging infrastructure available. Electric buses are an extension of the electric vehicle model, demonstrating how environmentally friendly urban mobility solutions can take center stage and fuel pollution reductions as well as expanding access to public transport. The case of electric cars in developing nations is different outright, calling for value-added policies to foster take-up. Electric two-wheelers also have potential due to lower front-end prices, but electric four-wheelers fail to break through affordability and infrastructure construction barriers. Policymakers should consider these hindrances in order to spur electric mobility effectively in such a situation. Policies such as subsidies for electric two-wheelers and investments in charging infrastructure will be most critical during the shift towards green transport options. Hence, in-depth understanding of the local drivers of electric vehicle uptake can aid in reducing emissions everywhere, and electric vehicles can be a cornerstone in combating climate change both in the developing and developed world (Figure 9.10).

Figure 9.10. Environmental impact of electric cars.

Source: https://www.buyev.my/images/Articles/EnvironmentalImpact.jpg.

9.3.2. Economic Advantages for Consumers

The transition from traditional gasoline cars to electric vehicles (EVs) has significant economic advantages for consumers, primarily in the form of gasoline and repair cost savings. Electric cars use electricity, which in most cases is less expensive than gasoline per mile. Such a shift allows consumers to reduce their monthly transportation expenses substantially. In addition, with progressing technology, the efficiency of batteries improves, leading to higher ranges and fewer recharges, again reducing operating costs. In urban areas, where air pollution caused by conventional vehicles has damaged the environment, the adoption of EVs reduces air pollution, and subsequently leads to reduced healthcare costs for pollution-related diseases. Therefore, while the initial price of electric cars will be greater, the long-term economic advantage surely benefits consumers, enhancing their own economic condition with rising fuel prices and maintenance (Dalton et al., 2013). Further, smart charging infrastructure maximizes the economic advantages of electric vehicle ownership. Smart charging technology allows consumers to charge their vehicles at the most favorable times, usually when electricity is lowest priced, i.e., off-peak. By applying time-of-use tariffs, customers are able to maximize changing energy prices throughout the day and, in the process, achieve extra savings on the energy bill. The integration of vehicle-to-grid (V2G) technology takes this economic model a step further by allowing consumers to sell excess energy from their automobile batteries back to the grid when the demand is high. This offers consumers a source of income, effectively making their automobiles economically productive. With energy distribution systems increasingly relying on sustainable practices, such practices not only bring economic gains for consumers but also enable a more integrated relationship between individual energy consumption and broader societal goals of sustainability (Ikpehai et al., 2017). Finally, economic gains for consumers extend beyond proximate cash savings. The evolving electric vehicle technology, made possible through smart charging solutions, positions the consumer well within the context of emerging mobility. City governments and governments are increasingly promoting EV usage through grants, tax credits, and rebates for the installation of charging stations. These incentives have the net result of reducing the entry point barrier for consumers who are considering an electric vehicle. Moreover, as the market for EVs expands, competition among producers will ensure prices come down, rendering electric vehicles even cheaper. In this new economy, consumers not just save on money but also invest in a cleaner future that, over time, can ensure a better quality of life and fiscal security as city mobility evolves (Ikpehai et al., 2017).

9.3.3. Energy Independence and Sustainability

The pursuit of energy autonomy and sustainability has become a hallmark issue in the recent discourse surrounding electric vehicles (EVs) and smart charging systems. As the transport sector continues to be among the most prominent causes of air pollution and energy usage, the transition away from traditional combustion engines to battery electric vehicles (BEVs) is vital. BEVs not only reduce the emissions of greenhouse gases but also reduce dependency on imported fossil fuels, thereby enhancing national energy security. In the use of alternative transportation with a focus on sustainability, policymakers are in a position to minimize the unwanted environmental effects. In addition, the growth in the BEV market is also evidence of shifting consumer patterns and technological advancement as a sign of transitioning towards sustainable means that can be adequately adopted within existing urban environments (Dalton et al., 2013). Central to e-vehicle adoption success is the implementation of smart charging systems, which play a significant role in making energy autonomy possible. These systems are an infrastructure framework for effectively balancing charging loads and reallocating energy from renewable sources further to meet sustainability objectives. New control strategies can maximize charging sessions in FEVs while maintaining balance in the generation profiles of DERs like wind and sun. A dual-level control system can be used to coordinate local Load Area Controllers (LACs) and Macro Load Area Aggregators (MLAAs) to increase the energy efficiency and economic efficiency of all involved parties. Such synergistic operation does not only favor the viability of EVs but also optimizes energy requirements being met through a renewable process, reducing reliance on conventional energy sources (Giorgio et al., 2013). With increasing integration of electric vehicle and smart charging technology, a clean energy future comes ever closer. The convergence of innovation in battery technology and dissemination of renewable energy sources makes energy independence a viable reality. By enabling individuals and communities to generate electricity for themselves, this paradigm shifts the flow of energy towards decentralized production and consumption. It is able to empower individuals to assist in making the environment sustainable and supporting the impending energy problems at a national level. Therefore, enhancing infrastructure in order to support both electric vehicle installations and smart charging infrastructure constructions is not a technical challenge, but a must for achieving an autonomous and sustainable energy future (Dalton et al., 2013; Giorgio et al., 2013) (Figure 9.11).

Figure 9.11. How electric vehicle (EV) can support energy independence.

*Source:*https://images.squarespace-cdn.com/content/v1/5eac87b47d1b210d7c0ab595/1588449904940-TEAPG6N3CJAUKFBO3IOX/V3-Vegas-e1563488073796.jpeg.

9.4. UNDERSTANDING SMART CHARGING

The transition to electric vehicles (EVs) is presenting several challenges that smart charging technologies are trying to surmount. Smart charging enables optimal distribution of energy during the process of vehicle charging so as to optimize the use of available energy resources and reduce the load on electrical grids. Using data analysis, smart charging systems are able to adjust dynamically charging rates against the need for electricity in real-time, price fluctuations, and grid stability. Not only does this responsiveness improve the user experience through cost savings, but it is also a built-in role of integrating renewable energy sources into the charging network. As such, smart charging can minimize the environmental impact of EVs by further decreasing reliance on fossil fuels and promoting sustainable energy habits. As cities and stakeholders recognize the importance of these technologies, they are well-equipped to handle the complexities of electrification, leaving space for a cleaner transportation future (Dalton et al., 2013).

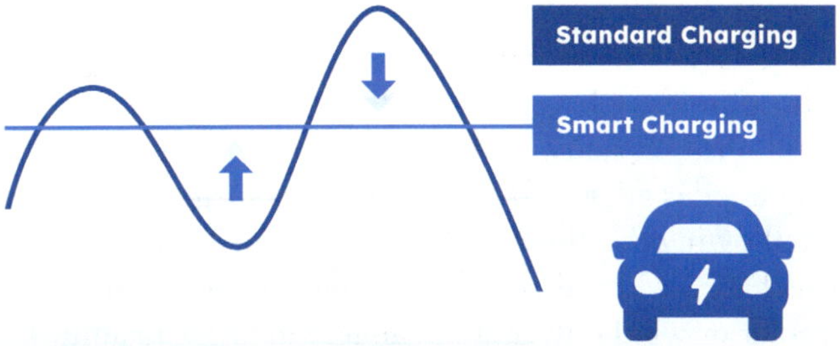

Figure 9.12. Understanding smart charging.

Source: https://www.ampeco.com/content/uploads/2022/10/smart-charging-types.png.

Cities are also now confronted with the challenge of the exploding demand for EVs, thus proving the urgent need for intelligent charging solutions. Urban planners and investors are shifting their focus to maximizing the supply and predictability of charging points. Smart charging systems bring this to life through guaranteeing that the energy grid supply can adapt to fluctuating demands without degrees of interference to services. Advanced grid integration and user-focused charging solutions make EVs more viable, allowing for expansion of the market into more diverse consumer demographics. In addition, new urban planning policies and infrastructure investments are increasingly becoming important, especially considering the cities' challenges presented by the expected surge in EV ownership. This new context necessitates a careful equilibrium between urban growth and environmental sustainability, thereby creating new collaborative platforms in the private sector and public policy (Fox-Penner et al., 2019). In acknowledging smart charging, its value to consumer behavior change and ability to stimulate driving style change should be viewed. Smart charging systems, through real-time charge status and energy consumption feedback, allow consumers to make choices regarding when and how to charge the vehicle. Such feedback could yield effective charging patterns, such as charging low-cost hours or off-peak charging for cost savings. Further, by linking cars to other types of energy, such as rooftop solar or neighborhood wind farms, smart chargers can enhance even more sustainability. The envisioned integration of advanced charging technology with user-centered design is a critical element of the changing transport system, evidence that being aware of smart charging is necessary in making a comprehensive shift towards electric mobility (Dalton et al., 2013; Fox-Penner et al., 2019) (Figure 9.12).

9.4.1. Definition and Functionality of Smart Charging

Smart charging is a critical infrastructure component for electric vehicles (EVs), which optimizes charging based on a variety of variables including energy demand, grid capacity, and renewable source availability. Essentially, smart charging applies advanced communication technologies to communicate with the grid, allowing dynamic load management. This feature allows electric vehicles to recharge when there is low demand or when renewable energy sources like solar or wind power are available in large quantities. As such, this system not only makes energy consumption more efficient but also reduces the strain on the electrical grid. By adopting smart charging systems, EV owners enjoy the option of programming when to charge their vehicles, thereby reducing the costs and contributing towards a cleaner energy environment. In addition to streamlining energy usage, smart charging assists in furthering the dependability of the grid. With the heightened utilization of electric cars in the coming days, exerting control over the combined charging processes of all the cars is important to prevent overloading the grid. Smart charging has the potential to facilitate Vehicle-to-Grid (V2G) technology, which can power EVs to supply electricity back to the grid when demand is peak. Two-way energy flow is a feature that can stabilize the voltage profiles and ease possible bottlenecks that result from increased penetration of renewable sources of energy. Research confirms that the effective utilization of smart charging will enhance the ability of distribution networks to cope with increased levels of both electric vehicles and renewable sources, i.e., photovoltaic (PV) systems, thus contributing to higher environmental goals as highlighted by current challenges in Voltage/VAR Optimization and Control technologies (Arghandeh et al., 2013). The implications of smart charging extend beyond operation efficiencies; they also influence consumer interest and behavior towards electricity consumption. Smart charging systems, by releasing real-time data and information about energy consumption, empower consumers to make informed decisions regarding energy usage. Such information can lead to reduced electricity bills and higher reliance on renewable energy sources, coinciding with consumer interest and sustainable purposes. As smart grid technologies continue to advance, the implementation of smart charging solutions becomes imperative in achieving regulatory objectives of renewable energy penetration. EV charging strategies that manage the grid constraints and are highly integrated with state plans, such as California's objective of 33% renewable penetration by 2020 (Arghandeh et al., 2013), become indispensable. Such pioneering approaches reflect the entwined interdependence of green energy solutions and electric vehicle technology innovation, paving the way for a more resilient and environmentally friendly future energy system (Figure 9.13).

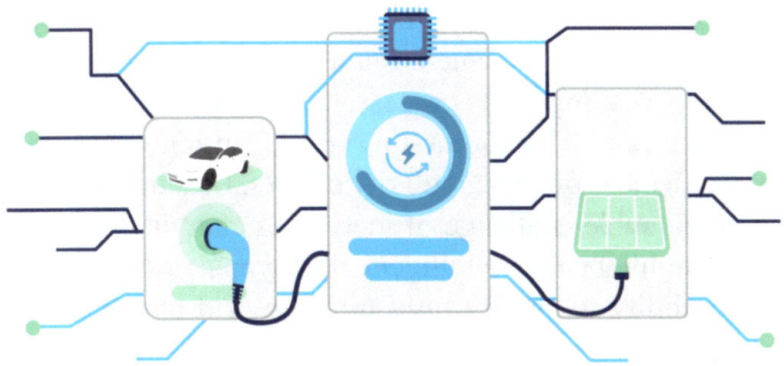

Figure 9.13. Smart electric vehicle (EV) charging.

Source: https://d2xqcz296oofyv.cloudfront.net/wp-content/uploads/smart-ev-charging-1.webp.

9.4.2. Types of Smart Charging Systems

Smart charging systems play a key role in facilitating the successful integration of electric vehicles (EVs) into existing power grids. These systems are mainly divided into two categories: smart grid charging and controlled charging. Controlled charging systems utilize rigid schedules of charging, based on pre-determined time slots, where the charge is executed during off-peak hours. This type is suitable for peak demand management and saves energy expenses for consumers to an enormous extent. On the other hand, smart grid charging systems possess the potential for dynamic charging such that EVs can vary their charging rates according to real-time data of the grid. These variations enhance the stability of the grid without depriving EV owners of the advantages of paying cheaper prices for energy during low-demand periods. With these means, smart charging systems render the energy system sustainable and efficient. Another significant classification of intelligent charging systems is in terms of the interaction with renewable energy resources. On one side, independent charging systems are designed to accommodate giving priority for the utilization of surplus renewable energy generated by a solar or a wind system. Not only does this make charging environmentally friendly, but it also takes the load off the power grid at peak usage hours. Conversely, vehicle-to-grid (V2G) technologies facilitate two-way interaction between the power grid and EVs to charge or send energy back to the grid at the peak demand. This type of flexibility creates a new source of income for EV owners along with increasing overall power grid resilience and dependability. Integration of renewable energy with smart charging systems can thus provide a greener, cleaner energy future for cities and citizens. Smart charging system adoption is not only a technological advancement but also a vital urban planning strategy and infrastructure development process. Cities worldwide are becoming increasingly conscious of incorporating these systems in their electrical infrastructures with the growing popularity of EVs. Urban planners and utilities have to tackle the double challenge of increasing charging points while being green. Scholars highlight that intelligent deployment of intelligent charging systems has the ability to significantly mitigate grid overload and make the system more efficient, as per the growing demand for electric vehicles. As indicated by present expert debate, cities that adopt smart charging within their structures are most likely to be the leaders in clean urban mobility solutions, ultimately paving the way to green cities and better environments (García Caro et al., 2019; Fox-Penner et al., 2019).

9.4.3. Role of Smart Grids in Electric Vehicle (EV) Charging

The integration of smart grids with EV charging systems is a significant move towards greater energy efficiency and management. Smart grids are based on advanced communication systems to create an integrated system that monitors and controls the movement of electricity in real-time. This aspect is immensely useful for EV charging because it offers dynamic responses to variable energy demands. With the help of the Connected Device Interface (CDI) Generic Enabler, intelligent grids are able to gather driver feedback on charging experience effectively. This is critical under situations when power to charge might be limited by grid requirements, e.g., when a Transmission System Operator must coordinate with the Distribution System Operator to maintain system stability and deal with contingencies. The flexibility of smart grids thus not only enhances consumer satisfaction but also enables operational integrity of the entire power grid to ensure reliable service delivery in a developing electric mobility scenario (Giorgio et al., 2014). Further, smart grids are an important enabler for managing simultaneous charging events to enable optimized allocation of resources across various Electric Vehicle Supply Equipment (EVSE) units. This optimization is realized by the Network Information and Controller (NetIC) Generic Enabler that manages the Load Area Controller in smart grid networks. By managing simultaneous charging processes efficiently, the NetIC GE accomplishes uniform demand allocation on the grid, reducing overloads during peak hours of usage. Increasing EV demands necessitate topology reconfiguration of distribution grid with an imperative for efficient communication among various nodes of the network. This coordination is facilitated by making it simple using the standardized model provided by the Smart Grid Architecture Model (SGAM) so that every stakeholder and process for charging EVs can be effectively coordinated for maximum efficiency (Giorgio et al., 2014). Furthermore, coordination of electric vehicle fleets with smart grids has also changed the consumption of energy and how the environment is maintained. By treating EVs as wandering loads within the power grid, smart grids make it easy to roll out novel charging priority techniques. For instance, the virtual power plant idea can be rolled out, which applies evolutionary computation algorithms in examining and optimizing charging policy. This optimizes energy efficiency through supply-demand balancing, in addition to curbing pollution through the prevention of grid overload. Thus, with the rise of electric vehicles, smart grids and smart charging solutions become essential to achieve environmental goals and create sustainable energy practices.

9.5. CHALLENGES OF ELECTRIC VEHICLES (EVS) AND SMART CHARGING

Electric vehicle transition has enormous challenges to be overcome in making this technology sustainable. Despite their potential to restrict air pollution and reduce the use of fossil fuels, electric vehicles are bedeviled by issues that are largely battery technology, infrastructural, and social acceptance. Battery durability, recharging time, and range anxiety are major issues that can fight mass uptake. Moreover, available charging stations are typically inadequate, particularly in city centers where there is high demand which can outstrip supply resulting in long waiting periods when charging. As highlighted in the study, while the battery electric vehicles (BEVs) are on the verge of growth, the complexities of the BEV system require a specific approach to address these issues properly (Dalton et al., 2013). Addressing these issues is crucial to enhance people's confidence and make it easier to transition towards an electric future. Smart charging systems also present their own challenges within the EV ecosystem, which are primarily associated with integrating the conventional power grids. As electric vehicles find greater utilization as dynamic loads on the grid, the load demand varies widely with the usage. The variability can potentially lead to grid overload and congestion unless handled appropriately. Researchers have also proposed a number of strategies, such as distributed charging priority and virtual power plant utilization, to optimize energy distribution and prevent infrastructure strain. Such strategies utilize advanced algorithms, such as genetic algorithms and particle swarm optimization, that attempt to balance energy needs and provide grid stability (García Caro et al., 2019). Although such technologies are sophisticated, there is a need to create effective smart charging protocols that can be employed to maximize the use of electric vehicles without detrimental connotations. Consumer sentiment is also a final key hurdle to widespread electric vehicle and smart charging system adoption. Although environmentally focused issues are now well known, many potential consumers are still to transition from traditional internal combustion engine vehicles to EVs. Factors such as perceived high initial costs, limited model availability, and skepticism over the adequacy of charging facilities dissuade buying decisions among potential customers. Moreover, awareness about the cost benefits of running electric vehicles—such as lower long-term cost and government subsidy—is a key influencer of buying behavior. Hence, targeted promotional campaigns and consumer awareness schemes are necessary to dispel wrong perceptions and highlight the virtues of electric vehicles and smart charging infrastructure. By transcending such informational and psychological boundaries, stakeholders can provide a more conducive environment for the deployment of electric mobility solutions.

9.5.1. Infrastructure Limitations and Charging Station Availability

The transition to electric vehicles (EVs) is naturally preceded by infrastructure limitations, primarily, the availability of charging stations. As urban environments evolve, the lack of existing charging networks continues to be a strong disincentive toward mass EV deployment. Spurts of growth in the EV market call for equally speedy development of charging stations, but most cities lag behind. Fresh research identifies that some cities are getting proactive to bridge those needs, developing end-to-end EV infrastructure plans to cope with increasing demands for electrical mobility. These interventions are critical because they provide a framework for urban planners and urban managers to facilitate a smooth transition to electrification, such that charging stations are strategically located in cities and made available for users, thus influencing the overall success of electric car initiatives (Fox-Penner et al., 2019). Various levels of infrastructure have a significant impact on the availability and effectiveness of charging points. Most of the existing systems provide equal charging capacity without efficient use of resources, which restricts their use in catering to EV users. Such a situation may cause traffic congestion at favored charging points and longer waiting times, which may deter prospective EV users from choosing to adopt EVs. An investigation looks into the advantages of implementing sophisticated management schemes for charging points, such as intelligent distribution of power resources and efficient customer routing. This approach not only enhances user satisfaction in terms of decreasing waiting times but also maximizes the utilization of the provided energy supply from the grid, addressing the important challenge of achieving efficiency in the energy-added system (Bayram et al., 2013). Integration between these systems stresses the need for innovative techniques to overcome the current limitations and build a more dependable charging infrastructure. Finally, addressing infrastructure limitations requires a multilayered approach that takes into account both the current status of charging stations and future demands. Charging network sustainable expansion needs to focus on areas most useful for users and local traffic patterns. In addition to expanding station presence as a whole, cities may also invest in intelligent charging technology that allows for maximum energy distribution optimization, which can assist in regulating demand on the power grid. This two-pronged emphasis on efficiency and accessibility is essential to the development of a charging infrastructure that can service a broad mix of electric vehicle users. By promoting the use of holistic planning methods that concentrate on the urban development patterns and energy usage, city governments can ensure that infrastructure emerges simultaneously with the increase in electric vehicle adoption, which further fosters a cleaner future (Bayram et al., 2013; Fox-Penner et al., 2019).

9.5.2. Range Anxiety and Battery Technology

Range anxiety is one of the biggest barriers to the widespread penetration of electric vehicles (EVs), primarily because of consumers' concerns over the battery capacity and perceived inadequacies in covering travel distance. It typically stems from the past experience of early electric models that failed to compete with traditional gasoline vehicles in both range and charging infrastructure. As a result, customers tend to resist migration to electric options for fear of being left stranded in the event there is no charging station. To mitigate this fear, battery technology must develop, with advances in more effective energy storage systems that will extend the driving range of EVs by a significant proportion. Therefore, enhancing the performance of batteries not only minimizes range anxiety but can also enhance the adoption of electric vehicles in the market. Developments in battery technology, including the creation of solid-state batteries and enhanced lithium-ion technology, are central to defeating range anxiety. Solid-state batteries, for example, are more energy-dense and safer than liquid ones, which could potentially increase the driving range of electric cars by a wide margin. More recent studies indicate that newer battery technologies would yield ranges far in excess of 300 miles per charge, making them somewhat comparable to conventional vehicles. Furthermore, newer charging technology, such as ultra-fast charging stations, can support these developments by reducing recharging time and anxiety over finding a charging station on extended trips. These technologies promise a future where range anxiety could much reduce, thereby promoting wider EV adoption and assimilation into the normal lifestyle. The social implications of the removal of range anxiety through the application of battery technology and smart charging networks are profound. As cities start to adopt the pivotal role of sustainable transportation, the role of urban planners and policymakers plays an important part in shaping the necessary infrastructure that EV rollouts will need. The application of intelligent charging systems enhances the user experience of EV users, as it relates to better charging schedules optimized with consideration for energy availability and grid requirements. In addition, fulfilling the need for urban society, in which charging is perhaps limited, becomes essential in dispelling concerns regarding the utilization of EVs. Scholarly literature suggests that solid city planning, together with enhanced battery technology, aligns with the change towards electric vehicles and also aligns with cities being able to change with this new transport wave without perpetuating existing conditions of pollution and energy consumption (Dalton et al., 2013; Fox-Penner et al., 2019) (Figure 9.14).

Figure 9.14. Electric vehicle (EV) range anxiety.

Source: https://www.buyev.my/images/Articles/RangeAnxiety.jpg.

9.5.3. Cost Barriers for Consumers and Manufacturers

The transition to electric vehicles (EVs) reflects greater sensitivity to the need for environmentally friendly transport solutions. There are, however, significant cost barriers that consumers and manufacturers cannot fully break in embracing this change. For consumers, the higher initial cost of electric vehicles compared to internal combustion engine vehicles can deter potential buyers, especially where affordability is an issue. This initial cost generally equates to a longer payback period, discouraging adoption despite lower operating costs and environmental benefits. As cities struggle to keep pace with increasing needs for EV facilities, the financial burden of doing so can be high, heavily taxing urban governments and private developers (Fox-Penner et al., 2019). Consequently, a proper comprehension of such economic constraints is essential to encourage wider acceptance and to ensure that the EV market will be viable. The cost of making and producing electric vehicles also poses a serious challenge to the manufacturers. The multifaceted supply chain employed to deliver fundamental parts such as batteries typically translates into increased costs of production that tend to cap scalability.

Also, the instability in raw material prices of lithium, cobalt, and other critical inputs poses an added financial uncertainty. The manufacturing firms also need to invest heavily in research and development in order to advance battery technology, which may result in a time lag in the recovery on investment. This situation is compounded by competition from traditional car manufacturers switching over to electric cars, therefore incurring higher costs as they seek to make the switch at a faster pace (Dalton et al., 2013). The synergy of design, cost of production, and market positioning creates a challenging situation for producers to remain competitive while meeting the demands of a rapidly changing marketplace. For reminiscing these cost implications on the consumers as well as manufacturers, coordination among all the stakeholders is a prime necessity. Public policies such as subsidy, tax rebate, and rebates can make electric vehicles more affordable for consumers and hence more accessible for the masses. Investments in EV infrastructure, i.e., charging stations, can also reduce the overall cost of EV ownership and boost consumer confidence towards making the switch. The manufacturers are able to benefit from collaborating with government institutions and research institutions in order to distribute innovation costs. The collaboration can lead to reduced production costs and faster technological improvement, providing a stronger market presence for electric vehicles (Fox-Penner et al., 2019). Lastly, eliminating these cost impediments will not only stimulate demand for EVs but also help lead us to a sustainable transportation future.

9.6. FUTURE TRENDS IN ELECTRIC VEHICLES (EVS) AND SMART CHARGING

The future of EVs will be transformed by advanced smart charging technologies of the future. As the demand for EVs keeps on increasing, new-fangled charging technology solutions will be required to keep up with the increasing demands placed on electrical grids. Among the trends is the development of Vehicle-to-Grid (V2G) technology, where EVs not only draw energy from but also supply energy to the power grid. The reverse flow not only allows supply-demand balancing but also promotes greater accommodation of renewable energy sources. Through the application of V2G technology, EV owners can potentially lower charging costs while supporting grid stability, particularly in peak periods of demand. The deployment of such smart charging solutions will involve collaboration between automobile manufacturers, utility firms, and regulators to create a seamless system that optimizes energy utilization without compromising the security of the electricity supply. With increasing usage of renewable energy sources, the relationship between electric vehicles and charging devices will become even more critical. States like California are leading the charge in greater penetration of renewables, with hopes such as 33% by 2020 as mandated by the Renewable Portfolio Standard (RPS). However, the growing utilization of electric cars is placing new constraints on existing distribution systems, particularly in voltage management (Arghandeh et al., 2013). The application of advanced Volt/VAR Optimization and Control (VVOC) technologies will be instrumental in solving the challenges.

Figure 9.15. Future of electric vehicles (EVs).

Source: https://i0.wp.com/evplugpros.com/wp-content/uploads/2024/06/blog-cover-photo-66.jpg?w=1366&ssl=1.

The new technologies will enhance the ability to manage voltage levels on distribution systems so that larger scales of adoption of EVs and solar power can be achieved without compromising the quality or reliability of service. The problems will require ongoing research and development to realize the full potential of smart charging and renewable energy. Aside from technological progress, policy regulations will drive the destiny of electric cars as well as smart charging systems. Government policies supporting the usage of EVs by offering tax credits and rebates, for example, are bound to stimulate consumers to opt out of traditional gasoline-powered vehicles. However, as the numbers of electric vehicles increase, the ensuing additional loads on the low-voltage power distribution grids must be efficiently managed (Butans et al., 2016). Uniform charging approaches that permit load redistribution minimize stress on local grids and thus minimize the need for costly upgrades in infrastructure. Policy-makers must ensure that the regulatory systems are adaptable to the rapid evolution of technology in this sector so that intelligent charging methods can flourish and support the realization of the ambitious goals of clean energy transition. This policy-tech-consumer synchronization will be required to initiate a future where EVs are a part of a clean, efficient and environmentally friendly transport system (Figure 9.15).

9.6.1. Innovations in Battery Technology

The fast-paced development of battery technology is a building block towards the evolution of electric vehicles (EVs) and their related intelligent charging systems. With the increasing demand for EVs, advances in battery chemistry, design, and management systems are key to improving energy density and efficiency. A handful of companies are producing lithium-silicon batteries, which are reported to have far greater energy storage capacity than lithium-ion cells available today. They do not just boost the driving range of electric vehicles but also reduce the frequency of charge sessions, hence addressing one of the largest concerns of potential users of EVs. With ongoing research and development, more is realized that improvement in battery technology will play a major role in a revolutionary transformation of mobility in urban areas and overall sustainability of our transportation systems, particularly in metropolitan cities with very severe challenges of EV infrastructure deployment (Fox-Penner et al., 2019). In addition to battery chemistry improvement, there has also been remarkable progress in the application of smart battery management systems (BMS). These BMS optimize the charge process, monitor battery condition, and ensure safety, thereby extending battery life. Integrated with algorithms rooted in machine learning, these BMS can predict best times for charging and optimize energy consumption, taking into account load from the grid as well as availability of green power. Beyond this, innovation in the form of solid-state batteries is on the cusp, offering enhanced safety as well as thermal stability. All these technologies work together to facilitate the establishment of a more efficient and robust EV ecosystem that reduces greenhouse gases and contributes to the development of a cleaner city. Enhanced battery management has a direct connection with improved smart charging solutions, hence facilitating urban planners to incorporate interventions which work in their favor for sustainable energy goals (Hussain et al., 2018). Moreover, the integration of battery technology with upcoming technologies such as blockchain is transforming the EV charging scenario. With electric vehicles being integral to the smart grid, new solutions are in demand currently for billing, charging, and audits. Proposals recently include the use of distributed ledger technology to implement a secure, efficient, and scalable charging system for supporting the growing number of autonomous electric vehicles. These techniques not only support real-time charging transactions but also ensure greater security and transparency in the charging process. Meeting these requirements, new models of smart charging can support a vast range of applications, from conventional driving to full-autonomous systems. The ongoing development of technology for batteries will thus be a core factor in the capability to seamlessly integrate electric vehicles into the wider energy system and continue urban infrastructure growth (Hussain et al., 2018).

9.6.2. Integration of Renewable Energy Sources

Integration of renewable energy into power systems is increasingly becoming necessary with the growing utilization of electric vehicles (EVs) and the rise in demand for renewable forms of energy. These developments require a dynamic and intelligent infrastructure that can efficiently manage shifting energy demand and supply. Smart grids spearhead this transformation by employing advanced technology to enable efficient incorporation of renewable energy sources. Through controllable loads and storage systems, smart grids enhance the stability of the electric grid, especially in rural communities where renewable energy production is localized. Literature highlights the significance of projects like “Well2Wheel” and “Smart Rural Grid” in the face of such challenges, showing the potential in innovative strategies to optimize the existing distribution system without any major investment in infrastructure (Boehner et al., 2016). Electric vehicle integration and renewable energy sources also present fresh avenues of energy efficiency as well as environmental sustainability. As more EVs are available, they can be utilized as mobile loads within the power grid, further enhancing the integration of renewables. This dual function means that EVs not only can absorb energy when charged but also, if it is necessary, can feed power back into the grid. In order to develop the full potential for this, researchers propose new methodologies for distributed charging prioritization with the virtual power plant concept. These systems make use of evolutionary computation methods to maximize charging periods and avoid grid overloading so that there can be symbiosis between green energy generation and EV infrastructure (García Caro et al., 2019). It not only supports grid stability through such synergy, but it also facilitates the optimum use of green energy. Moreover, operational efficiency thereby enabled by such technologies as intelligent distribution power routers (IDPR) enhances the integration of renewable energy significantly. IDPR enable dynamic reactive power and voltage management in distribution networks, enabling more levels of renewable generation to be hosted without significant upgrades to the infrastructure. Through facilitating controlled operation of renewable generators and stationary battery storage, these next-generation power routers show how innovative technology can solve traditional grid issues. Besides, the adoption of smart markets at a local scale enables distribution system operators to manage grid congestion effectively, thus incentivizing asset owners through intelligent pricing mechanisms that react to real grid conditions (Boehner et al., 2016). The deployment of these technological advancements places renewable energy integration on the agenda as a viable solution for electric mobility and consumption's sustainable future.

9.6.3. Policy Developments and Incentives for Electric Vehicle (EV) Adoption

In the recent years, there has been the central government policy-making playing a leading role in enacting the uptake of EVs. There have been expansive policy plans all over the world focused on reducing carbon emissions and embracing green transport. This is primarily triggered by the urgent demand to contain air pollution and environmental degradation caused by conventional internal combustion engine cars. Policymakers have realized increasingly that facilitating a mass transition to battery electric vehicles (BEVs) is essential in an attempt to achieve urban sustainability goals. The policies entail the provision of economic incentives through tax credits or rebates on purchasing EVs and investing in charging infrastructure, thus directly addressing consumers' barriers for adopting EVs, in the end making the market more affordable and shaping consumer demand for cleaner transport solutions (Dalton et al., 2013). Incentives for the adoption of EV extend beyond mere economic benefits, and the need arises to design full-scale infrastructure. As cities and municipalities transform to support EV demand, the implementation of smart charging technology becomes paramount. Smart charging technology maximizes energy usage as well as synchronizes grid demand with renewable energy production to create a sustainable energy system. Such systems are also able to effectively schedule vehicle charging times, with EVs charging when demand is low or at times of high renewable production. The availability of extensive charging stations, often subsidized through public-private partnerships or governments, also creates a place where EV ownership becomes more desirable and viable. Properly constructed incentives and solid infrastructure are thus key to policymakers seeking to encourage broader EV adoption (Fox-Penner et al., 2019). Apart from this, the role of public awareness campaigns and education cannot be overlooked in the context of EV policy development. Public awareness building through programs that raise awareness among potential consumers of the environmental benefits, cost savings, and technology in EVs has been effective. All stakeholders like manufacturers, urban planners, and government agencies need to collaborate in public-private partnership initiatives to craft an enduring narrative which emphasizes the necessity for electric mobility. This is required not just to boost the rate of adoption but also to ensure that every segment of the population is included in this transition. Policymakers must focus on inclusivity so that poor communities are offered access to EV incentives as well as charging installations. Doing so, they not only drive Smart Equity within the adoption of the sustainable technology but also optimize overall urban climate target efficiency (Dalton et al., 2013).

SUMMARY

Overall, the transition to EVs with support from smart charging systems is a watershed moment for the development of sustainable transport. As demonstrated in numerous researches, electric mobility not only offers tremendous savings in transport costs but also has potential to save the energy and harmful emissions. However, the means to utilize electric four-wheelers in developing countries represent a major hindrance because of their out-of-pocket cost. Electric two-wheelers, on the contrary, represent an affordable alternative, suggesting that targeted EV adoption strategies have the potential to reap maximum benefits to different regions. These variations must be acknowledged by policymakers so that there is equitable access to electric mobility solutions to enhance greater access to sustainable transport schemes by diverse types of communities (Albrecht et al., 2020). Moreover, incorporation of smart charging technologies is critical in order to gain complete advantages from renewable energy sources in the electric vehicle sector. Demand response algorithm design ensures coordinated charging, which allows for synchronization of energy usage with periods of high availability of renewables. This technique has been shown to significantly increase the proportion of renewable energy harnessed to meet EV charging demands along with minimizing $CO2$ emissions. In practical implementations, such practices have been proved to be very effective—understanding the supply of up to 68% of energy demand by wind power through smart coordination. These findings point towards the importance of using smart grid technologies, which can address current energy issues in addition to breaking reliance on fossil fuels (Turck et al., 2012). Ultimately, the combination of electric vehicles and smart charging technologies presents an opportunity to foster systemic change in both transportation and energy sectors. To ensure that transition towards electric mobility is not only a success but also equitable, stakeholders need to assess existing infrastructures, cost factors, and geographical needs. This will enable them to develop innovative policies which not only promote the adoption of EVs but also support the development of renewable energy and grid integration. This multi-pronged approach can give rise to a more sustainable future where electric vehicles are the cornerstone of an environmentally friendly transport system, lending support to better public health and environmental health in the process (Albrecht et al., 2020; Turck et al., 2012).

The discussion on EVs and smart charging has put in highlight some deep lessons on urban planning and infrastructure development. One observation of significance is the imperative for cities to get ready for the rising tide of demand for EVs, highlighting the issues and benefits involved in this shift. As reflected in existing studies, most cities are yet to address building the required infrastructure to embrace a bulk adoption of EV technology. Other cities, though, are not only meeting the challenge but are also stepping up to the leadership in increasing EV infrastructure. This is by adopting innovative strategies through collaborative coordination between city managers, private developers, and utilities to present strategic policies for optimizing sustainable urban mobility by establishing EV integration and smart charging solutions (Fox-Penner et al., 2019). The probable advantages of moving towards electric vehicle fleets are particularly evident to Small Island Developing States (SIDS), which are disproportionately vulnerable to climate change. They contribute very little to global greenhouse gas emissions but suffer full force from the effects of climate, and thus new methods of decarbonization are required. The specific geographical and economic conditions of SIDS—limited land space and limited populations—provide an opportunity for them to utilize effective practices of electrification. Further, leveraging the use of the tourism sector can facilitate shared mobility pilots that encourage awareness to tourists of sustainable travel modes. This targeted strategy has a multiplier effect since it acts as a catalyst for local communities to engage in zero-carbon practices. Therefore, investment in SIDS is beneficial to both climate mitigation and makes low-carbon living viable in the global environment (Blechinger et al., 2020). Moreover, the adoption of smart charging technologies and electric vehicles can possibly influence the energy pattern in both urban and rural regions. Smart charging technologies conserve energy by synchronizing the charging patterns with the availability of renewable energy resources such as solar or wind energy. Such coordination not only enhances grid stability but also facilitates the adoption of EVs by reducing charging expenses for consumers, hence enhancing the economic case for an electric mobility transition. Such synergy between consumer behavior and technology is driving this transformation to the cutting edge. As energy suppliers and utilities begin implementing demand-response interventions to balance energy loads, the role of smart charging becomes increasingly imperative. By these innovations, the way will be set for a sustainable and connected transportation system that will be a major turning point in the connection between energy and mobility in our day-to-day lives.

REFERENCES

1. Alahakoon, S., Jayasinghe, S., Pashajavid, E., Ramraj, P., et al. (2023). Quality of service and associated communication infrastructure for electric vehicles †. https://core.ac.uk/download/613908013.pdf (accessed on 30th September 2025).
2. Albrecht, J., & Rajper, S. Z. (2020). Prospects of electric vehicles in the developing countries: A literature review. https://core.ac.uk/download/322829320.pdf (accessed on 30th September 2025).
3. Ali, M. M., Fuad, A., Lipu, H., Molla, M., Islam, M., et al. (2024). Empowering fuel cell electric vehicles towards sustainable transportation: An analytical assessment, emerging energy management, key issues, and future research opportunities. https://core.ac.uk/download/643801310.pdf (accessed on 30th September 2025).
4. Arghandeh, R., Brown, M., Cibulka, L., Miller, S., et al. (2013). Distribution system voltage management and optimization for integration of renewables and electric vehicles: Research gap analysis. https://core.ac.uk/download/288432751.pdf (accessed on 30th September 2025).
5. Arroyo, V. (2019). Investing in America's surface transportation infrastructure: The need for a multi-year reauthorization bill: Hearing before the S. Comm. on Env't & Pub. Works, 116th Cong., July 10, 2019. https://core.ac.uk/download/225012733.pdf (accessed on 30th September 2025).
6. Bayram, I. S., Devetsikiotis, M., Granelli, F., Michailidis, G., et al. (2013). Electric power allocation in a network of fast charging stations. https://core.ac.uk/download/265813830.pdf (accessed on 30th September 2025).
7. Bezruchonak, A. (2019). Geographic features of zero-emissions urban mobility: The case of electric buses in Europe and Belarus. https://core.ac.uk/download/225150005.pdf (accessed on 30th September 2025).
8. Blechinger, P., Creutzig, F., & Soomauroo, Z. (2020). Unique opportunities of island states to transition to a low-carbon mobility system. https://core.ac.uk/download/322792155.pdf (accessed on 30th September 2025).
9. Boehner, V., Franz, P., Gallart, R., & Girbau Llistuella, A., et al. (2016). Smart grids for rural conditions and e-mobility–Applying power routers, batteries and virtual power plants. https://core.ac.uk/download/81571789.pdf (accessed on 30th September 2025).
10. Bunch, D., Chakraborty, D., Jenn, A., et al. (2020). Factors affecting demand for plug-in charging infrastructure: An analysis of plug-in electric vehicle commuters. https://core.ac.uk/download/287624006.pdf

(accessed on 30th September 2025).

11. Butans, E., Nieto-Martin, J., & Zafred, K. (2016). Electric vehicles–Effects on domestic low voltage networks. https://core.ac.uk/download/74409962.pdf (accessed on 30th September 2025).
12. Dadlani, A., Kim, K., Mehrabidavoodabadi, A., Moon, Y., et al. (2020). Decentralized greedy-based algorithm for smart energy management in plug-in electric vehicle energy distribution systems. https://core.ac.uk/download/323307094.pdf (accessed on 30th September 2025).
13. Dalton, R., ElBanhawy, E., & Thompson, E. M. (2013). Eras of electric vehicles: Electric mobility on the verge. *Focus Attention Scale*. https://core.ac.uk/download/17299059.pdf (accessed on 30th September 2025).
14. De Turck, F., Develder, C., & Mets, K. (2012). Distributed smart charging of electric vehicles for balancing wind energy. https://core.ac.uk/download/55773539.pdf (accessed on 30th September 2025).
15. Develder, C., Haerick, W., & Mets, K. (2010). A simulator for the control network of smart grid architectures. https://core.ac.uk/download/55823207.pdf (accessed on 30th September 2025).
16. Di Giorgio, A., Lanna, A., Liberati, F., Zuccaro, M., et al. (2014). Electric vehicles charging control based on future internet generic enablers. http://arxiv.org/abs/1503.01267 (accessed on 30th September 2025).
17. Di Giorgio, A., Mercurio, A., & Purificato, F. (2013). Optimal fully electric vehicle load balancing with an ADMM algorithm in smart grids. http://arxiv.org/abs/1305.1044 (accessed on 30th September 2025).
18. Fox-Penner, P., Hao, J. (R.), Hatch, J., Helveston, J., et al. (2019). The critical role of public charging infrastructure. https://open.bu.edu/bitstream/2144/39112/1/MeltingIceBook_Web_FINAL.pdf (accessed on 30th September 2025).
19. García Caro, S., García Delgado, A., Guerrero Alonso, J. I., León de Mora, M., et al. (2019). Forecasting recharging demand to integrate electric vehicle fleets in smart grids. https://core.ac.uk/download/322845464.pdf (accessed on 30th September 2025).
20. Graff, R. (2011). *Ready to Roll? Overview of Challenges and Opportunities*. https://core.ac.uk/download/71346025.pdf (accessed on 30th September 2025).
21. Hussain, R., Lee, J., Mazzara, M., Mustafin, R., et al. (2018). On M2M micropayments: A case study of electric autonomous vehicles. https://core.ac.uk/download/492594510.pdf (accessed on 30th September 2025).
22. Ikpehai, A., Adebisi, B., Cluzel, F., Conn, D., Rabie, K., Maciejowski,

J., et al. (2017). Bi-directional coordination of plug-in electric vehicles with economic model predictive control. https://core.ac.uk/download/187716664.pdf (accessed on 30th September 2025).

23. Klitkou, A., Bolwig, S., Coenen, L., Solér, C., et al. (2015). *Technology Opportunities in Nordic Energy System Transitions (TOP-NEST): Final Report*. https://core.ac.uk/download/pdf/154665071.pdf (accessed on 30th September 2025).

CHAPTER 10

CYBERSECURITY IN SMART GRIDS

Contents

INTRODUCTION

Rapid evolution of electrical power grids into smart grids demands in-depth knowledge of cybersecurity concerns inherent in these new infrastructures. As the United States is transitioning to an integrated grid founded on new digital technologies, susceptibilities connected with such a transition come into vivid picture. Smart grids integrate renewable energy sources that, while being favorable to sustainability, carry with them complexity not envisioned by traditional grid infrastructures. Smart technology expansion, as noted by Millis et al. (2019), brings new, untested security risks that can potentially significantly impact grid operations. Greater reliance on information technology to handle energy increases the attack surface, making these systems ideal targets for attackers seeking to exploit weaknesses. Attaining resilience and reliability, therefore, means addressing cybersecurity without compromising innovative capabilities of smart grid technologies.

Figure 10.1. Cybersecurity in smart grids.

Source: https://www.researchgate.net/publication/387649510/figure/fig4/AS:1143 1281391682268@1745330877241/Cybersecurity-in-smart-grids-19.tif.

The addition of distributed energy resources and advanced communications adds functionality to the smart grid but in the process adds to the need for addressing cybersecurity. The new rural distribution grid design, for instance, is evolving to incorporate a decentralized approach in management that makes use of new telecommunication synergies and electrical storage, as discussed in Díaz González et al. (2018). This strategic change not only optimizes energy supply but also raises security concerns regarding decentralized operational frameworks. As these new frameworks are being adopted by operators, the danger of cyberattacks that not only disrupt the energy supply but also critical services increases. The consequence of such a disruption cannot be ignored; there should be a framework for developing robust cybersecurity policies that secure such systems without undermining innovation and operational efficiency. In order to effectively manage the complexities of cybersecurity in smart grids, all stakeholders are required to participate in multidimensional strategies that also involve both technical and regulatory aspects. Comprehensive vulnerability assessments, as indicated in Millis et al. (2019), can yield useful information regarding areas of improvement for grid elements. Second, it is essential to bring together a mix of sectors including government agencies, energy corporations, and security specialists to create an environment that makes security a core component of smart grid development. Real-time monitoring and response schemes, along with constant evaluations of potential new threats, become increasingly crucial as technology proliferates across grid networks. Lastly, their solution not only emboldens the physical infrastructure, but it also creates public trust in the security and reliability of smart energy systems (Figure 10.1).

Definition of Smart Grids

The concept of smart grids has developed as an essential step within the evolution of electrical energy systems, suggestive of an integration of digital technology with power distribution and management. Smart grids are defined by their implementation of smart meters, sensors, and advanced communication networks to enhance the efficiency, reliability, and sustainability of energy consumption and production. The advanced architecture enables real-time monitoring and control of electricity flow, enabling utilities to respond quickly to variations in demand, accommodate renewable energy sources, and optimize grid performance. Therefore, smart grids promote not only the distribution of energy but also greater consumer engagement through data-driven decision-making, balancing production levels with consumption levels. Emphasis on the pivotal role of technology in infrastructural transformation is necessary, particularly where frictionless operation forms a vital part of national security and economic equilibrium. Understanding about the operation of smart grids and potential areas of weakness is critical in the area of cybersecurity. Interconnectedness of the elements such as advanced metering infrastructure (AMI) and supervisory control and data acquisition (SCADA) systems creates a vast cyber environment which becomes vulnerable to attacks by malicious entities if not well secured. Over the past few years, successful cyberattacks on power grids have validated the catastrophic consequences following from weak spots in smart grid infrastructure. For instance, the Ukraine cyberattack revealed the effects of simultaneous attacks on core cyber infrastructure, symbolizing the extreme need for proper security measures in safeguarding such systems (Ginter et al., 2018). As more trends towards smart grids emerge, knowledge of and action against their inherent vulnerabilities is critical to maintaining the integrity of national energy systems. Besides, integration of big data analytics in smart grids enhances their efficiencies of operation while simultaneously introducing new security risks. Through robust streams of data formed from diverse sources, power systems can achieve intelligent decision-making and predictive analysis, making grid operation streamlined and facilitating new uses. Grid performance monitoring, detection of anomalies, and ultimately system stability improvement are significant roles of big data analytics. However, the reliance on networked Internet of Things (IoT) and cloud-based infrastructure is a significant threat that must be carefully worked against (Gharehpetian et al., 2019). These analytics platforms must be accompanied by robust cybersecurity controls to protect against data breaches and ensure the smart grid operation's resiliency. As the energy environment evolves, a dual focus on security and innovation will be vital to guide the complexities of modern energy systems (Figure 10.2).

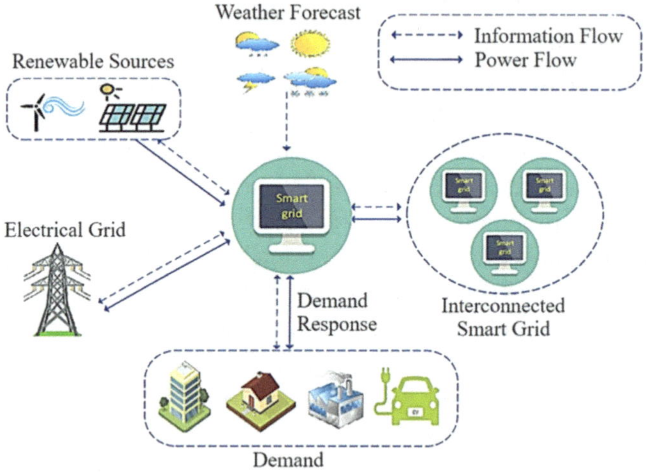

Figure 10.2. Typical block diagram of a smart grid.

*Source:*https://advanced.onlinelibrary.wiley.com/cms/asset/5263afa3-da3c-4cfe-8d62-a8d462eda858/aesr202400088-fig-0008-m.jpg.

Importance of Cybersecurity in Smart Grids

Incorporation of cybersecurity into smart grids is basically imperative due to the increasing application of digital infrastructures characterizing modern energy systems. Smart grids use advanced communication technologies to ensure efficiency and dependability in the delivery of electricity but are also vulnerable to being hacked because the integration makes them hackable. With the capabilities of harvesting and transmitting bulk data, these networks would become potential points of attack for malicious actors who wish to disrupt energy supplies or compromise operational integrity. As mentioned in rural distribution grid conversations, the new architecture of upgrading these networks places particular focus on sufficient cybersecurity to protect not just the power and management planes of the system but also to ensure that distributed energy resource operators can operate distributed energy resources securely without risking themselves to cyberattacks (Díaz González et al., 2018). Furthermore, the application of big data analytics in smart grids reinforces the importance of cybersecurity. With energy management increasingly adopting the use of advanced algorithms and artificial intelligence to process and analyze large volumes of data, the threat of vulnerability also increases in tandem. Decision-makers for the power system face the critical task of ensuring data security with the widespread number of cyber-attacks seeking to undermine the effectiveness of traditional security measures. Reliance on big data analysis also brings with it the threat of unauthorized use as well as data integrity. Cybersecurity will thus need to evolve to respond to these threats by providing multi-layered defense mechanisms for protecting sensitive information and enabling smart grids to keep running because creative steps must be taken to deflect threats and enhance resilience against prospective cyber threats (Gharehpetian et al., 2019). Moreover, the ramifications of inadequate cybersecurity measures can be wide-ranging, impacting not only utility companies but also end-users and the economy overall. A successful cyber-attack on a smart grid can potentially lead to widespread power disruptions, economic impacts, and declining public trust in energy providers. Such is an underlining note that underscores the relevance of regulatory actions and industry standards putting a priority on cybersecurity in the planning and implementation of smart grids. With cooperative efforts involving the stakeholders of utility operators, cybersecurity experts, and policymakers, a worldwide cybersecurity blueprint can be adopted. This would enhance the robustness of smart grids against cyber threats and safeguard important infrastructures so that they can meet the needs of the new energy paradigm without compromising operational security and consumer trust.

10.1. OVERVIEW OF SMART GRID TECHNOLOGY

The evolution of Smart Grid technology is an intrinsic transformation in the control and distribution of electric energy, eventually giving rise to the enhanced reliability, efficiency, and sustainability of the energy system. Smart Grids employ higher-end information and communication technologies to create an integrated network that enables real-time data management and decision-making processes within electrical grids. Through the integration of two-way communication between utilities, consumers, and various energy resources, Smart Grids enable the optimum allocation of power and expanded utilization of renewable energy resources. Technologies such as automated demand response and intelligent monitoring system capabilities mark a shift away from the traditional grid paradigms in which manual interventions were the rule and thus an entry into a more responsive and resilient energy infrastructure imperative in today's energy landscape. One of intriguing advancements in Smart Grid technology is the deployment of decentralized management architectures tailored to rural distribution networks. The installation of agents that assist legacy Supervisory Control and Data Acquisition (SCADA) systems is a demonstration of how Smart Grid installations can increase operational efficiency in poorer system configurations. In practice, the new architecture promotes the increased integration of distributed energy sources on the basis of better communications links and power electronics.

Figure 10.3. Overview of smart grid technology.

Source: https://yourelectricianwinnipeg.ca/wp-content/uploads/2023/12/view-bio-engineering-advance-with-human-hand.jpg.

As illustrated in real-world deployments in pilot networks throughout Europe, these deployments not only improve efficiency and quality of service but also benefit the growing reliance on climate change combatants that are sustainable energy technologies. Overall, these advances indicate the foremost importance of Smart Grid design in solving contemporary energy challenges and showing the increasing imperative for robust cybersecurity practices to protect these advanced networks from emergent threats (Díaz González et al., 2018). While Smart Grids thrive with exponential benefits, their reliance on interconnected technologies carries monumental risks that demand comprehensive cybersecurity efforts. The development towards a more decentralized and digitized electrical system results in complicated interdependencies that can be exploited by malicious actors to create means of possible interference and data compromise. Further, the shift towards renewable resources, as environmentally friendly as it is, reveals existing security problems that were not addressed prior to now. A vulnerability analysis, highlighted by recent studies, shows that while energy storage helps in better resilience to disruption, the operation of such systems depends on the careful implementation of cybersecurity measures (Millis et al., 2019). While the cyber world keeps evolving with Smart Grid technology, foresight is required to ensure that all necessary countermeasures are given priority to safeguard the integrity of these systems so that there can be a secure and reliable energy future (Figure 10.3).

10.1.1. Components of Smart Grids

Smart grid entails a collection of networked components designed to enhance efficiency and reliability of the supply of electricity. At its core are advanced metering infrastructure (AMI) systems, which enable real-time data communication between customers and utilities. AMI includes two-way communicating smart meters, enabling dynamic pricing as well as consumption metering. Besides, the systems also provide electricity companies with precious data on energy usage patterns, making it possible to balance supply and demand better. But the interconnectivity of the components also presents severe cybersecurity risks. Each piece of equipment, ranging from smart meters to grid management systems, is a point of vulnerability for cyber-attacks, and thus the need to protect such technologies in the overall context of grid security (Ogunmoye et al., 2025). Distributed Energy Resources (DERs) are another vital component of smart grids, integrating renewable energy like solar and wind into the power system. The distributed energy sources add flexibility and resilience to the grid but also introduce vulnerabilities with their digital interfaces and control systems. Increased reliance on Internet of Things (IoT) devices to control and monitor DERs complicate the cybersecurity. The attackers can exploit weaknesses in the systems that control these resources, risking disrupting energy delivery or grid stability. Strategically implementing cybersecurity solutions, like incident response controls and system segmentation, can mitigate these risks while developing an efficient energy ecosystem. Safe DER integration will be the core of future grid operations and will need an active response to overcome their associated cybersecurity challenges (Simonthomas et al., 2024). Lastly, SCADA systems play a critical role in the management and automation of smart grids. The systems enable real-time monitoring and control of grid operations, ranging from generation to delivery. But the integration of SCADA systems into the smart grid system exposes them to a variety of cyber-attacks. A successful intrusion on SCADA would trigger significant disruption, not only in the operational efficiency of the grid, but also trigger cascading failures on interdependent systems. Thus, it becomes important to utilize cutting-edge cybersecurity technologies, such as intrusion detection systems and artificial intelligence-powered threat analytics, to safeguard these critical components.

With SCADA resilience enhancement, utilities can make smart grids not only yield the expected benefits of modern energy management but also uphold integrity and security required to secure consumers as well as infrastructure (Ogunmoye et al., 2025) (Figure 10.4).

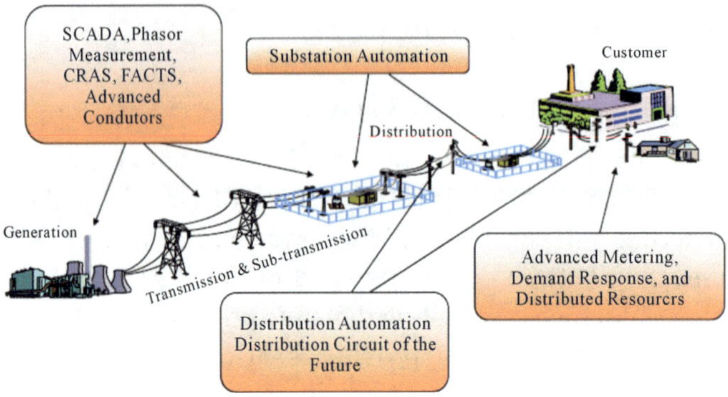

Figure 10.4. Components of smart grid.

Source: https://www.scirp.org/html/1-6401052/189bc43d-6bf4-4e2c-af07-3f2e476cc82a.jpg.

10.1.2. Benefits of Smart Grids

The transition to smart grids introduces a number of benefits that largely complement the increase in cybersecurity. At the most basic level, smart grids offer enhanced monitoring and management of electrical networks and thereby support the integration of renewable energy sources. Smart grids offer a decentralized system that enhances resilience against outages through the facility for local energy generation and storage facilities. As identified in recent architectural advancements, the emerging smart grid architecture leverages telecommunications and power electronics to extend rural network capabilities, thereby creating increased operating efficiency as well as facilitating increased management of distributed energy resources. The confluence of such technologies facilitates a more dynamic power distribution system, crucial for adaptation to alterations in energy production and consumption, leading finally to a more secure and stable electricity supply system (Díaz González et al., 2018). Beyond efficiencies of operation, the application of smart grids significantly boosts energy security by facilitating the possibility of having state-of-the-art cybersecurity capability. With the increased reliance on digital and networked infrastructure, smart grids leverage advanced data analytics and real-time monitoring capabilities to identify and react to potential threats. These technologies not only improve the overall security of the electrical system but also provide operators with actionable intelligence on areas of vulnerability. With the potential for disruptions having far-reaching economic impact, the integration of resilience measures—such as energy storage solutions—is prized in optimizing network performance under difficult circumstances. A thorough vulnerability assessment of the electrical grid operations has shown that storage is particularly beneficial in grids with renewable energy overcapacity, thereby further increasing security through diversified sources and improved management practices (Millis et al., 2019). Smart grids also increase a pro-activeness in addressing energy issues, both for security and sustainability. The integration of smart technologies with renewable resources makes electricity use and generation more sustainable. Through an optimization of the energy efficiency of systems and facilitation of the integration of variable renewable sources, smart grids contribute to reducing fossil fuels and carbon emissions dependence. As demonstrated in pilot projects, such grids allow operators to make intelligent decisions about the supply of energy and demand response, generating a more environmentally friendly energy environment that by its nature integrates cybersecurity issues. Thus, the benefits of smart grids extend beyond quantitative gains in operations to encompass longer-term effects on resilience, sustainability, and energy security in the context of modern electric infrastructures (Díaz González et al., 2018) (Figure 10.5).

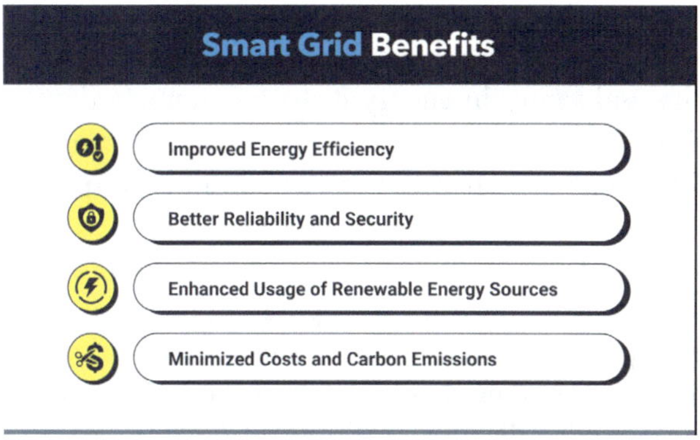

Figure 10.5. Smart grids improve energy efficiency, reliability, and renewable energy usage while minimizing costs and carbon emissions.

Source: https://assets.eweek.com/uploads/2024/11/EW_SmartGrids_2024_MEL_rnd2-03-1024x683.png.

10.1.3. Current Trends in Smart Grid Development

As the energy paradigm shifts, recent trends in smart grid development reflect a growing focus on incorporating advanced technologies, including artificial intelligence (AI) and machine learning. These technologies aim to enhance operational efficiency while overcoming the unique challenges of dynamic energy loads and cyber-attacks. Adaptive project management processes are becoming more prevalent in the smart grid, promoting iterative development and collaborative working, needed to keep pace with rapidly changing technology. The combination of smart grid technology with agile principles not only enables more effective utilization of resources but also increases capabilities in terms of response to emerging security risks. Prioritizing user engagement and flexibility, stakeholders can enhance the resilience of energy systems and create a foundation for future expansion in smart grid innovation and security integration (Adegboyega et al., 2024). Apart from agile methodologies, the use of AI-based anomaly detection systems is also emerging as a dominant trend in smart grid evolution. These systems utilize sophisticated algorithms to identify anomalies in power grid operations, filling gaps that can be exploited by attackers. The complexity of modern power grids also necessitates such intelligent detection systems, which are capable of identifying various types of anomalies, such as data integrity breaches and unusual consumption patterns. Anomalies and cyber-attacks related to network infrastructure are top priority areas of critical significance in current research with high needs for efficient security measures. As machine and deep learning algorithms increasingly find applications, smarter grids are evolving into next-generation grids that not only react but also predictively anticipate prospective security threats, promoting overall reliability and safety in energy distribution networks (Burgos MFG et al., 2024). High-profile trends in smart grid investment also highlight the need for cybersecurity frameworks that can serve to diminish threats from increasing digitalization. Connected devices and systems lead to an amplified attack surface, and the choice is thus an important consideration. The current strategy prefers a multi-layered framework, incorporating physical security, enhanced encryption techniques, and real-time monitoring to detect and respond to threats. With this, constant innovation of hybrid solutions, combining traditional security with cutting-edge detection techniques, becomes a necessity. Through joint collaboration between technology innovators and cybersecurity experts, the smart grid market can come up with more resilient systems that are well positioned to adapt to various threats, enabling energy infrastructures to be secure in spite of rising cyber threats (Adegboyega et al., 2024; Burgos et al., 2024).

10.2. CYBERSECURITY THREATS TO SMART GRIDS

The implementation of smart grids in modern energy networks has not only significantly raised operational efficiency but also facilitated the integration of renewable resources. This innovation in technology has, however, introduced a range of cybersecurity threats that have the potential to jeopardize the reliability and stability of power systems. Among the most critical examples is the cyber-attack on the power grid of Ukraines, which indicated the possibility of disaster by means of hacking into significant infrastructure through administrative credentials. The hackers discovered vulnerability in supervisory control and data acquisition (SCADA) systems, leading to coordinated attacks on a number of substations. Such incidents reinforce the necessity for multi-faceted approaches in order to curb the cyber threats. As these types of systems become increasingly interconnected, the threat of coordinated attacks grows, and this illustrates the pressing need for upgraded cybersecurity infrastructure across the grid network in order to ward off cascading failure that could impact a broader interconnected system (Ginter et al., 2018). The industrial internet of things (IoT) has proven to be a double-edged sword when it comes to smart grids, providing new opportunities for efficiency but at the same time enormously expanding pre-existing security concerns. Power grid DDoS attacks have increased, increased vulnerabilities within systems reliant on continuous connectivity. Current legislations, the EU Network and Information Security Directive and General Data Protection Regulation, have begun to address these novel threats, but gaps exist between those regulations in practice and their enforcement. The complexity of smart grid infrastructure amplifies these security challenges because traditional cybersecurity measures might be inadequate to the complexities of interdependent systems. Therefore, it is important not just to recognize the shift from offline to an online infrastructure but also to address the unique temporal dimension and complexity in addressing cybersecurity for smart grids with a focus on ensuring that regulatory support keeps up with technological advancement (Accenture Technology et al., 2018). Also, when the stakeholders and utilities transition their technology platforms, proactive cyber security is the hour of need. Continuous risk assessment, increased capability for real-time monitoring, and implementation of advanced anomaly detection systems can effectively make the smart grids more robust. An intersectoral strategy where government and private sectors collaborate becomes the determinant of establishing best practices unique to the individual challenges of the smart energy supply chain. By focusing on a combination of proactive and reactive strategies, asset owners are better positioned to predict possible cyber threats. A formalized risk profile analysis, as discussed in research regarding catastrophic impacts on grid reliability, will enable utilities to focus resources most efficiently on securing critical cyberinfrastructure. Lastly, these steps are critical to the establishment of a secure and robust future for energy that is less vulnerable to the ever-changing world of cyber threats (Ginter et al., 2018) (Figure 10.6).

Figure 10.6. Cybersecurity threats.

Source: https://icssindia.in/blog/wp-content/uploads/2024/10/Cybersecurity-Threats.png.

10.2.1. Types of Cyber Threats (e.g., Malware and Phishing)

The integration of smart grids into modern energy systems has introduced unprecedented efficiency and connectivity. But in the process, this increased dependence also exposes these infrastructures to a world of cyber assaults. The most prevalent are malware and ransomware, which infiltrate systems to sabotage or siphon off sensitive information. Malware exists in various forms, including Trojan horses and viruses, often employed to break down the integrity of systems by taking control of operations protocols. In smart grids, where control and real-time data flow are of prime importance, such kinds of threats can lead to complete failure and power shutdown, demonstrating the extreme need for suitable cybersecurity measures on every operational level to decrease the risks, as advocated by latest research in cyber defense frameworks (Anglano et al., 2018). Phishing is another large category of cyber threats involving fraudulent efforts to acquire sensitive information by masquerading as authentic sources. Phishing in smart grids can target utility employees or consumers and result in unauthorized access to the control systems of critical infrastructure. Through expertly crafted emails or messages, cybercriminals attempt to take advantage of human vulnerability, employing social engineering strategies capable of breaching even the most sophisticated technological defenses. Given the potential for such attacks to be exploited to facilitate even more widespread threats, such as direct penetration of grid control systems, intensive awareness and training of staff against phishing techniques become essential components of an effective cybersecurity framework, as recognized in more recent evaluations of cyber threats (Hamim et al., 2023). The cybersecurity landscape for smart grids changes on a daily basis, necessitating an aggressive posture towards cybersecurity given the increasing sophistication of attackers. Besides malware and phishing, other types of cybersecurity attacks such as Denial of Service (DoS) attacks and Zero-Day exploits make it difficult to safeguard these vital services. DoS attacks, for instance, can fill system resources, hence paralyzing operations and service delivery. Zero-Day exploits exploit yet-to-be-known vulnerabilities in software, offering cybercriminals a chance to infiltrate networks prior to the potential of building defenses. This tangled array of threats demands a multi-faceted approach to cybersecurity—a one that incorporates not only technical countermeasures but also education, risk management, and incident response protocols tailored to the challenges presented by smart grid systems (Anglano et al., 2018) (Figure 10.7).

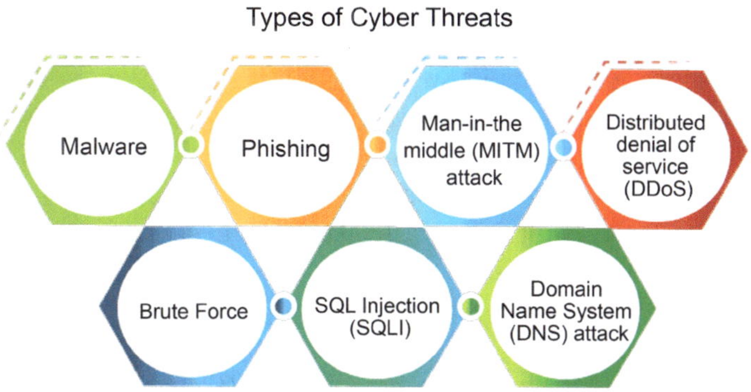

Figure 10.7. Types of cyber threats.

Source: https://www.acte.in/wp-content/uploads/2024/11/what-is-cyber-security3.png.

10.2.2. Potential Impact of Cyber Attacks on Smart Grids

The susceptibility of smart grids to cyber-attacks can be tremendous since the systems rely on advanced technologies aimed at improving performance and reliability. However, their reliance on standardized protocols such as the IEC 61850 communication standard unwittingly increases vulnerability. This is due to the fact that attackers use the standard semantics of such protocols to launch false data injection (FDI) attacks that may mislead system data as well as interleave them to undermine grid control. To counter such attacks, testbeds of cybersecurity that simulate these kinds of attacks must be developed. Research emphasizes the use of hybrid solutions, e.g., the proposed sequence content resolver, to be able to counter FDI attacks efficiently. Application of these measures can diminish the risks of cyber-attacks significantly, ultimately making the smart grid more resilient to potential weaknesses (Alammari et al.,

2023). Also, field events emphasize the devastating impact of cyber-attacks on power grid operation. The early cyberattacks on Ukraine power grids showed how intrusions into vital cyber assets could have catastrophic consequences. By compromising supervisory control and data acquisition (SCADA) networks, cyber attackers introduced interference switching commands between substations and created widespread instability throughout the interconnected grid. Such high-impact incidents highlight the importance of preventive interventions in cybersecurity solutions to protect against targeted attacks that have the potential to become system crashes. Asset owners need to prioritize routine maintenance while considering investing in cybersecurity tools with strong security measures to prevent these risks (Ginter et al., 2018). Finally, the issue of managing cyber threats in smart grids calls for a holistic assessment framework considering both steady-state and dynamic stability factors. Detailed analysis of potential "nightmare" scenarios can be its weight in gold for energy providers in terms of optimizing cybersecurity investments. Real-time monitoring and anomaly detection enable stakeholders to analyze the impact of outages and sequential switching orders in their grids. This systematic approach to risk assessment comes through critical and noncritical component verification, cascade confirmations, and re-assessment of combinations to enable a more effective response to cyber-based contingencies. Finally, prioritizing the development of rigorous mitigating strategies is the key to safeguarding smart grids amidst an elevated level of interconnectedness and exposure (Ginter et al., 2018).

10.2.3. Case Studies of Notable Cyber Incidents

Smart grid interconnectivity gives rise to new cybersecurity threats, as verified through major cyber incidents that have highlighted vulnerabilities within these systems. One of the most impactful case studies is the Stuxnet attack on Iran's nuclear facilities, illustrating the devastating level to which sophisticated malware can sabotage critical infrastructure. The attack made apparent the need for enhanced cybersecurity measures not only for nuclear equipment but also for other important utilities like electricity grids. A strike of this magnitude makes evident how cyberattacks can exploit inherent flaws in smart systems to bring tremendous disruption in operations. Since smart grids start adopting advanced digital communication technologies, the experience of such an incident emphasizes the necessity of developing strong security frameworks with the ability to guard against both external vulnerabilities and internal flaws. Another illustrative case is the 2015 cyberattack on Ukraines power grid when hackers could control operational controls, leading to a blackout that affected thousands of residents. This incident provided practical demonstrations of the cyber vulnerabilities of smart grid, as the attack employed basic phishing techniques and unpatched vulnerabilities to gain unauthorized access. This incident not only demonstrated that the cyberattacks revealed the potential to cause physical harm but also provided the need for utilities to place cybersecurity at the top of their operational practices. It emphasizes the utmost significance of secure communication in smart grid systems and ongoing testing of potential vulnerabilities that may be introduced while updating the system since they could lead to devastating failures affecting utilities and consumers (Qian et al., 2012). Last, the 2021 Colonial Pipeline ransomware attack serves as a stark reminder of the danger cyber events pose to energy infrastructure, including smart grids. Even largely a fuel supply chain disruption, the attack served to highlight the interdependencies between the physical and cyberspace operations of critical infrastructures. The resulting response generated a national conversation on the need for better cybersecurity practices across a variety of sectors, including energy. This incident is a case of the financial impact and operational problems that ensue from cyberattacks, urging policymakers and business leaders to implement more stringent security solutions. The implications extend beyond immediate operational impacts, as they entail full familiarity with evolving cyber environments in an attempt to enhance resilience against future attacks, underscoring the need for security for smart grid technology from future cyberattacks (Botezatu et al., 2023).

10.3. REGULATORY FRAMEWORK FOR CYBERSECURITY

The regulatory framework for cybersecurity is pivotal in safeguarding smart grids where digitalization of energy infrastructure exposes systems to threats of various cyber threats. Effective regulations not only establish a minimum level of security but also ensure resilience against possible attacks. For instance, following the likes of significant cyber assaults on large enterprises such as in the Ukraine power grid hack where the SCADA system vulnerability was targeted by cyber attackers, there has been increased emphasis on regulation through control. This is because adaptive frameworks that can keep pace with the evolving threat environment are needed, with checks on dominant protocols being stringent. It is then the responsibility of developing a robust regulatory system that makes asset owners invest heavily in cybersecurity and safeguard essential infrastructure from potential cyber intrusion, as stipulated by the assessment and management standards proposed in recent research (Ginter et al., 2018). Implementation of cybersecurity regulations into smart grid system operation strategies must also emphasize innovative technological steps. The development of a new architecture to upgrade rural distribution networks is an example of how regulatory environments can direct the evolution of cybersecurity-conscious systems. By supporting complementarities between telecommunications and power management technologies, such architectures enable improved resilience while at the same time enhancing system efficiencies. Decentralized management systems as delineated in recent proposals allow for more efficient monitoring and response capacities, empowering operators to address vulnerabilities preemptively. Therefore, regulatory frameworks that facilitate such innovations constitute an essential element in guiding traditional power infrastructure into robust smart grids that can mitigate cybersecurity threats, thereby improving enhanced overall energy management (Díaz González et al., 2018). In addition, the evolving nature of cybersecurity in smart grids necessitates a joint regulatory framework consisting of players from various industries. This multi-faceted effort must aim for something more than mere compliance; it must instill a culture of continuous improvement and knowledge sharing among utilities, regulators, and technology providers. By embracing frameworks that will foster collaborations and sharing of information, the sector can build an enhanced understanding of cyber threats and develop effective counter-measures. Periodic review and updates of such regulatory mechanisms ensure that the cybersecurity strategies stay current with respect to the rapidly changing threat vectors. The request for systemic evaluations and real-time surveillance mechanisms as discussed in the literature serves to highlight the significance of proactive and collaborative approaches to cybersecurity in ensuring the stability and reliability of smart grids (Ginter et al., 2018).

10.3.1. Overview of Existing Regulations (e.g., North American Electric Reliability Corporations Critical Infrastructure Protection (NERC CIP))

The regulatory framework that governs cybersecurity in smart grids has evolved to address the growing complexity and exposure that accompany sophisticated energy infrastructure. Among the most prominent regulations are NERC CIP standards, whose purpose is to enhance the security and reliability of the bulk power system. NERC CIP creates mandatory guidelines in terms of physical and cybersecurity controls for utilities to implement. It puts emphasis on risk management methods along with ongoing improvement to ensure the resilience of key infrastructure (Sandoval et al., 2019). As more reliance comes to be put on interdependent technologies, NERC CIP compliance has become a priority for all organizations that handle energy distribution, which influences their operation processes as well as response process. Aside from NERC CIP, the regulatory environment is influenced by federal and state policy to reduce risks of cybersecurity breaches. Federal Energy Regulatory Commission (FERC) is responsible for enforcing standards and works in cooperation with NERC to evaluate compliance. However, Federal Communications Commission (FCC) rollback of net neutrality introduced new vulnerabilities to the scene that are typically overlooked in traditional regulatory frameworks (Sandoval et al., 2019). Without robust net neutrality protection, Internet Service Providers (ISPs) can potentially traffic-manipulate, endangering the integrity of vital energy communications networks. This synergy between cyber defense and regulatory monitoring requires an examination of existing legislation and further emphasizes the requirement for sophisticated policies that address both traditional threats and emerging issues confronting the energy sector. As computer technology reliance increases in smart grids, the existing regulatory framework must adapt to reflect the changing nature of cyber threats. It is necessary that regulators conduct proactive analyses which include modeling the probable impact of communications delays imposed by ISPs on electric reliability. Recent discussions suggest that federal and state regulators need to drive a revival of net neutrality values to safeguard against increasing cybersecurity threats (Sandoval et al., 2019). In addition, the integration of stress tests and simulations within regulatory processes can provide insights into the energy system's inherent risks. By closing gaps in current regulations and integrating cybersecurity as a central element of grid operation, government agencies can help enhance smart grids against any probable threats and render the energy market sustainable and secure in the long term (Sandoval et al., 2019).

10.3.2. Role of Government Agencies in Cybersecurity

With advancing digital technologies to form smart grids, the role of government agencies in the context of cybersecurity becomes increasingly crucial. They are tasked with developing regulatory guidelines that set down rules of cybersecurity practice to neutralize smart grid infrastructure-specific attacks. By framing guidelines that cover data privacy and security features, the agencies help counter risks related to the huge volume of data in smart grids. For instance, as quoted in the smart grid implementation debate, data privacy concerns require strict regulation that safeguards user data from interception but also allows for authentic analysis by authentic vendors (Bartholomew et al., 2016). This is imperative in terms of upholding consumer trust and promoting the adoption of smart grid technology. Additionally, there is the role of the government agencies as drivers of coordination among the various stakeholders of smart grid cybersecurity. Coordination in this case includes public-private partnerships to enable the sharing of best practices, threat intelligence, and technological innovations. An effective cybersecurity strategy cannot be devised in isolation; therefore, agencies shall play the role of a conduit to facilitate open conversation between the utility corporations, the technology vendors, and the general public. Through investments in research activities and promotion of the development of technologies for increasing cybersecurity protection, government agencies have an important role in establishing a safer smart grid infrastructure that is resilient against upcoming cyber-attacks. As highlighted in literature, strategic initiatives and frameworks of national research communities can provide important information for building robust cyber defense systems (Anglano et al., 2018). In addition to developing regulatory standards and encouraging collaboration, government agencies are responsible for educating and training smart grid personnel. Comprehensive training programs ensure that employees are equipped with the skills needed to detect, deter, and respond to cyber-attacks effectively. As the threats to cybersecurity continue to change, ongoing education is essential to keep industry professionals informed of new threats and countermeasures. Further, educating consumers about the importance of cybersecurity and how their data can be protected can enhance overall system resilience. By prioritizing these education initiatives, government agencies can foster a culture of cybersecurity, where every stakeholder takes proactive measures to secure smart grid facilities and advance national security interests.

10.3.3. International Standards and Best Practices

With the rapid pace of smart grid technology, international standards and best practices are essential to create further cybersecurity initiatives. These standards form a basis for addressing the many vulnerabilities inherent in interdependent energy systems. Creating protocols is essential in designing secure interfaces between various components, like those facilitating communication and data transmission between power producers and consumers. As businesses all over the globe try to implement smart grid technology, a consolidated framework of guidelines can prevent cyber-attacks, ensuring a more reliable energy infrastructure. By adhering to globally recognized models of cybersecurity, stakeholders can co-ordinate more effectively in safeguarding critical assets, thus enhancing reliability and stakeholder confidence in smart grid operations. In addition, the inclusion of cybersecurity standards in utility operations enhances the ability to manage and regulate risks effectively. New architectures such as those seen in cutting-edge studies focus on the use of decentralized management systems and robust Supervisory Control and Data Acquisition (SCADA) systems (Díaz González et al., 2018). Apart from boosting the supply of electricity, these systems also enhance the security mechanisms against cyber-attack. By adopting such sophisticated architectures, operators can optimize energy flow and render their systems strong enough to be resistant to external attacks. The transition to such sophisticated frameworks aligns with international best practices and enables the introduction of new technologies for enhancing overall cyberinfrastructure in smart grids. Moreover, growing threats to power systems across the globe signal that a proactive approach to cybersecurity is critical. Losing the lesson of actual incidents, such as the Ukraine cyberattack against the electric grid that took advantage of SCADA system flaws, emphasizes the necessity for continuous evaluation and improvement of security procedures (Ginter et al., 2018). Ensuring real-time monitoring and anomaly detection mechanisms to identify and counter threats prior to their escalation is critical. This framework not only enhances near-term response capability but also facilitates longer-term strategy for maintaining operations in stable mode in smart grids. Forefront positioning of international standards and best practices enables energy suppliers to support a culture of awareness and preparedness for cybersecurity, ultimately securing the future of power distribution worldwide.

10.4. RISK ASSESSMENT AND MANAGEMENT

Risk assessment and management have become the most important factors in the evolving environment of smart grids to safeguard critical infrastructure from cyber-attacks. Incorporating emerging technologies demands extensive knowledge regarding potential vulnerabilities that can be exploited by cyber criminals. The efficient risk assessment process begins with identifying asset dependencies, particularly on supervisory control and data acquisition (SCADA) systems. The cyberattack targeting the Ukraine electrical grid emphasizes the potentially catastrophic consequences that can arise when systems are not properly protected. Studying planned cyberattacks that disrupt substation networks by circumventing SCADA capabilities assists stakeholders in placing their investments in the best cybersecurity systems. A robust risk management system that incorporates system-level worst-case scenario analyses, as in Ginter et al. (2018), allows organizations to anticipate cascading failures, thus enhancing overall grid resilience. The role of big data analytics in risk assessment and management in smart grids highlights the necessity of utilizing large datasets for predictive analysis and decision-making. Traditional methods of power system management are insufficient since they do not have the ability to handle the complexity and volumes of data in today's era. With the utilization of artificial intelligence and machine learning as proposed in Gharehpetian et al. (2019), organizations are able to improve their risk understanding and automate real-time operations anomaly detection. This transition towards data-driven operations facilitates anticipatory risk management, which enables power system operators to anticipate potential threats and take necessary measures. The ability to utilize big data not only optimizes operational performance but also considerably enhances the security position against cyber-attack, leading to more secure smart grid infrastructures. Moreover, continuous enhancement of risk management practices requires the development of adaptive frameworks that address new issues in cybersecurity. The interdependent nature of intelligent grid systems demands periodic reviews and consolidation of evolving technologies and practices. Re-examination of must-have and nice-to-have combinations in the event of any potential incidents should be part of strong protocols of whole risk assessment. Flexibility here is critical with respect to vulnerabilities as they evolve and responding to the ongoing evolution of cyber threats. Stakeholders must remain vigilant through the establishment of a culture of cybersecurity and investing in training programs aimed at reinforcing risk management skills. Lessons learned from exhaustive research within the industry, as captured in the discussions included in Gharehpetian et al. (2019) and Ginter et al. (2018), will have a significant role to play in shaping strategies to ensure the integrity of smart grids in the future.

10.4.1. Identifying Vulnerabilities in Smart Grid Systems

The increased reliance on smart grid infrastructures to ensure successful electrical power distribution offers an incredible amount of vulnerabilities that must be detected and examined very carefully. Smart grids apply several different technologies, such as intelligent sensors, communication platforms, and sophisticated data analytics, that can significantly evolve their capabilities. The evolving capabilities also create new planes of attack for cyber threats. It is important to realize that the intricacies of such systems may bring unforeseen vulnerabilities, especially at their earliest phases of design. Early-stage vulnerability analysis is crucial since it enables system architects to apply countermeasures that reduce the risk associated with potential exploits. The development of an extended model representing the features of a system can facilitate the identification of vulnerabilities and their corresponding attack vectors, thereby improving security mechanisms throughout the life cycle of smart grid systems (Bakirtzis et al., 2018). The design of smart grid systems also plays a vital role in preventing cyberattacks. Through the use of decentralized management systems and optimizing telecommunications and power management technology integration, further reduction in the vulnerabilities is possible. For instance, the proposed architecture focuses on rural distribution networks transforming them into smart grids through the addition of Intelligent Distribution Power Routers as well as enhanced Supervisory Control And Data Acquisition (SCADA) systems. This architecture promotes better distributed use of energy resources and automatically enhances weaknesses by means of more secure operational frameworks. By examining case cases in which similar architectures have been implemented, stakeholders can learn from potential vulnerabilities and measures that have proven to enhance cybersecurity (Díaz González et al., 2018). Smart grid network vulnerability identification also entails holistic understanding of the interdependencies among the various components and of the cascading effects of the security breach. Smart grid cybersecurity entails assessing the whole ecosystem, not only technological but also human factors and organizational policy. An effective cybersecurity plan should incorporate risk analysis, continuous monitoring, and harsh testing of system defenses. This integrated approach guarantees that the weaknesses are constantly detected, analyzed, and remitted, maintaining the overall security and resistance of smart grid systems. With the continued expansion of smart grids, creating a mutual framework that involves all the stakeholders—government agencies, private sectors, and end-users—is important in creating a culture of security awareness and preparedness in the face of persistently evolving cyber threats.

10.4.2. Risk Assessment Methodologies

In the avoidance of the cyber security attack menace in smart grids, there have to be robust risk assessment strategies that quantify potential vulnerabilities and the consequent effects. These strategies are developed to the end that threats are identified systematically, the probability of occurrence of such threats is analyzed, and the potential impacts of successful attacks are analyzed. A general risk assessment not only increases the understanding of existing weaknesses within the grid but enables improved prioritization of resources for protective measures. This proactive approach is particularly critical due to the interdependent nature of smart grid technologies, where one compromised sector can cascade across the rest of the infrastructure, posing system safety and reliability threats. The increasing complexity of these systems necessitates stakeholders adopting an integrated vision for risk management, integrating various elements of risk within both cyber and physical spaces to create a resilient operational configuration. From the foundation of risk assessment, it is necessary to adopt an integrated approach that prioritizes technical as much as procedural solutions to recognized threats. RAIM methodology consisting of real-time monitoring, anomaly detection, impact analysis, and mitigation strategies is an indispensable part of asset owners who are ready to face the realities of smart grid operation. With its application, organizations can examine nightmare attack scenarios emulating the effect of simultaneous multiple attacks such as those experienced in the Ukraine power grid attack where SCADA systems were compromised. The comprehensive analysis afforded by the RAIM framework supports critical decision-making regarding asset prioritization, maintenance spending, and long-term cybersecurity strategy. The framework further ensures continuous reevaluation of risk indicators, keeping organizations responsive and attuned to responding to evolving threats as emerging vulnerabilities emerge, ultimately safeguarding the integrity of the smart grid. Besides, the application of system theoretic method promotes the development of secure Cyber Physical Systems (CPSs) within smart grid infrastructure. Analyzing the cyber and physical integration, the stakeholders are able to see interdependencies and vulnerabilities within smart grid operation. Research has shown that analyzing these vulnerabilities at the design stage can lead to the identification of potential attack vectors, hence preventing situations like the Stuxnet attack that brought down critical infrastructure. Incorporating system safety practices enables it to perform a complete risk assessment that not only identifies risks but also provides actionable advice for designers to actually mitigate them. This forward-looking design strategy can potentially increase the resilience of CPSs significantly and be instrumental in the creation of a secure environmental framework for the future smart grid evolution. Ultimately, the integration of risk assessment methods and new design practices will be the basis for the realization of effective cybersecurity measures in the future smart grid environment.

10.4.3. Strategies for Risk Mitigation

The integration of artificial intelligence (AI) into smart grids is a two-edged sword, both enhancing operational efficiency as well as exposing the systems to more cybersecurity threats. Hence, it is important that effective risk mitigation methods are employed to safeguard these advanced infrastructures. AI technologies, such as predictive analytics and machine learning, can be employed to anticipate possible attacks and system vulnerabilities. For instance, through patterns of data traffic observation, AI can identify aberrant patterns as indicative of cyber-attacks. Consequently, this real-time observation allows for tailored interventions having the impact of diminishing likelihoods of major breakdowns. The relentless development of AI validates continuous change in cybersecurity measures to guarantee that they keep abreast of the technologies they seek to protect, underscoring the necessity of creating robust frameworks for securing smart grid systems against many cyber-attacks (AbuDawas, 2025). Furthermore, the incorporation of blockchain technology into smart grids presents an exciting opportunity for enriching cybersecurity measures. The decentralized nature of blockchains underlying platform ensures authenticity and integrity of the data, rendering it much more difficult for hackers to manipulate critical information. Transactions and communications among system entities can be enforced and authenticated automatically through the utilization of smart contracts in the blockchain platform, further installing security layers in the grid. These smart contracts also enable greater transparency and accountability since all the transactions are tracked on an unalterable ledger, which can be accessed by authorized entities. This transparency not only provides assurance to stakeholders but serves as another protection against would-be attackers as well.

Figure 10.8. Mitigating the risk of cyberattack on smart grid systems.

Source: https://www.mdpi.com/sustainability/sustainability-13-03196/article_deploy/html/images/sustainability-13-03196-g004.png.

The integration of blockchain technology with AI-based tools presents a multi-dimensional risk mitigation approach to both operational vulnerabilities and the constantly evolving nature of cyber-attacks in smart grids (Betouil et al., 2025). Lastly, a multi-dimensional risk mitigation approach must integrate technological innovation and human factors to create an actually secure smart grid system. Cybersecurity best practices should be trained to employees because human error remains the prime reason for data breaches and system failure. Furthermore, a well-designed incident response plan can mitigate the impact of a cyber-attack by enabling quick recovery and business continuation. Regular vulnerability scans and audits should be conducted to find and close security loopholes before they are being targeted. Lastly, integration of blockchain and AI technologies with hardened training and readiness procedures provides a robust setting to react to the changing threat landscape. Multidisciplinary integration not only enhances the security of the smart grid but also facilitates achieving the overall vision of energy resilience and sustainability in the increasingly digital future (AbuDawas, 2025; Betouil et al., 2025) (Figure 10.8).

10.5. TECHNOLOGIES FOR ENHANCING CYBERSECURITY

The incorporation of advanced technologies in smart grids has not only revolutionized their efficiency and dependability but also their vulnerability to cyber-attacks. Traditional cybersecurity measures often fail in the dynamic world of cyber threats; therefore, the development of special frameworks is required. A dedicated testbed for energy system cybersecurity is proposed for measuring threat vectors and encouraging resilience. By applying interdisciplinary methods with software-defined networking and intrusion detection, researchers can simulate real attack scenarios. Through this empirical tool, actionable insights on how security testing of the grid can be enhanced are generated. These attempts highlight the necessity of strict methods, which can be followed to counter sophisticated cyber threats, ultimately closing the gap between practice and theory in the field of cybersecurity for smart grids (Elbez et al., 2025). The incorporation of innovative solutions such as blockchain technology and artificial intelligence (AI) provides an effective paradigm for enhancing cybersecurity for smart grids. Emerging developments put into the spotlight a hybrid optimization approach that incorporates blockchain and the real-time anomaly detection capability of AI. The combination, in addition to boosting data integrity, also improves decision-making mechanisms in times of cyber-attacks. With the integration of emerging algorithms like the Adaptive Vulture Optimization Algorithm (AVOA) alongside Convolutional Neural Networks (CNN), researchers have made substantial improvement in detection accuracy coupled with computation expenses. These advancements reflect the ability of AI-Blockchain models to provide a scalable and effective way of securing smart grids against increasing cyber-attacks, reflecting an improved advantage over the traditional models (Al-bermani et al., 2025). Additionally, the overall effectiveness of such systems in enhancing the security of smart grids is reflected through in-depth performance evaluations. Hybrid optimization architecture, in comparison to current models, has reflected impressive results. Over 99% detection rate signals a significant advancement in energy system security as timely identification of potential threats is the most important factor in preventing risks. Additionally, the reduction in computation time enhances overall security protocol efficiency, ensuring protection mechanisms do not slow down the operational performance of smart grids. This intersection of pioneering technologies not only secures the current networks but also lays a solid ground for future developments, rendering the systems more resistant to even more sophisticated cyber-attacks on key energy infrastructures (Al-bermani et al., 2025) (Figure 10.9).

Figure 10.9. Artificial intelligence (AI) in cybersecurity: Enhancing threat detection.

Source: https://ccoe.dsci.in/storage/blogs/July2023/Gfzv5wjNgeDVTw1GMDSW.jpeg.

10.5.1. Encryption and Data Protection Techniques

In the field of smart grid cybersecurity, encryption is one of the most important methods to protect data. Smart grids are inherently networked systems that exchange data among parties such as consumers, utilities, and third-party services. Such integration makes them more susceptible to cyber-attacks, hence the need for strong encryption methodologies to safeguard critical information. Encryption methods, such as using Advanced Encryption Standard (AES) and RSA algorithms, provide strategies that encrypt data into imperceptible patterns to intruders. By deploying such strategies, smart grid operators can protect critical infrastructure data from confidentiality violations and integrity loss, hence mitigating the risk of data breach that can jeopardize operations and lead to massive financial loss. With the changing nature of the smart grid world comes a change in data protection methodology. With new technologies such as distributed ledger technology (DLT) and quantum encryption come more security components. DLT provides secure, decentralized transactions with protection against data tampering and enhanced transparency across stakeholders. Quantum computing also introduces a higher degree of security through quantum key distribution (QKD), which allows for the creation of unbreakable cryptographic keys. These advanced techniques are particularly relevant given the increasing demand for fast and safe data communications in smart grids with the looming migration towards 6G networks. The integration of such next-generation data security and encryption techniques will not only enhance the cybersecurity infrastructure of smart grids but also pave the way to more resilient and adaptive cybersecurity techniques. In short, as smart grids develop into more advanced networks founded on complex technologies and greater connectivity, the requirement for robust encryption and methods of data security cannot be overemphasized. Existing cyber-attacks, coupled with anticipated 6G network challenges, necessitate proactive strategies in cybersecurity. The foreseen 6G features, i.e., enhanced data rates and reduced latency, come with their respective vulnerabilities that require countermeasures through extensive security measures. By emphasizing encryption and state-of-the-art data protection technologies, stakeholders can increase smart grid resilience to future threats while maintaining the integrity of essential services and public trust in the cybersecurity ecosystem. As the field grows, constant research into potential solutions and standardization processes will be crucial to ascertain the future of cybersecurity in smart grids (Figure 10.10).

Figure 10.10. Data encryption.

Source: https://kinsta.com/wp-content/uploads/2023/07/what-is-encryption.jpg.

10.5.2. Intrusion Detection and Prevention Systems

In smart grid cybersecurity, Intrusion Detection and Prevention Systems (IDPS) are a crucial element that assists in shielding critical infrastructure from a plethora of cyber-attacks. Since smart grids use sophisticated technologies such as Supervisory Control and Data Acquisition (SCADA) systems for regulating electricity supply, the requirement for a robust cybersecurity becomes increasingly urgent by the minute. Cyber-attacks on such systems have the potential to exploit vulnerabilities, leading to severe operational disruption and long-term economic repercussions. Use of IDPS facilitates monitoring of network traffic in real time to detect suspicious activity and potential threats before they can cause serious issues.

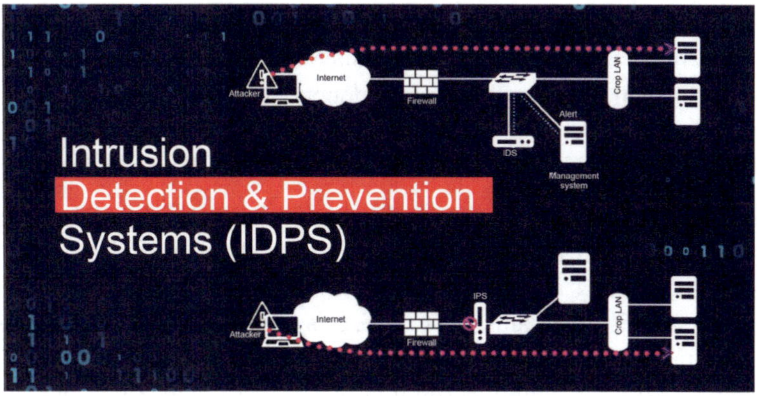

Figure 10.11. Intrusion detection and prevention systems.

Source: https://www.goallsecure.com/wp-content/uploads/2024/05/Intrusion-Detection-and-Prevention-System.webp.

Methodical evaluation and continuous refining of IDPS are the secrets to making these systems remain effective against ever-evolving cyber threats that compromise the integrity of smart grids. IDPS effectiveness is contingent upon the use of sophisticated detection methods capable of distinguishing between malicious activity and normal user actions. Such differentiation is even more essential in smart grid contexts where high-speed connectivity of numerous components expands the attack footprint of possible intrusions. IDPS achieves this discrimination using anomaly detection and signature-based approaches that pick up on abnormal patterns that could indicate a possible attempt at breaching or compromising systems of interest. Moreover, combining these techniques with sharing threat intelligence will further enhance the pre-emptive capability of an organization against weaknesses. Practice shows that holistic security models integrating IDPS with continuous risk assessment goes a great way in securing the overall resilience of smart grid operations against unprecedented cyber threats (Ginter et al., 2018). Continuous evolution in cybersecurity threats necessitates a proactive response by the asset owners and operators of the smart grids. Adopting an adversarial mindset towards system security will enable weaknesses in IDPS to be easily detected early on. With such a mindset, organizations can detect blind spots and implement appropriate countermeasures to enhance their defenses against intrusion threats. Furthermore, continuous training for personnel running IDPS is necessary as human errors are usually exploited by attackers. As emphasized in recent research, the interdependency of SCADA systems presents unique operational challenges that may escalate if not properly addressed (Almalawi et al., 2017). Therefore, embracing an end-to-end strategy that combines IDPS, internal staff awareness, and regular security audits will enhance the cybersecurity stance of smart grids significantly and render such critical infrastructures more dependable (Figure 10.11).

10.5.3. Role of Artificial Intelligence (AI) in Cybersecurity

The use of AI within cybersecurity mechanisms is increasingly vital, particularly within intelligent grids. Such grids are heavily based on interlinking devices and real-time data, and thus are among the top targets for cyberattack. AI technology can enhance security mechanisms because it is capable of identifying anomalies in network traffic that otherwise would indicate breach or attack. For instance, algorithms based on machine learning are capable of analyzing huge amounts of data generated due to the operations of smart grids and can recognize patterns and identify potential threats much more quickly than traditional methods. This is not only a guarantee for a more effective response to threats but also identifies potential future vulnerabilities, hence guaranteeing a more proactive approach to cybersecurity. Moreover, big data analytics is also a significant component of AI-based cybersecurity of smart grids. The sheer amount of data from thousands of devices poses enormous challenges for conventional systems; conventional methods generally lack the capability to manage such scale economically. With the application of AI methods such as data mining and time-series forecasting, business organizations are in a position to derive significant insights from big data, thereby enhancing decision-making potential. This becomes all the more relevant with smart grids utilizing renewable energy and electrical vehicles, necessitating intelligent management systems. The successful deployment of big data analytics can lead to increased system resilience, operational efficiency, and ultimately more secure energy supply with reduced vulnerability to cyberattacks (Gharehpetian et al., 2019). Finally, additional research emphasizes the necessity for building a robust cybersecurity framework to be able to effectively protect smart grids. The debate now across the cybersecurity industry is all about identifying and adopting essential controls and technologies for strengthening defenses (Anglano et al., 2018). The initiative needs to be addressed with a multi-pronged approach that addresses not only technical capability but also training and awareness among stakeholders. As AI is at the forefront of these initiatives, the present vulnerabilities can be minimized by continuous learning and adaptation. The professionals in the industry must prioritize the design of AI-based security solutions, which will ultimately create a secure and reliable smart grid infrastructure that will withstand future cyber-attacks (Figure 10.12).

Figure 10.12. The role of artificial intelligence (AI) in cybersecurity.

Source: https://www.ducatindia.com/_next/image?url=https%3A%2F%2Fadmin.ducatindia.com%2Fblog%2F1735997971316The%20Role%20of%20Artificial%20Intelligence%20in%20Cybersecurity.jpeg&w=750&q=75.

10.6. CHALLENGES IN IMPLEMENTING CYBERSECURITY MEASURES

The transition to smart grids, though promising extensive advantages in power management and efficiency, simultaneously also creates serious challenges to cybersecurity. Smart grids are founded on a large system of interrelated devices that may be able to optimize operations but also create vulnerability points. The integration of Internet of Energy (IoE) technologies, as seen in the literature, brings with it functionalities like real-time demand response and automation. But these improvements also come with the threat of safeguarding sensitive data against cyber-attacks. In particular, the dependencies among different components within a network require stringent cybersecurity mechanisms to seal off unauthorized access and maintain data integrity. The intricacies brought about by sophisticated communication protocols and real-time analysis of data enlarge these vulnerabilities, and it becomes exceedingly important that there be holistic mechanisms to make smart grids more resistant to possible attacks (Meslouhi et al., 2025). Aside from technical weaknesses in smart grid infrastructure, interoperability and regulatory complexity complicate instituting effective cybersecurity. The diversity of smart grid technologies used in a grid makes the establishment of uniform cybersecurity standards that can be applied across all complicated. A lack of a common regulatory framework enables jurisdictions and actors to apply different security practices. Moreover, the relationship between blockchain and machine learning offers promising solutions to enhancing the efficiency and security of energy transactions but is hampered by issues such as insecure communication channels and potential vulnerabilities. Such a multilateral setting not only demands sustained collaboration among the stakeholders but also highlights the imperative need for innovations in cybersecurity measures to counteract both current and potential threats (Fatima et al., 2025). Finally, the adaptive mode of cybersecurity threats demands adaptive solutions that can keep pace with the dynamism of the smart grid environment. Traditional approaches to cybersecurity are not adequate when the future threats embodied by sophisticated attackers targeting strategic energy infrastructure are being confronted. Smart grids evolving, so too must protective methodologies. The adoption of new technologies, such as light blockchain technologies and advanced machine learning models, presents new possibilities for building robust cybersecurity routines. The technologies enhance anomaly detection and risk mitigation and allow for more adaptive responses to potential breaches. Yet, integration of these new systems is not complication-free, and additional work must be accomplished in addressing how to best align technological advancements with functional security measures to provide an efficient and secure energy future (Fatima et al., 2025).

10.6.1. Cost Implications for Utility Companies

Smart grid transformation will come at a high cost to utility companies primarily due to the necessity of upgrading existing infrastructure to enable new technologies. These improvements also involve significant investments in telecommunications, smart meters, and data management systems. Utility firms also need to invest in cybersecurity measures to protect this infrastructure from attacks. As attacks on critical infrastructure become more sophisticated, the cost of installing sound security procedures can be expensive. These costs are not only required to safeguard sensitive consumer data but also to maintain the overall integrity of the power supply. Failing to have an active emphasis on cybersecurity will leave utility companies vulnerable to even higher costs through potential hacking that can create disruptions to service and reputational loss. Secondly, the integration of Distributed Energy Resources (DERs) into smart grids poses certain financial challenges for utility companies. While DERs promise cost savings and improved efficiency, they are also placing strain on existing electricity infrastructure. The fact that utilities are forced to incur expense on advanced control systems and grid management technologies needed in order to facilitate coordination of the resources comes at a hefty price in terms of capital expenditure. Utilities are also required to deal with regulatory reforms that aim at creating a more even playing field for energy producers but are reacting to security concerns presented by DERs. As highlighted by recent research, such challenges need a general framework addressing price design and lowering barriers to market participation so that any source of energy can play its role efficiently without compromising system safety and reliability (Gómez Román et al., 2017). Lastly, the cost of cybersecurity threats extends beyond overt technology and infrastructure investments; it also involves latent liabilities due to data breaches and operational disruptions. For utilities, customer data protection has become the hub of trust and privacy regulation compliance. The sheer volume of information generated by smart grids mandates not just security but also transparent policies for data use and protection, the twin imperative of utilizing data to realize operational efficiencies while upholding consumer interests. The trade-off is critical because consumer anxiety about data privacy continues to escalate. Having proper regulations in position, such as those suggested for Massachusetts, can allow the utilities to minimize risks while inducing consumer engagement and acceptance of smart grid technologies (Bartholomew et al., 2016).

10.6.2. Balancing Security with Operational Efficiency

The integration of cybersecurity into smart grids has a curious challenge: striking a balance between security and operational efficiency. With growing decentralization and interconnectedness of smart grids, the integration of differing technologies is critical for convenient energy dispersion and management. This complexity, however, is often accompanied by vulnerabilities that can be exploited by attackers. The architecture proposed in the European Smart Rural Grid initiative shows the prospect of innovation for improving system resilience and efficiency. Upgrading rural distribution networks using Intelligent Distribution Power Routers and other state-of-the-art technologies enables operators to better manage weak rural distribution systems. This balance is crucial since, along with the provision of an enabling environment for the integration of distributed energy resources, it also speaks to security. Failure to achieve this delicate balance will result in catastrophic service disruptions, exposing these systems to potential cybersecurity breaches that will compromise operational goals and grid stability in general. Operational efficiency in smart grids is increasingly reliant on the use of big data analytics, which maximizes decision-making processes critical to maintaining a high level of system performance. Big data is crucial to the processing of large amounts of data generated by smart devices, thereby enabling utilities to optimize energy distribution and predict demand fluctuations. However, the incorporation of big data analytics must occur in conjunction with sound cybersecurity practices to prevent sensitive data from potential breaches. Quoted, big data analytics use cases encompass enhancing microgrid performance up to better detecting electricity theft. Combining such analytics capability with good cybersecurity practices maintains operational integrity in the smart grid while achieving maximum value from new technology. Thus, businesses must invest in developing and deploying analytics tools and keeping their systems secure against potential emerging cyber threats. Besides, security verses operational efficiency on smart grids is an evolving process of continuous evaluation and adjustment. The dynamic nature of cyber threats demands a forward-looking approach towards cybersecurity countering the threat before it occurs while allowing for operational goals. With advancing technology, utilities must rethink their cybersecurity paradigms to accommodate emerging vulnerabilities introduced by innovation such as IoT devices and increased interconnectivity. In addition, ongoing cooperation among stakeholders—e.g., information technology and operational technology professionals—is paramount to ensure that security should not impair operations. An integrated paradigm such as this results in a robust infrastructure that not only enhances security frameworks but also enhances overall system efficiency. As described in more recent activity, finding this balance will eventually lead to a more secure, stable, and more efficient smart grid infrastructure that meets the growing needs of today's energy consumers and safeguards critical infrastructure.

10.6.3. Human Factors and Insider Threats

Human factors in cybersecurity, notably insider threats in smart grids, need to be understood. Insider threats are caused by individuals within an organization who utilize their authorized access to information systems for malicious purposes, either accidentally or intentionally. The advanced level of the smart grid infrastructure, whereby it relies on connected networks and mixed technologies, increases the risk of human error and malicious behavior. Factors of inadequate training, lack of knowledge, and organizational culture are a major factor in rendering the smart grid systems susceptible. Individuals who poorly comprehend the implications of their actions or those who are non-compliant with stipulated security protocols can inadvertently facilitate breaches, which leave vulnerable infrastructure open to external attack and internal abuse. Therefore, organizations have to highly prioritize large-scale training initiatives, creating a culture of security that emphasizes individual accountability as being paramount to safeguarding critical infrastructures. Apart from human factors, the regulatory framework also has a significant role to play in mitigating insider threats to smart grids. The EU's NIS Directive and GDPR set out the security controls organizations must implement to protect critical infrastructure from cyberattacks and data breaches. Such regulations compel organizations to adopt best practices in security management, thus making them more susceptible to insider attacks. With increasingly sophisticated cyber threats, compliance is insufficient; organizations must go beyond compliance by continuously evaluating and updating their security controls to counter evolving threats. This includes audits and threat analysis, and the deployment of sophisticated security technology. The incorporation of new methodologies, e.g., security architecture orthogonality, allows organizations to implement several layers of control. These controls not only enhance security against insider threats but also enhance the integrity and reliability of smart grid systems in an era characterized by emerging cyber challenges (Accenture Technology et al., 2018; Al-Khateeb et al., 2019). Finally, addressing the human aspects of cybersecurity requires a comprehensive approach that combines technology, policy, and business culture. The implementation and setup of reliable cybersecurity practices must be supplemented with building a work culture where workers recognize the significance of their role in safeguarding systems. Training sessions and awareness for staff members continuously enhance their ability to identify possible security threats both from themselves and other staff members. Moreover, creating a culture of reporting encourages employees to report issues without fear of backlash, hence empowering positive action against weaknesses. There is an urgent need for interventional collaboration among cybersecurity experts, organizational management, and human resource departments to create policy that successfully eliminates insider threats. By combining technological safeguards with human considerations, organizations can significantly improve their defense against insider threats, safeguarding the integrity and reliability of smart grid operations in an interconnected and increasingly vulnerable world.

SUMMARY

The implications of smart grid cybersecurity are far-reaching, as such systems represent an integrated step forward in energy management and sustainability. Smart grids enhance the efficiency and reliability of power delivery, but in the process introduce unprecedented vulnerabilities due to their reliance on advanced technologies and interconnected systems. While the move toward renewable energy sources is beneficial in terms of lowering carbon emissions, it introduces unprecedented risks to grid security. Increased dependency on digital infrastructure necessitates a strict vulnerability analysis of the operations of electric grids. This examination not only encompasses assessing the potential threats but also studying the resilience and flexibility of energy grids in varying situations, underscoring the need for end-to-end cybersecurity strategies to protect such strategic infrastructures (Millis et al., 2019). Furthermore, the role of big data analytics cannot be overstated in the cyber security of smart grids. As large volumes of operational data are being generated, advanced analytical techniques have become the only way to identify threats and optimize systems. Traditional methods are found wanting in dealing with such volumes of information, and this generates inefficiencies and potential security vulnerabilities. By implementing advanced applications of big data analytics, energy managers can leverage artificial intelligence and machine learning capabilities to enhance response capabilities by allowing them to detect anomalous activities faster and allocate resources more effectively. Besides enhancing smart grid resilience, this forward-thinking approach also fosters a culture of continuous improvement in cybersecurity practice such that adaptive responses are prepared in anticipation of evolving threats (Gharehpetian et al., 2019). In brief, countering the cybersecurity challenges in smart grids needs an integrated solution that integrates advanced analytical capabilities, resilience-building strategies, and ongoing communication between stakeholders. The interplay between escalating reliance on renewable energy sources and escalating digital interconnectivity creates an intricate context for decision-makers. In advancing through this context, a clear vision of vulnerabilities, along with a commitment to leveraging new technologies, will be critical. Finally, the escalation of cybersecurity controls on smart grids not only guarantees the security of energy resources but also supports sustainable development that is crucial for future generations. The paramount importance of electricity supply networks in today's world calls for constant innovation and advancement in cybersecurity measures, which is essential to ensuring the dependability and integrity of these life-support services (Gharehpetian et al., 2019; Millis et al., 2019).

With developing smart grids, the importance of cybersecurity for this critical infrastructure cannot be overemphasized. The interconnection and reliance on IoT devices create vulnerabilities that necessitate strict security measures. Key points are the realization of the requirement for extensive datasets such as the Edge-IIoTset, by which one can design machine learning-driven intrusion detection systems. This particular data set is designed to enhance understanding of potential attack vectors by simulating real-world scenarios by employing an array of IoT devices. By identifying and analyzing various security threats, for instance, DoS attacks and malware, researchers are able to develop more effective mitigation strategies and ultimately protect the integrity of smart grids that are essential to modern energy management. Moreover, the nexus of cybersecurity and sustainability is increasingly significant if put on smart grids. The imperative to address climate change was placed in the spotlight by the UNs COP26 summit alongside highlighting how technology was fitting into this process. Technology for digital and information systems is a significant part of monitoring energy consumption and emissions and thus sustainability targets. This must be balanced against the risk of digital innovation, however.

As the stakeholders are managing the impact of responsible digitalization, the implementation of smart technologies has to be ensured so that there are no negative environmental implications stemming from it, such as e-waste. The points underscore the necessity of an interdisciplinary approach towards securing smart grids with technical, environmental, and societal elements. Moreover, the approaches adopted for enhancing cybersecurity in smart grids need to be dynamic with respect to the rapidly changing nature of threats. Continuous upgrade through data processing and machine learning is needed to stay ahead of potential cyber-attacks. The information gained from large datasets such as the Edge-IIoTset allows one to gain an in-depth knowledge of the changing strategies employed by the cyber hackers. Additionally, as technologies become more integrated into the smart grid system, periodic evaluation of their security characteristics and vulnerabilities has to be prioritized. A proactive approach to cybersecurity can not only safeguard not only the operational integrity of smart grids but also assist in buttressing public trust in the transition to more intelligent and efficient energy infrastructures, suggesting a broader commitment to resilience in the face of future adversity.

REFERENCES

1. AbuDawas, M. (2025). The role of artificial intelligence and cybersecurity in energy management and optimization. *Journal of Information Systems Engineering and Management*. https://www.semanticscholar.org/paper/68d4807150f92faafa43362946710dbd0621db8f (accessed on 30th September 2025).
2. Adams, A., Alexander, A., Alperovitch, D., Anastasi, D., Anderson, R., Bilge, L., et al. (2018). Avoiding the internet of insecure industrial things. https://core.ac.uk/download/141198917.pdf (accessed on 30th September 2025).
3. Adegboyega, A. W., Oladunni, S. O., Pramanik, A. S., Ani, E. C., & Oyesanya, A. (2024). Agile project management in smart grid design: Comparative insights from Africa and the United States. *International Journal of Science and Research Archive*. https://www.semanticscholar.org/paper/625fecd3233748d22b50ee27162a77ae21d551b2 (accessed on 30th September 2025).
4. Alammari, R., Hussain, S. M. S., Hussain, S., Iqbal, M., et al. (2023). A novel hybrid methodology to secure GOOSE messages against cyberattacks in smart grids. https://core.ac.uk/download/558772196.pdf (accessed on 30th September 2025).
5. Al-bermani, N., Bermani, A. K., Raad, A., & Manaa, M. (2025). AI-driven cybersecurity-based hybrid approach using blockchain for smart grids. *Journal of Discrete Mathematical Sciences and Cryptography*. https://www.semanticscholar.org/paper/66a0edf989d474a1cfd8b88db3a353c89fd467eb (accessed on 30th September 2025).
6. Al-Khateeb, H., Burnham, K. J., Epiphaniou, G., Hammoudeh, M., et al. (2019). Preliminaries of orthogonal layered defense using functional and assurance controls in industrial control systems. https://core.ac.uk/download/196185875.pdf (accessed on 30th September 2025).
7. Almalawi, A., Baecher, G., Boldea, B., Bouchti, A., Brand, S., Byres, E., et al. (2017). Assessing and augmenting SCADA cyber security: A survey of techniques. https://core.ac.uk/download/293885085.pdf (accessed on 30th September 2025).
8. Anglano, C., Aniello, L., Antinori, A., Armando, A., et al. (2018). The future of cybersecurity in Italy: Strategic focus area. https://core.ac.uk/download/188830442.pdf (accessed on 30th September 2025).
9. Bakirtzis, G., Carter, B. T., Elks, C. R., Fleming, S. D., et al. (2018). A model-based approach to security analysis for cyber-physical systems. http://arxiv.org/abs/1710.11442 (accessed on 30th September 2025).

10. Bartholomew, A. (2016). *The Smart Grid in Massachusetts: A Proposal for A Consumer Data Privacy Policy*. https://core.ac.uk/download/71468025.pdf (accessed on 30th September 2025).
11. Betouil, A., El Haddouti, S., & Chaoui, H. (2025). Global research trends in AI and blockchain for smart grids: A bibliometric analysis with a focus on Morocco (2014–2024). *Electronics*. https://www.semanticscholar.org/paper/556ad8ebbba6d36f72e2d64940a3ac12cfe6dde6 (accessed on 30th September 2025).
12. Botezatu, U. E. (2023). Smart cities: Linking cyber resilience to outer space security. https://core.ac.uk/download/655344073.pdf (accessed on 30th September 2025).
13. De Alwis, C., Kalla, A., Pham, Q. V., Kumar, P., Dev, K., Hwang, W. J., & Liyanage, M. (2021). Survey on 6G frontiers: Trends, applications, requirements, technologies and future research. *IEEE Open Journal of the Communications Society, 2*, 836–886. https://doi.org/10.1109/ojcoms.2021.3071496 (accessed on 30th September 2025).
14. Díaz González, F., Gallart, R., Girbau Llistuella, F., Heredero Peris, F., et al. (2018). Smart grid architecture for rural distribution networks: Application to a Spanish pilot network. https://core.ac.uk/download/157811722.pdf (accessed on 30th September 2025).
15. Dwivedi, Y. K., Hughes, L., Kar, A. K., Baabdullah, A. M., Grover, P., Abbas, R., Andreini, D., et al. (2021). Climate change and COP26: Are digital technologies and information management part of the problem or the solution? An editorial reflection and call to action. *International Journal of Information Management, 63*, 102456. https://doi.org/10.1016/j.ijinfomgt.2021.102456 (accessed on 30th September 2025).
16. El Meslouhi, S., El Fadil, H., Lassioui, A., Hasni, A., & Fhail, A. (2025). Internet of energy in microgrids and smart grids: State-of-the-art. *2025 5th International Conference on Innovative Research in Applied Science, Engineering and Technology (IRASET)*. https://www.semanticscholar.org/paper/87f4b7cfa687494302bad50b103f8fdfbf38dbb5 (accessed on 30th September 2025).
17. Elbez, G., Sánchez, G., Canbolat, S., Corallo, S., Fruböse, C., Lanzinger, F., Kellerer, N., et al. (2025). Insights and lessons learned from a realistic smart grid testbed for cybersecurity research. *Proceedings of the 16th ACM International Conference on Future and Sustainable Energy Systems*. https://www.semanticscholar.org/paper/5ee5e709b6f436afc7a5867fcc738b47bf2129b1 (accessed on 30th September 2025).

18. Fatima, S., & Arshad, M. J. (2025). A comprehensive review of blockchain and machine learning integration for peer-to-peer energy trading in smart grids. *IEEE Access, 13*, 92756–92782. https://www.semanticscholar.org/paper/3b987f26dec7cacec454b192621c60c986ab7e55 (accessed on 30th September 2025).
19. Ferrag, M. A., Friha, O., Hamouda, D., Maglaras, L., & Janicke, H. (2022). Edge-IIoTset: A new comprehensive realistic cyber security dataset of IoT and IIoT applications for centralized and federated learning. *IEEE Access, 10*, 40281–40306. https://doi.org/10.1109/access.2022.3165809 (accessed on 30th September 2025).
20. Gharehpetian, G. B., Marzband, M., Moradi, J., Nafisi, V., et al. (2019). Attributes of big data analytics for data-driven decision making in cyber-physical power systems. https://core.ac.uk/download/305121394.pdf (accessed on 30th September 2025).
21. Ginter, A., Ten, C. W., Vasilakos, A. V., Yamashita, K., et al. (2018). Impact assessment of hypothesized cyberattacks on interconnected bulk power systems. http://arxiv.org/abs/1801.01048 (accessed on 30th September 2025).
22. Gómez San Román, T. (2017). Integration of DERs on power systems: Challenges and opportunities. https://core.ac.uk/download/83830035.pdf (accessed on 30th September 2025).
23. Guato Burgos, M. F., Morato, J., & Vizcaino Imacaña, F. P. (2024). A review of smart grid anomaly detection approaches pertaining to artificial intelligence. *Applied Sciences*. https://www.semanticscholar.org/paper/7de62c78020cc2bf1e56e0186d0356d34fb77e6c (accessed on 30th September 2025).
24. Hamim, S. A., & Jony, A. I. (2023). Navigating the cyber threat landscape: A comprehensive analysis of attacks and security in the digital age. https://core.ac.uk/download/599374058.pdf (accessed on 30th September 2025).
25. Madnick, S. E., & Nourian, A. (2015). A systems theoretic approach to the security threats in cyber physical systems applied to Stuxnet. https://core.ac.uk/download/83235017.pdf (accessed on 30th September 2025).
26. Millis, M. (2019). Vulnerability analysis of modern electric grids: A mathematical optimization approach. https://core.ac.uk/download/215464251.pdf (accessed on 30th September 2025).
27. Ogunmoye, K. A., Agupugo, C. P., Ejichukwu, E., Barros, P., & Hayden, M. D. (2025). Protecting the grid: Cybersecurity strategies for renewable energy integration. *World Journal of Advanced Research and Reviews*. https://www.semanticscholar.org/paper/753f3ac06cf8556be51950e04367fd0c9c60d4ff (accessed on 30th September 2025).

28. Porambage, P., Gür, G., Moya Osorio, D. P., Liyanage, M., Gurtov, A., & Ylianttila, M. (2021). The roadmap to 6G security and privacy. *IEEE Open Journal of the Communications Society, 2*, 1094–1122. https://doi.org/10.1109/ojcoms.2021.3078081 (accessed on 30th September 2025).
29. Qian, Y., Sharif, H., Tipper, D., Yan, Y., et al. (2012). A survey on cyber security for smart grid communications. https://core.ac.uk/download/12211131.pdf (accessed on 30th September 2025).
30. Sandoval, C. J. K. (2019). Cybersecurity paradigm shift, the risk of net neutrality repeal to energy reliability, public safety, and climate change solutions. https://core.ac.uk/download/269681085.pdf (accessed on 30th September 2025).
31. Simonthomas, S., Subramanian, R., & Mathiew, S. A. (2024). A survey of enhancing cyber physical system security in smart grid. *2024 International Conference on Communication, Computing and Internet of Things (IC3IoT)*. https://www.semanticscholar.org/paper/6e92a0b3399d920c618bc270f0fe6a620d57cc09 (accessed on 30th September 2025).

INDEX

D

E

F

G

Q

R

S

T

U

V

W